D0929628

MYSTERIES AND SECRETS REVEALED

FROM ORACLES AT DELPHI TO SPIRITUALISM IN AMERICA

Loren Pankratz

Prometheus Books

Lanham • Boulder • New York • London

Prometheus Books

An imprint of The Rowman & Littlefield Publishing Group, Inc.
4501 Forbes Boulevard, Suite 200, Lanham, Maryland 20706
www.rowman.com

Distributed by NATIONAL BOOK NETWORK

Copyright © 2021 by Loren Pankratz

All rights reserved. No part of this book may be reproduced in any form or by any electronic or mechanical means, including information storage and retrieval systems, without written permission from the publisher, except by a reviewer who may quote passages in a review.

British Library Cataloguing in Publication Information Available

Library of Congress Cataloging-in-Publication Data

Names: Pankratz, Loren, author.
Title: Mysteries and secrets revealed : from oracles at Delphi to spiritualism in America / Loren Pankratz.
Description: Lanham, MD : Prometheus Books, [2021] | Includes bibliographical references and index. | Summary: "This book uncovers the science behind mysteries of nature and secrets of frauds who have been fooling us for centuries. Beginning at the Greek oracle in Delphi, author Loren Pankratz, PhD. guides us through the mysteries of the ancient world, the rituals of the Renaissance Church, and the readings of early mystics and spiritualists of the modern world to expose the deception of those claiming to tap into supernatural realms"— Provided by publisher.
Identifiers: LCCN 2020034780 (print) | LCCN 2020034781 (ebook) | ISBN 9781633886681 (cloth) | ISBN 9781633886698 (ebook)
Subjects: LCSH: Occultism and science—History. | Supernatural—History. | Deception—History.
Classification: LCC BF1409.5 .P36 2021 (print) | LCC BF1409.5 (ebook) | DDC 130—dc23
LC record available at https://lccn.loc.gov/2020034780
LC ebook record available at https://lccn.loc.gov/2020034781

∞™ The paper used in this publication meets the minimum requirements of American National Standard for Information Sciences—Permanence of Paper for Printed Library Materials, ANSI/NISO Z39.48-1992.

To Ray Hyman

CONTENTS

INTRODUCTION   vii

PART I: SECRETS OF THE ANCIENT GREEK ORACLES

1   HIDING A BLUNDER   3

2   DOUBTING THE ORACLES   10

3   CHAMBER OF COMMERCE
     PROMOTIONAL SECRETS   18

4   EXPLOITING THE MYSTERIES OF NATURE   30

PART II: REVISING AND REVISITING THE PAST

5   UNCOVERING A FORGERY   40

6   ASTROLOGY AND PROBABILITY   54

7   NATURAL MAGICK   66

PART III: REVISING THE WORLD

8   EXTRAORDINARY VISION   84

9   FIGHTING FOR MEANING   102

10   HIDDEN AGENDA   109

PART IV: REVISING THE UNIVERSE

11   SOME ENCHANTED EVENING   120

12   MEN ON THE MOON   127

13   SPACE RACE   133

PART V: SECRETS OF CHEATS

14   POSTER BOY FOR CLAIRVOYANTS   148

15   MIRACLES OR ENTERTAINING TRICKS?   164

16   REMOTE TRAVEL   179

17   WHO IS BEING TRICKED?   189

18   MESMERISM   204

19   MODERN HYPNOSIS   218

**PART VI: THE MYSTERIES OF NATURE AND
SECRETS OF SPIRITUALISTS**

20   RAP IF YOU BELIEVE IN SPIRITS   234

21   THE TURING TEST   251

22   TABLE TURNING   263

23   PALLADINO: YOU JUST HAD TO BE THERE   280

24   PALLADINO: CHEATING IN AMERICA   289

25   SLATE WRITING   306

CONCLUSION   320

ACKNOWLEDGMENTS   323

NOTES   325

BIBLIOGRAPHY   395

INDEX   457

INTRODUCTION

The opposite of truth has a hundred thousand shapes, and a limitless field.

—Montaigne

Have you ever lost sleep because you were confronted with some uncomfortable new fact? Sometimes the truth becomes too explicit to ignore. Time to face a painful reality.

Have you ever wondered whether a belief you have traveled with is a worthy companion? Beliefs are easy to acquire but difficult to discard. We grow up with many ideas we never question because they define our family and culture.

This book focuses on some individuals who asked uncomfortable questions. They were curious about cherished beliefs and decided to think, investigate, and discover. Their questions created discomfort; their explorations were variously construed as intrusive, disrespectful, unpatriotic, or blasphemous. Sometimes the truth was blocked by lying rascals, but often good people who resisted change to traditional beliefs opposed it.

We discover here that conflict and opposition can lead to more thoughtful searching. How do you know if someone is holding back secrets? How do you know if you are listening only to answers that confirm your own cherished beliefs?

The characters we visit explore these questions. They emerge from different times and explore diverse topics. However, all experienced an unconquerable desire to uncover reliable information. They were willing to accommodate to the reality of what they discovered no matter the consequences.

Our journey of discovery begins in ancient Delphi, where oracle priests ran a bait-and-switch scam, promoting their own ambiguous messages as pronouncements from Apollo. Unfortunately, one of their proclamations resulted in a war that cost King Croesus his life. This blunder caused some individuals to question the whole institution, which was deeply embedded in Greek life.

The focus then shifts from ancient Greece to men of the Italian Renaissance who attempted to direct the thinking of the people toward a more natural way of understanding the world. Their ideas aroused the disdain of academics and the anger of religious authorities. The master among such troublemakers was, of course, Galileo. He was determined to describe the world as it existed, not as it best suited expectations. His telescope rearranged the furniture of the heavens, and some had difficulty adjusting to the new living situation. Just as his telescope showed that the world was larger than anyone imagined, his microscope demonstrated it was also populated with things smaller than the naked eye could see. The secret was now out: direct observations best describe the world, not the pronouncements of ancient authorities.

Galileo's emphasis on observation as a test of theory sowed the seeds for modern science, but extraordinary claims about the natural world continued to attract attention. Some argued that the universe was populated by extraterrestrials who were visible to clairvoyants. Others claimed that subjects in a deep mesmeric trance could perform psychic feats that demonstrated mysterious forces. The most convincing clairvoyant of the time, Alexis Didier, regularly displayed a variety of psychic feats, like reading words inside sealed envelopes. The secrets of his techniques are revealed here through the process of working backward from the effect to the method, somewhat like reverse engineering.

Spiritualists declared they were in contact with departed spirits and mysterious forces. Among those considered here, one of the most convincing was an Italian peasant woman, Eusapia Palladino. Undoubtedly, she fooled more scientists and academics than any other person. After testing in New York by individuals acquainted with trickery, she sailed back to Italy with her reputation in tatters and her secrets exposed. In a grim admission, one professor said that the massive scientific literature on Palladino was a monument to groveling, imbecilic judgment.

The following pages are intended to help readers avoid contributing their judgment to similar monuments and to recognize groveling.

I

SECRETS OF THE ANCIENT GREEK ORACLES

1

HIDING A BLUNDER

If you abolished mythology throughout Greece, all the official guides would starve to death, for foreign tourists have no wish to hear the truth about anything.

—Lucian

The most popular oracle of ancient times was located at Delphi. Athenians trekked more than one hundred miles, ascending the magnificent slopes of Mount Parnassus that Apollo himself had once traversed. Upon arriving at the oracle, a guide introduced first-time visitors to the local attractions while preparing them for their consultations with priests. Looking back, we see our guide walking his small group past unending rows of statues. Then he stops in front of a golden lion.

"This statue was a gift from Croesus," he informs his attentive audience. He tells them that Croesus, once king of Lydia, wanted to know which oracle of all those scattered throughout the region could be trusted. So he devised a challenge. He deputized loyal attendants, sending each to a different sanctuary. Then on an appointed day and hour, his ambassadors were to ask the priests what Croesus was doing at that exact moment.

Once back in Lydia, each told Croesus what his assigned priest had revealed. The priest at Delphi, consulting Apollo, proclaimed:

I count the sand, I measure out the sea,
The silent and the dumb are heard by me;
Even now the odors to my sense that rise,

A tortoise boiling with a lamb supplies;
While brass below and brass above it lies.[1]

Only the priests at Delphi correctly identified the actions of Croesus. He was boiling a stew of tortoise and lamb in a large brass cauldron with a brass lid.[2]

"In gratitude, Croesus sent us this solid gold statue that weighs a quarter of a ton. It's sitting on 117 silver ingots. Count 'em yourself." Here, standing in front of the pilgrims was proof of the oracle's success.

Those early oracle seekers did not need to be reminded that sibyls and priests at Delphi provided divine guidance from Apollo, but the story of Croesus was pleasant persuasion. The guide also mentioned, in an offhand manner, that Croesus additionally gave the two large urns prominently displayed at the entrance to Apollo's temple, one silver and the other gold.[3]

"And he sent some other bric-a-brac as well, by the way. And now, moving along, if you will follow me over here. . . ."

The guide's story contrasted the success of Delphi with the failure of other oracles, but this bragging was not designed to outshine the competition. The priests were hiding a blunder by rewriting an embarrassing piece of history. They encouraged Croesus into a war against Persia that badly failed.

They did what always directs attention away from uncomfortable facts. They told a good story. Throughout history, nothing has been better for hiding the truth. In this instance, they said they had told Croesus that if he went to war, a great nation would be destroyed. One was. His.

For centuries, the Greeks and their philosophers trusted the oracles. Previous mistakes were easily ignored, but the magnitude of their failure with Croesus created the first troubling cracks in the foundation of a trusted institution. The defeat of Croesus had a profound impact on Greek thinking. The revision of history by the priests tore the veil of secrecy sufficiently to allow a peek behind the curtain. Some concluded it was time to acknowledge that priests were dispensing fallible human opinions that they promoted as guidance from Apollo. When looking behind that curtain, other deceptions were obvious as well.

THOSE GIDDY GOATS

The word "oracle" commonly refers to (1) the *message* of an individual who delivers a divine or uncommonly wise proclamation, (2) the *shrine* or *sanctuary* where such an utterance is given, or (3) a *person* who delivers an important message. The context conveys the meaning. The oldest known shrine is believed to have been established in Egypt during the fifteenth century BCE, although most Greek oracles probably began in the eighth century BCE. The locations of more than three hundred shrine sites have been identified, mostly in Greece, and they remained active until about the fifth century CE.

Oracles were often built near unusual geographic formations like caves or underground springs to instill a sense of wonder or fear.[4] The temple at Delphi is said to have originated at a spot where farmers noticed their goats bleating in a strange manner.[5] On further investigation, peasants discovered that they could get high by breathing the fumes emanating from a fissure in the ground. Some were so overcome that they fell to their deaths. Thus, attendants were assigned

View of Mount Parnassus at Delphi, Greece. Lithograph by Salvatore Puglia, 1842. *Getty Images*

to protect the stream of visitors, and rituals were introduced to inform the curious. Ordinary people were no longer permitted to inhale. Young women from the nearby region of Pytho performed rituals over the fumes, issuing divinations. They were given the title of Pythoness or Pythia, and they were known as sibyls in their role of providing divination.[6]

Various gods were honored at the oracle until a temple to honor Apollo was built in the mid-seventh century BCE. The temple and other structures were subsequently reconstructed after wars, fires, lightning strikes, mudslides, and earthquakes.[7] A seventeen-meter-high statue of Apollo was dedicated in 346 BCE. After about twelve centuries of popularity, the oracle was in decline when closed by the edict of Theodosius the Great, Roman emperor from 379 to 395 CE.[8] Its location was then lost for about a thousand years. When it was rediscovered in modern times, there was considerable skepticism about the existence of euphoric fumes. In 1891, for example, French investigators found no evidence of fissures or fumes, which led scholars to conclude that the descriptions of vapors were myths. Historians concluded that the ecstasy of sibyls resulted from satanic influence, a poisonous drug, autosuggestion, mesmerism, hysteria, chewing laurel leaves, or inhaling smoke.[9] Some historians claimed that seekers were merely enthralled by rituals devised to impress the superstitious; still others denied that frenzy was ever a part of the prophetic process. Nevertheless, those early reports were difficult to dismiss, like that of Plutarch, who was a priest at Delphi for some years. He said that at least one Pythia died when overcome by fumes.[10]

Then in the 1980s, a geologist noticed a fault line nearby that he followed to the exact location of the temple. By the 1990s, an interdisciplinary team of scientists confirmed an east-west and north-south intersection beneath the area where the temple once existed.[11] A local resident directed the scientists to a "wind hole" that emitted strange odors. The area was fenced off, just as locals had done thousands of years earlier. Nearby gulls acted drunk, which recalled the images of those giddy goats. The fissure at the temple had likely closed during one of the commonly occurring earthquakes in the area. This new finding provided an opportunity to study the emanating vapors.

The team concluded that the underlying stratum contained bituminous limestone, seeping water, and organic deposits that were producing petrochemical gases, namely ethane, methane, and ethylene. All three can produce psychogenic effects, but ethylene in particular is characterized by a sweet-smelling odor that can create an immediate euphoric state of excitement, confusion, and even frenzy.

Gases like ethylene and nitrous oxide commonly give a sense of euphoria associated with metaphysical illumination, the impression of being in touch with otherwise unobtainable knowledge. In 1799, chemist Humphry Davy "applied himself with the greatest assiduity and zeal to the investigation of the effect of gases in respiration."[12] The psychedelic effects of nitrous oxide were so startling that the experience became popular among philosophical and literary men of the day. William James, the famous Harvard psychologist and philosopher, was amazed by the wisdom laid open before him with "blinding evidence." When his mind cleared, however, he viewed his revelations with astonishment. His most coherent sentence read: "There are no differences but differences of degree between different degrees of difference and no difference."[13]

The psychogenic gases are only part of the mystery of Greek oracles.[14] But it appears that sometimes the voice of Apollo was conflated with the utterings of someone with a confused brain.

MEANING IN A TROUBLED WORLD

Ancient Greeks took their religion seriously. Priests in the Athens Assembly consulted collections of written oracles kept by the city, which they interpreted for guidance. The city spent considerable funds on religious festivals in an effort to keep the gods happy, and military campaigns involved a continual process of divination.[15] The big questions of individuals and nations were brought to Delphi. Sparta, Athens, and other cities built embassies there so that consultations could be conducted smoothly, just as later governments established relationships with the Vatican.[16] Delphi also became a financial clearinghouse for the Greek world.[17]

The oracle provided a spiritual sanctuary for contemplation. Above the doors to the temple of Apollo were the words: "Nothing in excess" and "Know thyself."[18] To know oneself was not an invitation for self-analysis or a suggestion to figure out your own problems, as might be currently interpreted. Instead, the two words were an invitation to remember one's vulnerabilities and mortality.[19] Or, alternatively, as some wag suggested, these might have been an injunction against excessive consumption of wine—know your own capacity and avoid getting drunk.[20]

The early Greek religion may seem primitive to us today because we view their gods as mythical, anthropomorphic creations. However, early Christian churches were often devoted to specific saints who were similarly venerated and invoked for favors. Greek gods were national heroes worthy of respect. Plato was distrustful of individual soothsayers and charlatans, but he expressed confidence in the oracles of Delphi, Dodona, and Ammon. He suggested that the religious life of his ideal city would be based on the revelations stemming from these institutions, integrated with a philosophical system based on reason.[21] The oracles would motivate people to live a moral life in a chaotic world.[22] Freedom of thought in religious matters, on the other hand, could threaten social order and cohesion. Myths kept alive the nation's history and values, uniting citizens through a common bond in times of crisis.[23]

The ancient Greek system was operative for hundreds of years because the oracle provided services that we should not be quick to criticize or dismiss.

GETTING THE MESSAGE

The messages delivered at Delphi, where we have our best knowledge, are still a matter of scholarly dispute. The methods and procedures varied over time, as did the interactions between priests and the Pythia. It is generally believed that priests transcribed the utterings of the Pythia into hexameter verse, but they probably prepared their responses in advance.[24] Most proclamations were probably never written down, and many were answered with simple yes or no responses. Oracles considered successful were later memorialized in verse, giving them a grandeur that was not evident in the original presentation.

For the poor, itinerant charlatans wandered the streets of Greek cities offering potions, incantations, healing balms, amulets, and the interpretation of dreams.[25] And for surefire success, some sold lethal poisons on the sly. These oracle mongers were mocked in plays, and philosophers said they degraded true religion like televangelists. Nevertheless, these itinerant charlatans gleaned what money they could from the superstitious who, like themselves, were surviving on the edge of social and economic life.

Belief was encouraged by legends about the consequences of denying or resisting an oracle's decree. These legends usually featured a protagonist who goes out of his way to avoid the oracle's instruction only to encounter his inevitable fate with a dramatic, unexpected, or tragic ending that punishes his disbelief or disregard. These early urban legends served as a warning against doubting the oracle edicts.[26] This theme of "punishment for doubt" returns during other periods of history, for example, those who deny the Christian message, doubt the existence of demons, reveal the secrets of alchemy, scoff at the existence of ghosts, reject the efficacy of relics, inquire into forbidden things, or doubt the power of black magic.[27] Like urban legends, these cautionary tales of retribution are propagated from person to person because they carry a message that encourages conformity or teaches a moral lesson for society.[28]

The story of Croesus is a good example. The Delphi priests turned their blunder with Croesus into a moral lesson that cast the blame on him. Shaken by their miscalculation, they quickly devised a revision of what they told him. They said he was informed that a great nation would be destroyed if he went to war. If he did not understand their warning, he should have asked for clarification.

Their post hoc construction of an ambiguous response (a great nation being destroyed) was a brilliant solution. They were well acquainted with obscure responses, but they decided that questions of this magnitude should henceforth be answered in a way that could always imply fulfillment no matter what the outcome.[29] This change was noticed by scoffers who gave Apollo the nickname of Loxias, the ambiguous one. Cicero said, "Apollo, thy responses are sometimes true, sometimes false, according to chance, in part doubtful and obscure, so much so that the expositor has need of another expositor."[30]

Croesus sent Delphi that golden lion, which was exceedingly generous, but probably not much more than his gifts to other shrines.[31] He completely rebuilt the sanctuary at Ephesus, and his name was inscribed on their temple column, which now resides in the British Museum.[32] Croesus was not spreading charity as much as seeking endorsements to inspire his troops. He did not need to remind the priests at Delphi that the Persians would eventually turn their attention to Greek lands. The fierce Persian fighters were known for pillage and rape, so the priests were eager to send Croesus out to fight—better him than us. So they told him, "Just do it." After his defeat, the priests told the guides to inform visitors that the gold lion was an acknowledgment of their spectacular distant vision of him cooking stew in a brass pot.

The oracle's endorsement of war was nothing unusual. Decade after decade, Delphi encouraged Greek expansionism. The Greeks were successful, however, not because the gods were smiling on Greece but because Greek warriors had better training and equipment.[33]

Croesus learned the hard way that believing what you want to hear can hurt you. Thus, we turn now to consider more about the power of belief.

2

DOUBTING
THE ORACLES

When we argue from what is said in history, what assurances have we that these historians were not prejudiced, nor credulous, nor misinformed, nor negligent?

—Fontenelle, *The History of Oracles*, 1753

We think of ancient Greek society as democratic, but historian Hugh Bowden reminds us that in contrast to popular belief, their civilization was not liberal, individualistic, or secular.[1] Ancient Athens spent considerable money on religious festivals to keep the gods happy. Any dissent was considered an affront to the gods who protected their society. The ancient Greek Enlightenment encouraged variant opinions, but that freedom decreased as society became more insecure. Disbelief in the gods became an indictable offense, and educated individuals became more alienated from the general populace. Philosophers were the first to be charged with heresy. The most notable was Socrates (469–399 BCE), who drank poison rather than flee as had other accused philosophers. He was charged with atheism and corrupting the youth, but his real crimes were pestering people to be virtuous and needling them to examine their personal belief systems.

> But I shall be asked, Why do [the youth] delight in continually conversing with you? I have told you already, Athenians, the whole truth about this: They like to hear the cross examination of the pretenders to wisdom; there is amusement in this.[2]

Here Socrates was practically testifying against himself. Citizens did not laugh when their beliefs were challenged. As William James said centuries later, "We find ourselves believing, but we hardly know how or why."[3]

BELIEF SYSTEMS

If we think of *belief* as information held to be "true" with some degree of conviction, then a *belief system* denotes interconnected beliefs that impose a framework on our view of the world. These systems provide a structure for understanding and organizing life. Belief systems, which we consider throughout this book, are mostly unnoticeable because they are like looking through a tinted window. We are scarcely aware that our preferences and expectations have been colored by former experiences.

Moreover, we are not aware how easily our belief systems can lead us astray because our brains are hardwired to have a strong preference for information that confirms our preconceived conclusions. This "confirmation bias" is so strong that it may feel as if we are thinking and reasoning when we are actually spending effort justifying our belief. Unfortunately, "reasoning" is mostly designed to seek justification for belief, not to find truth.[4] As a result, ideas inconsistent with our belief system are often not even noticeable—they remain hidden.

Being bright does not protect us from bias; educated people are better at articulating reasons to confirm their beliefs while ignoring things that contradict their theories. It takes careful planning and attention to overcome these serious limitations. Helpful reasoning involves searching for alternatives that better match the world outside our brain. Scholars across a wide range of disciplines have conducted fascinating research on this topic.[5]

A belief system has the advantage of allowing one to evaluate new ideas quickly so that time is not wasted in processing and analyzing information. The absence of a belief system would slow life to a crawl by making every situation a challenging process of evaluation. A belief system does not need to be rational in order to be successful.[6] And even the best of belief systems gets us into trouble from time to time. For example, someone's guiding belief about solving interpersonal problems might be highly successful at work but destructive in one's domestic relationship. Unfortunately, belief systems are rapidly acquired but difficult to reconsider and replace.

Because Greeks had a long history of consulting oracles, their trust did not weaken despite the cracks that continually appeared in the supporting structure.

Belief systems tend to support the status quo so that problematic information is not always apparent. Outsiders or those from different historical perspectives often see the tricks and traps that entangle otherwise intelligent people. The secrets and lies not apparent to one group might be blatantly obvious to others.

Ancient Greeks maintained their belief in oracles for centuries when life was reasonably stable. No one thought about whether advice came indirectly from the gods or directly from crafty priests. Once the stability of life was seriously threatened by the Peloponnesian War, the citizens of Athens began questioning the value of consulting priests who had not warned them about a dreadful and unexplainable plague (430–429 BCE) that killed about one-quarter of their land army.[7] People usually died after seven days of terrible symptoms, and survivors were left with necrotic fingers, toes, and genitals; some lost eyesight or memory.[8] The priests of Delphi were unable to protect themselves or the populace from its ravages. Athenian losses at sea were equally disastrous. A battle near Sicily in 413 BCE resulted in the loss of more than two hundred military vessels and about forty-five thousand men killed or enslaved.[9] Sparta eventually starved the Athenians inside their city to end this senseless war.

As society devolved into lawlessness, some of the secrets of the priests became more obvious to the Athenians. Herodotus (484–425 BCE) tried to prop up the faltering structure of belief in the traditional myths of his culture.

MORAL LESSONS

Herodotus began his famous *Histories* by informing readers that his purpose was to preserve the memory of the great Greek accomplishments. Additionally, he wanted to provide an understanding about the rise and fall of leaders and empires. Events unfold, according to the belief system of Herodotus, in ways consistent with moral rules that inevitably produce consequences. Noble intentions result in a just outcome, and pride must be punished.[10] As the first case example in his book, Herodotus opened with the story of Croesus. He denied that the priests had blundered with their advice. The Lydian king made his first mistake, Herodotus said, by testing the authority of the Delphic oracle. He asked the priests to say what he was doing at a specific time (when he was cooking that tortoise and lamb stew). He should have, instead, sought Delphian council with more appropriate humility. Therefore, his transgression would be ultimately punished, and he would be forced to admit that the consequences were his own fault.[11]

Croesus asked about going to war with the Persians, who were encroaching on the far side of his territory. According to Herodotus, the priests told Croesus:

When Croesus shall o'er Halys River go
He will a mighty kingdom overthrow.

And it happened, but not in the way he expected. Herodotus presented this narrative as an illustration of how the world invariably works. Over the centuries, those who believe in oracles and supernatural predictions have cited Herodotus, who is known as the "father of history." Alternatively, he has also been labeled "father of lies."[12] And if we read his book, we discover why he received this latter appellation.

HIDDEN PARTS OF THE HERODOTUS STORY

As mentioned, those who promote miracles often cite Herodotus's story about Croesus to promote their belief in prognostications. However, they fail to tell the rest of the story. Herodotus reports that Croesus was taken prisoner by the Persian king Cyrus, who chained him to a pyre to be burned alive. But, observing the dignity of Croesus in this perilous situation, Cyrus rescinded his sentence and demanded his servants quench the flames.[13] Unfortunately, they had no water. Croesus, now knowing that Cyrus intended to save his life, called out for Apollo to come to his aid. Suddenly, out of a clear and windless day, clouds appeared with violent rain that extinguished the flames.[14]

Herodotus said that after the defeat of Croesus the Persians attacked Greek cities including Delphi, just as the priests feared might happen. The marauding army was specifically intending to loot those golden gifts of Croesus.[15] When word got to Delphi, the people asked their god if they should bury the city treasures or transport them out of town. Apollo reassured them that they could leave everything in place because he would provide protection. Nevertheless, the women fled to other cities and the men headed into the caves of surrounding mountains. Herodotus claimed that as the enemy marched toward Delphi, Apollo unleashed lightning that split off two boulders from the upper crags, crushing many and terrifying the remainder, who fled in disarray. The emboldened Delphi militia chased the surviving soldiers down the mountain, and two statues near the temple came to life and successfully routed the enemy warriors.

This delightful but absurd narrative by Herodotus is never included by writers who support supernatural explanations. The complete story is understandable only as a myth for teaching those moral lessons that Herodotus intended. He was not as interested in communicating the actual events. Evolutionary biologist Robert Trivers calls such myths "false historical narratives" that permit national self-deception.[16] All nations have similar narratives, and they are often fiercely defended.

We can easily read right over the glaring absurdities of Herodotus because of our admiration for the historical importance of his writing. We instinctively understand that the war with Persia actually occurred,[17] but the details Herodotus gives are myths that we accept as part of the culture of ancient Greece. We do not conclude that Herodotus was an intentional liar, even if Lucian was not so certain. Legends, myths, and false historical narratives become integrated into the belief system of all cultures, enduring without any awareness or concern about their falsity.

For example, English theologian Conyers Middleton reluctantly pointed out that a considerable number of declarations of early church patriarchs were obvious fables,[18] and early church historians were eager to insert miracles into the *Golden Legend*, a popular Medieval review of saints. However, these myths inadvertently exposed "the gross immorality of many of them, the ridiculous absurdity of others, and the shameless fraud."[19] Yet the early church fathers and saints were so revered that no one ever commented about their flights into fantasy and the bizarre stories they perpetrated.

Similarly, Bertrand Russell noted that most commentators on the great philosophers have politely ignored their silly remarks,[20] and this was especially true for followers of Aristotle, who suppressed his fallacies and retained only what seemed plausible and consistent.[21] As we learn in later chapters, these mistakes of Aristotle, like the blunders at Delphi, eventually became too obvious or constricting to ignore.

THE END OF ORACLES

Many factors contributed to the slow collapse of the Greek oracles. After the big lie about Croesus, Delphi loaned money to Sparta for its conflict against Athens in the Peloponnesian War (431–404 BCE).[22] Athenians felt betrayed by Delphi, and they were ultimately defeated by Sparta and decimated by a plague. But all of Greece was psychologically damaged by the loss of leadership and the subsequent inattention to arts, culture, and social needs. As a result, wealth and power shifted to Rome. Romans visited oracles more as tourists or curiosity seekers because their questions about the future were more easily answered by astrology. Historian E. R. Dodds said that astrology weakened the traditional moral values of Greeks, infecting them like a new disease that consumes those newly exposed.[23]

Cicero (106–43 BCE) mocked the oracles because he was a Roman who had not grown up with belief in their value.[24] He sarcastically noted that the oracle trade secrets worked best for the credulous and naive. He ridiculed his brother for believing that a bird's entrails could possibly have any meaningful relationship to future events. He recommended that citizens give up their superstitions because events previously attributed to the actions of the gods were now understood as natural phenomena. Lightning, for example, was an ordinary event of weather. Jupiter had no reason to throw thunderbolts in uninhabited mountains or deserts. Anyway, if the gods were serious about communicating with us, they would send straightforward messages, not frightening tribulations, unintelligible verses, or ambiguous dreams that occur in our most disorganized state. If some dreams are true and some false, Cicero asked his brother, how do I know which is which? He was unwilling to rely on someone to tell him which ones were significant because he noted that every soothsayer with one prescient interpretation was responsible for multiple failures. He said he would not attribute to the gods that which could just as likely be chance or lucky guesses.

Similarly, Plutarch (46–120 CE), a Greek biographer who later became a Roman citizen, complained that the oracles had fallen on hard times. Delphi's glory days of dispensing wisdom had sadly declined. But Plutarch was comparing the oracles of his day with events of the past that were inflated into historical legend. The quality of the oracles was probably never very good, then or now.

As an example, Oenomaus of Gadara was impelled to write a pamphlet entitled *Exposure of Cheats* after he visited the oracles at Klaros.[25] When he asked the priests about the future of his business enterprise, he was told "There is in the land of Trachis a garden of Herakles, full blooming, well watered, where fruit is gathered every day without diminishing."[26]

The reference to Trachis suggested hard times, but the oracle seemed to promise ultimate success despite obstacles. As he was leaving, Oenomaus discovered that others had been given this same message. Oenomaus concluded that his presumed personal response was only one of many scripts drawn from a stack of boilerplate options. Knowing this, how could anyone believe they had received special attention from the gods?

This experience prompted Oenomaus to evaluate the quality of the messages at Delphi. Like any responsible investigative reporter, he condemned the oracle for deceit and for giving advice that was nonsensical and utterly worthless. He labeled the priests as cheats and the oracle as practicing institutional duplicity.

Reports about the decline of oracles was probably justified. As Greek society became more secular, answers degenerated into quick-response, self-help consultations rather than searches for the right path and enlightenment.

Statesmen and war heroes constructed their statues and monuments for self-aggrandizement instead of gratitude.[27] Oracles imploded from internal difficulties as well. Rumors about sexual improprieties of priests were so common that Herodotus did not even attempt to deny them. Plutarch informs us through the mouth of cynic Didymous that the interactions between priests and supplicants had degenerated so much that divine providence "has gathered up its oracles and departed from every place!"[28] Priests were in conflict over rituals, and they treated the rich and politically connected with deference. Bribery hastened procedures that ordinarily required waiting for days.

Nero confiscated more than five hundred brass statues from Delphi for his own palace, although a Spartan declared that he could not name and describe the statues that remained because they were so numerous.[29] Then Constantine, the first Christian emperor, looted treasures from Delphi to decorate his new capital at Constantinople.[30] Christian churches created competition with oracles when they developed their own rituals, hymns, healing ceremonies, and prayers to saints.[31] Oracles had spas and springs for healing, but one could always hope for an instantaneous miracle at a church.[32] Then in 392 CE, emperor Theodosius banned pagan practices and closed Greek temples. That was the last year of any known oracle delivered at Delphi, although its Olympic-style games were still celebrated as late as 424 CE.[33]

We must conclude by admiring the success of oracles. The practice lasted because they satisfied supplicants century after century. They provided a mutually beneficial exchange between consumer demand and service provider until that equation broke down. In a modern analysis of business practices, A. A. Leff proposed that it is ethical for sellers to puff their products and entice customers, but it is not ethical to bait and switch.[34] The core of the oracle was a bait-and-switch swindle.

The oracle at Delphi, in particular, implicitly told seekers that they would receive the opinion of a deity, but they got advice from fallible priests. The seeker may have obtained worthy counsel, but the answer was a deceptive switch. Priests could have presented their opinions as "wisdom from the best philosophers and holy men of the kingdom" but they did not use that advertisement on their billboards. Customers would have kept walking north to the oracle at Dodona. The Delphi Chamber of Commerce decided that honesty was not the best policy.

Most of us are surprisingly talented in the fine art of ignoring things that conflict with our belief or faith. Typically, the occurrence of a crisis causes a reinterpretation of that event rather than creating doubt in our beliefs. Most people clutch their persuasions even tighter during times of trial.[35] And we see here a relentless theme as we note that many ancient Greeks persisted in their belief despite Delphi's abandonment of political neutrality, the massive loss of life from plague and war, internal conflicts among priests, and looting. And finally, the decree of Theodosius ended consultations, and the last priest went back to herding goats.

That is where matters stood until the seventeenth century when a French scholar raised some additional accusations against the Greek priests. Bernard de Fontenelle described the oracle service industry as one of exploitation. His larger, more comprehensive descriptions revealed a better understanding of how and why the oracle system held together and lasted so long. Priests were hiding secrets and exploiting the mysteries of nature.

3

CHAMBER OF COMMERCE PROMOTIONAL SECRETS

The secret of business is to know something that nobody else knows.

—Aristotle Onassis

When he was thirty years old, Bernard de Fontenelle (1657–1757) experienced an existential crisis. Fontenelle was born a couple of decades after Galileo had asserted that the cosmos was immeasurably larger than anyone expected, and Fontenelle felt lost in the vastness of this new universe. However, he was unwilling to abandon his Christian faith because he considered its larger truth still valid. His problems were with the dogmatic theology and obsolete doctrines that mired the church in superstition. Society was infused with belief in spirits, devils, and angels. In order to comprehend the natural world, people needed to disentangle their everyday lives from belief in demon manifestations.[1] Fontenelle decided to abandon the fantasy that his church had all the answers; he was ready to embrace Galileo's method of uncovering the mysteries of nature, even if it left him untethered in a limitless galaxy.

Fontenelle was simmering in this toxic cultural brew when he discovered a small book written by a Dutch scholar named Anthony Van Dale. Van Dale challenged the prevailing view that the birth of Jesus terminated the power of demons who controlled Greek oracles.[2] This gave Fontenelle an idea: Christians then believed that demons were responsible for the success of those ancient oracles. If he could persuade people to doubt that old idea, then maybe he could cast doubt on the responsibility of demons in connection with events and activities in his own time. An indirect approach suited his style, because in contrast to Voltaire, a later French skeptic, Fontenelle was more private and less confrontational. With a sense of vigor, he took the theme of his Dutch muse and boldly asserted two conclusions: (1) oracles were not delivered by demons, and (2) the presence of Jesus did not stop the oracles.[3]

To set the stage, we begin by considering how the church got ensnared in such a robust belief in demons in the first place. This belief diverted humans away from better understanding the world. Similarly, the deceptions of priests discussed here created false beliefs that diverted people from seeking better ways of making decisions.

SPEAK OF THE DEVIL

Plato believed in personal "daemons" that operated as intermediaries between God and man. These beings were like messengers: winged creatures with human passions that primarily accomplished good deeds. Plato said these demons were the interpreters of all divinations and soothsaying, which were activities that Plato did not consider evil. He even claimed that a "genius" demon

Bernard de Fontenelle (1657–1757). *Musée d'art et d'histoire, Genf. Getty Images*

attached itself to Socrates from the days of his infancy.[4] Historian Valerie Flint reminds us that the extent of Greek belief in demons is a matter of enormous philosophical and textual complexity.[5] Nevertheless, Plato gave early Christian writers a basis for believing that demons were responsible for oracles.

As Christianity gained popularity, early church fathers promoted their faith by claiming that the advent of Jesus disempowered those sinister demons that assisted the Greek priests in their oracle proclamations. Without the help of demons, the priests were no longer effective, and the oracles ceased. Unfortunately, the idle hands of the devil found new workshops within the Christian faith. For example, Satan was presumed to have supervisory responsibility for thunder and lightning, which was previously the lash of Greek gods. He and his associates managed hail, plague, fire, drought, floods, and other such harassments,[6] although sometimes these were acts of God when he became provoked by the sins of man.[7]

Augustine appropriated Plato's good demons and made them into angels, consistent with his tendency to view the world in terms of good and evil. By the fourth century, people had filled the air with cosmic battles. They believed the devil could not produce miracles, but he could exploit the deep mysteries of nature in ways that enabled him to create *apparent* miracles. Thomas Aquinas said that demons had superior knowledge of nature, and soon thereafter books appeared that revealed the secrets of how magicians compelled them to reveal knowledge and locate hidden treasures.[8]

Anyone interested in exploring the mysteries of nature was asking for trouble from the church. Curiosity was dangerous because seeking information inherently involved a risk that the devil might intrude to offer his services.[9] Thus, the very process of seeking hidden knowledge was suspect, and curiosity was considered a snare of the devil. Therefore, all magic was labeled dangerous and condemned in the twelfth century by Gratian's Decretum, which was canon law of the church for eight centuries.[10] Pope John XXII (1316–1334) was particularly obsessed with the workings of magicians, and he vigorously persecuted them.[11] Pope Urban VIII, in contrast, accepted the magic rituals of Campanella in 1629 for protection against predictions of his impending death.[12] Horoscopes suggesting future unpleasantness were so disturbing to Urban that he decreed anyone guilty of magical threats against a pope would be punished by death.[13]

Fontenelle asserted that Christian theologians could save themselves some serious embarrassment by rejecting these old belief systems about magic and demons. In the past, he said, clergy concocted fables, forged false quotations, and fell for hoaxes.[14] Christianity should not be propped up by bogus ideas, and a belief in demons controlling oracles was another pitfall to avoid. To make his point, Fontenelle was ready to reveal the secrets that Greek priests devised to create the appearance of miracles.

THE DELPHI CHAMBER OF COMMERCE PROMOTION PLAN

Philosophers and historians repeatedly suggested that early Greek priests at oracles issued vague proclamations that were later considered uncanny if not supernatural predictions. But Fontenelle believed there was more to the game than lucky guesses and ambiguous responses. Fontenelle painted a vivid picture of the oracle as a successful service industry. Oracles were an interaction between priests and pilgrims—providers and consumers—that took place in a larger psychosocial, political, and economic environment. Delphi survived because the whole city adopted a successful business plan that promoted its primary economic enterprise. The city was dependent on the success of oracles, so everyone benefitted when the priests appeared to create miracles. Armies and nations often gave 10 percent or more of their spoils of war to Delphi. In fact, priests sometimes stipulated that a portion of booty should be dedicated to the oracle.[15] Therefore, local citizens had a powerful incentive to keep the oracle vibrant.[16]

Delphi lies about one hundred miles northwest of Athens on Mount Parnassus, which is surrounded by the natural beauty of mountains, streams, and caves. The path was arduous, whether coming from Athens or by ship into the Gulf of Corinth. To make the adventure worthwhile for ancient pilgrims, Delphi became a destination resort with spas, healing centers, and obligatory gift shops. ("My grandpa went to Delphi, and all I got was this stupid toga!") The mountain air was especially refreshing to those who lived in lower swampy areas. About a million visitors a year still visit Delphi because of its appeal and beauty.[17]

Delphi had something for everyone. Its citizens sponsored games every four years that rivaled the Olympics.[18] There were lute- and lyre-playing contests, recitations by poets and writers, and performances by acrobats and conjurors.[19] The outdoor amphitheater offered entertainment, and pictures of Hades on the temple walls provided opportunity for meditation. The city was filled with memorials, elaborately ornate buildings, and hundreds of statues, including that gold lion given by Croesus.[20] The visitors' warm welcome from tour guides was part of a larger plan. Seekers could scarcely contain their hopes, fears, and expectations. Guides and innkeepers gleaned information that they forwarded to priests.

Moreover, the priests employed spies in the major cities who reported the political winds back to Delphi.[21] Porters carried more than luggage and revealed more than the pathway by conveying aspirations and political intrigue. The secret Delphi Chamber of Commerce Plan worked successfully for centuries by employing the latest technologies of industrial espionage. This plan helped the oracle business become an integral part of Greek cultural life, positioned to maintain enormous power over citizens and cities across a wide geographic region, even over outlying barbarians.

The conclusions were obvious to Fontenelle: the Delphi priests manipulated belief through comprehensive deceptions.

CRAFTY PRIESTS

Fontenelle said that statues and idols were not animated by the voices of demons. He checked this out, he said, because he had an obligation to protect himself against false belief.

> God is only obligated, by the laws of his goodness, to protect me from those impositions from which I cannot defend myself; as for other things, it is incumbent on [me to use my] reason.[22]

Fontenelle noted that talking statues were always situated in caves, near holes, or strategically placed within buildings. Indeed, a secret passageway at Delphi led to a small room where the sibyl proclaimed her message. Sounds and fragrances were transmitted upward through an opening in front of the assembled oracle seekers.

Statues talked because of speaking tubes that priests exploited. The whole process was a performance by well-informed priests. After a question was placed on the altar of Apollo, that room was locked and sealed. Then priests obtained access through secret passages, completely undetectable. Or a priest might tell a seeker he needed to sleep on a petition to obtain an answer in a dream. The priest thereby secured additional time to open and reseal the message.

ANGRY RESPONSES TO FONTENELLE

Despite his persuasive arguments, Fontenelle's denial of demons stirred up many Christian scholars. A response by "A Priest of the Church of England" was longer than Fontenelle's original work.[23] This alleged English priest was later identified as a French Dominican monk, Jean François Baltus.[24] His angry review was presented as a "translation," but there was no original English manuscript. Baltus repeatedly decried Fontenelle's rejection of the traditional view of the church fathers. In distress, he wondered how Fontenelle could justify such an attack on the Christian faith.

> You account as nothing the most ancient and most constant tradition . . . all the opinions of the Church clearly expressed in the scriptures. You think yourself at liberty to believe nothing of them; nay sufficiently warranted to reject them.[25]

Baltus had captured the critical issue: that was exactly Fontenelle's point. Fontenelle championed empirically demonstrated truth over religious dogma, ancient texts, and superstition. He was not anti-religious but an advocate for a "purified Catholicism."[26] He considered himself entitled to reject old superstitious beliefs.

Baltus accused Fontenelle of citing only the opinions of heathens and Christians who agreed with him and of misrepresenting the ancients by reporting only the parts that supported his thesis. Baltus provided his own translation of Porphyry to enlighten Fontenelle about Porphyry's intended meaning, and Baltus instructed Fontenelle about his misunderstanding of Augustine. On and on he went, telling Fontenelle that the experts all opposed his opinion.

Baltus was correct that Fontenelle had cherry-picked passages from those older authors, just as Baltus himself ignored passages that contradicted his beliefs. Baltus argued in the tradition of the scholastics who cited authorities, jabbed with reason, and turned clever phrases to catch the opponent off guard. However, Fontenelle trusted that his audience of ordinary citizens would be more interested in what really happened at Delphi and care less about the opinion of Porphyry, a dead Greek philosopher.

Baltus was living in the demon-infested world that Fontenelle was trying to clean up, and Baltus was desperate to preserve his version of reality. What really happened at Delphi, Baltus insisted, was the result of demon intrusion. Sure, maybe there were underground passages, but those tunnels were constructed to provide access for devils or for priests who had been informed by devils. The idea that priests animated speaking statues was so far-fetched, he dismissively asserted, that the question did not deserve discussion. Furthermore, Baltus proclaimed that no debate was required because oracles were never given through speaking statues.

Baltus was wrong, and he knew better because he mentioned a Greek charlatan who got caught giving proclamations through a speaking tube. We know of Alexander of Abonuteichos through the writings of Lucian and Eusebius. They described Alexander as a charismatic religious charlatan whose personality disorder converged all his talents into powerful lies, perjury, blackmail, and successful conjuring.[27] He grew rich by roaming from city to city "fleecing the fatsoes."[28] He hired servants, secretaries, spies, and interpreters to promote his oracle pronouncements. Alexander was famous for his lifelike figure of a snake, which he controlled like a puppet under his arm, and displayed only in dim light. The mouth and forked tongue moved by the manipulation of undetectable horsehairs. The windpipes of cranes were attached to the snake's mouth and ran back to a hidden confederate.[29] Baltus cited this piece of history as evidence that priests who used deception would be caught like Alexander. But Alexander was highly successful for a long period of time despite the harassment and exposure by Epicureans and lampooning by Lucian.

The blunders of Baltus kept coming. He identified Samuel Morland (1625–1695) as the inventor of the speaking trumpet, a device to amplify sound for the hearing impaired or for broadcasting sound over a farther distance. But Morland was, instead, the first person to conduct scientific research with the device.[30] When someone speaks, sound is ordinarily transmitted spherically (in all directions), but sound spoken into a trumpet is constrained and transmitted without dispersal.[31] The entirety of the voice is concentrated at the opening of the tube and all that acoustic energy is focused in one direction, which deceives the uninformed.

Baltus asserted that crafty priests would not be able to fool intelligent Greeks. But he was wrong again.

SPEAKING TUBES

In 1890, a Greek bust of Epicurus was examined by a Danish sculptor who discovered alterations from its original construction.[32] An ancient stonecutter had drilled a hole in the mouth that led to a hollow section inside. This modification was not designed to create a fountain because that would have resulted in a residue of minerals from the water on the marble. The only feasible explanation was that the statue had been altered to install a speaking tube.

As a test, archaeologists attached a twelve-meter bronze tube to the statue, running it into an adjoining room. When a person spoke into the tube, his voice emerged from the mouth with a mysterious sound that these investigators concluded would have created a powerful effect on believers. To illustrate how this would have worked, David Abbott, an early twentieth-century magician and psychic investigator, said that he installed hundreds of feet of half-inch iron pipe underground. Those passing near the opening thought they were being addressed by someone nearby and looked around in a foolish manner in search of the speaker.[33] Abbott tested speaking tubes because they were used by hidden observers who passed information to stage clairvoyants of the nineteenth century.[34] Cutting-edge, battery-powered communication devices replaced tubes when that technology became available.[35] A similar megaphone funnel was discovered at the temple in Corinth. At the door of the passageway, an inscription from about the fifth century BCE warned any potential trespasser of dire consequences for intrusive curiosity.

Hippolytus, bishop of Rome (d. 249 CE), described a pagan priest who had his supplicants lie down near a speaking tube made from brass or the windpipes of cranes.[36] When the priest entreated a demon to reveal himself, a frightening voice emerged. Similarly, speaking tubes were placed inside skulls. Hippolytus also revealed the technical details for creating a speaking statue made of asphalt, gypsum, and cowhide.

When early Christian priests tore down Greek idols, they discovered speaking tubes. Not willing to let a good secret escape their use, priests used the idea to construct speaking statues of their own. Some statues wept or spoke in coordination with a moving jaw like that of a ventriloquist's dummy.[37]

THE INVISIBLE GIRL

Baltus dismissed speaking tubes, but generations of sophisticated people have been amused and baffled by this secret method of transmitting sound.[38] At the end of the eighteenth century, a former priest created a sensation in an old French Capuchin convent where he displayed the "Invisible Girl."[39] A small box was suspended by four brass chains in front of a window. Glass panes in the front and rear of the box showed that the box was empty. A speaking horn emerged from the front of the box, extending out toward visitors. Questions spoken into this horn were answered with surprising clarity by an enchanting voice, and any object displayed nearby was described. When the editor of the *Gazette de France* learned about this exhibition, he immediately visited the convent to conduct a personal investigation.[40] He was sufficiently cautious to make certain his host was not tricking him with ventriloquism. After the attendant left, "we spoke so low to the Invisible, that it was altogether impossible that any other person should hear what we said." He concluded that "electrical virtues" could explain the voice, but he could not account for how it described objects and people in the room. Thus, he was reduced to considering the possibility of an invisible dwarf or perhaps a dwarf hidden behind the horn where he could not see. But he was wise enough not to pretend that his guesses provided the proper solution of the mystery.

The explanation that evaded the editor was subsequently revealed in a couple of pamphlets. One was undoubtedly written by Etienne-Gaspard Robertson, the former priest who created the illusion to promote science, like Fontenelle, as an alternative to superstition and religious fallacy.[41] Later he staged a phantasmagoria (an early version of motion pictures) of ghosts in a cloistered chapel surrounded by the tombs of monks.[42] A playbill professed his intention "to expose the practices of artful impostors and pretended exorcists, and to open the eyes of those who still foster an absurd belief in ghosts or disembodied spirits."[43]

The Invisible Girl worked because the speaking horn projected sound from the back of the box, through the air, and into another tube that ran to an upper level where a young girl could hear and respond. She could see what was happening below through an opening in the floor that was cleverly disguised by a lamp on the ceiling.[44] Robinson wrote his exposure after another "Invisible"

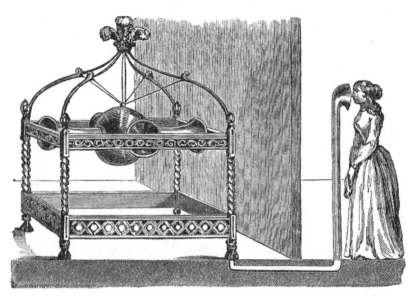

Invisible girl illusion. J. H. Pepper, *Cyclopaedic Science Simplified* (London: Frederick Warne/Scribner, Welford, 1869). *From the author's collection*

was displayed in Paris.[45] Surprisingly, his revelations did not dampen interest in the illusion, which continued to be popular for three decades.[46]

OPENING SEALED MESSAGES

When Fontenelle claimed that priests could open sealed messages, his critics were certain such skills required the assistance of demons. Baltus again declared that Fontenelle presumed extraordinary cunning of priests and foolishness of inquirers.

> They who consulted Oracles by sealed letters were mistrustful persons, who took this way only to avoid being imposed upon, and to try even to impose upon the oracle, if they could.[47]

Baltus insisted that even devils had difficulty reading sealed messages, so they often provided ambiguous responses because they could not see the future.

> The Devils therefore being thus ignorant of what is to come, in order to conceal their ignorance, were obliged to involve their oracles in affected obscurities and ambiguities, by means of which they might be accommodated to several quite different, and sometimes even contrary events. Thereby, as the [Christian] fathers have observed, they made sport with the credulity of the heathen, they miserably seduced them; and whatever might happen, they always appeared to have foretold the truth.[48]

Baltus said that despite sometimes relying on trickery, devils were usually clear in their messages because they could describe in one country what was seen in another. This was how they discerned that Croesus was boiling a turtle in his brass cauldron.[49]

Henry Beaumont, an English scholar, insisted that opening sealed letters would be impossible.[50] Even some modern historians, like H. W. Parke, have doubted the feasibility of opening sealed messages. He echoed Baltus by saying it was unlikely that "magistrates of a Hellenistic city lent themselves to such a naive imposture."[51] But both Parke and Baltus knew about Alexander, the wandering prophet who used the windpipes of cranes to speak through his snake. Alexander *requested* that questions be sent to him in sealed envelopes.[52] He took an impression of the seal with a soft mastic, which he then used to re-form the seal. Other fortune-tellers of that time similarly instructed those seeking opinions to seal their questions.[53]

Was Fontenelle right about priests reading sealed messages? Hippolytus provides a convincing answer that was not available to Fontenelle or Baltus.[54]

Hippolytus, an early bishop of Rome mentioned previously, challenged the philosophies and religions of his day, considering them unsatisfactory in comparison to Christianity. He debated with himself about whether to reveal the secrets of heathen priests because he feared some knave would attempt to practice these juggleries. However, he decided that exposure provided protection against trickery.

Hippolytus revealed how priests created the sound of thunder, demonstrated resistance to fire, performed tricks with candles, and established their authority by devising secret methods for killing goats and sheep at a distance. He said the deeds of pagan priests included "innumerable other such tricks" that fooled their followers.

Hippolytus's manuscript was lost until a Greek scholar found it in an ancient convent in 1842, and the first English translation appeared in 1868, about two centuries after Fontenelle first published his work. Of importance to us here, Hippolytus described several ways that early priests obtained messages from inquirers. In one method, the priest asked the supplicant to write his question on paper using water instead of ink. He then folded the paper so that nothing was seen by the priest, who burned the paper in a candle flame, sending the message up in smoke to the gods.

Here's what really happened. The priest secretly switched the paper to be burned while walking to the candle. The original paper was then exposed to heat to retrieve the message that was written in water infused with milk, urine, brine, or fig juice—better oracles through chemistry.

The priest was not suspected of cheating because he had removed himself two steps from discovery. First, the inquirer did not know that the "water" was invisible ink, and second, he saw the priest take his question directly to the flame. This method of deception illustrates the thinking of a talented conjuror, and Hippolytus shows us that ancient priests used crafty cheating for temple trickery.

Renaissance author Giovanni della Porta wrote an extensive essay on invisible inks followed by this attention-grabbing offer:

> It may be of great use when places are besieged, and in armies, and affairs of great men, to know how to open letters that are sealed with the general's seal, and signed with his name, to know what is contained within, and to seal them again, writing others that are contrary to them, and the like. I will show how.[55]

If the inquirer brought a sealed envelope, Hippolytus said the seal was removed by exposure to heat. The seal could be protected against breaking or melting if first encased in a concoction of hog's lard, hair, and wax. Similarly, a letter to Lucian of Samosata (125 CE–ca. 180 CE) mentions two methods of opening sealed letters,[56] and a recently discovered Italian manuscript from the seventeenth century explains this procedure in detail.[57]

Porta explained that a letter could be opened by sliding a horsehair under the heated seal.[58] He also revealed how to make a cast of the seal so that it could be duplicated and replaced on the letter or on a different document, thus thwarting the intended purpose of secrecy.

Opening sealed messages and the forgery of seals became increasingly problematic. By the fifteenth century, this political hacking was so common that governments were forced to protect themselves by frequently modifying (usually in small detail) their sealing stamps.[59] Seals with subtle alterations were earmarked for different purposes. William Pybus[60] recommended gluing different parts of a seal together so that if someone tried to remove it by heat, the different colors would melt together.

Finally, there is a small monograph written in 1920 by Burling Hull, *Twelve Sealed Message Reading Methods As Employed in Stage Mind-reading Acts*.[61] Hull and Hippolytus have revealed sufficient evidence to conclude that the content of sealed messages can be surreptitiously obtained in a way that fools the most intelligent and perceptive skeptic.

A few years after Hull published that 1920 monograph, he published a hardbound book entitled, *Fifty Sealed Message Reading Methods*.[62]

Fifty.

EXPLOITING THE
MYSTERIES OF NATURE

*Salting may be defined as the act of fraudulently increasing the
value of a sample of ore for purposes of deception, although it
may also mean any addition of valuable mineral in the sample
beyond what legitimately belongs there.*

—Herzig, *Mine Sampling and Valuing,* 1914

In addition to talking statues, we now know that the ancient Greeks also created statues that moved, usually as an answer to a question.[1] For example, priests directed questions to a statue of Apollo that moved backward if he dissented or forward to signify approval.[2] Early English natural philosopher John Wilkins said Aristotle described a wooden statue of Venus that was set in motion by the action of quicksilver (mercury), but Wilkins was disappointed that no further explanation of the mechanism was provided.[3] Most likely, quicksilver flowed through a hidden hourglass tube, slowly changing the center of gravity and thus precipitating movement.[4]

We know that another ancient statue moved after a small fire was kindled on an altar. Differences in air temperatures rotated the base on which stood human images in the posture of dancing. Whereas Egyptian statues had their eyes shut, arms hanging down, and feet joined together, Greek statues displayed open eyes with extremities suggesting animation.

It was Hero of Alexandria, though, who designed even more complex movement of statues.

THE WONDERS OF HERO[5]

Hero of Alexandria (10–70 CE) provides amazing insight into the wondrous devices employed at early oracles. His importance here is that he illustrates that things are not always what they seem, which can result in false beliefs and misunderstandings about the world around us. Hero was the first to harness the power of wind on land and the first to build a steam-powered engine, although it was used only for making puppets dance and a ball spin.[6] Hero designed clocks, engines of war, parallel gearwheels, automatons, and pneumatic marvels.[7] One of his inventions showed a knife slicing completely through the neck of a horse, without the head falling off. The mechanism was not duplicated again until French magician Robert-Houdin re-created it some two thousand years later.[8] Similarly, Hero constructed a coin-operated machine that dispensed holy water at shrines.[9] Nothing similar was seen until a coin-operated perfume dispenser was patented in 1889.[10] Hero is of particular interest to us because many of his inventions were designed to "produce amazement and alarm."[11] He brought together new technology and the finest craftsmanship to enhance the mysteries of temples and oracle sites.[12]

Hero's writings were lost until Renaissance scholars discovered sections saved by Arabs,[13] and Renaissance engineers soon installed pneumatic fountains described by Hero in the ostentatious gardens of the rich.[14] Hero's work pushed science forward all over Europe, and a book about his inventions, *The Mysteries of Arts and Nature*,[15] inspired Isaac Newton to study science.

Reconstruction of the Temple of Serapis at Alexandria revealed extensive secret underground tunnels that permitted activation of mechanisms described by Hero.[16] One of his miraculous fountains was discovered as far away as the Dodona oracle in northern Greece.[17]

Hero's "Shrine of Bacchus" was a marvel of ancient engineering, incorporating hydraulic, pneumatic, acoustic, and mechanical actions into one mechanism. Hero mounted a statue of Bacchus on a large pedestal.[18] In one hand he placed a staff that spurted milk, and the other hand grasped a decanter that poured wine. When the wine splashed on an altar, it spontaneously burst into flame. Bacchus himself was within a six-columned cylindrical temple with a conical roof topped with a statue of Victory. Four nymphs adorned the outside of the temple pillars. Once the apparatus was set in motion, the shrine moved by itself to a designated place and stopped. When the fire on the front altar ignited, drums and cymbals sounded as the nymphs danced around the temple columns. When they moved, garland wreaths unexpectedly encircled their heads. When the dancing ceased, Bacchus and Victory both turned around to face the opposite direction. Again,

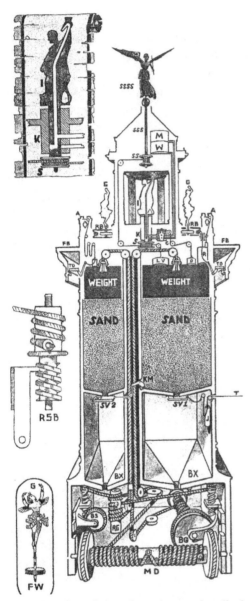

A moving altar, Shrine of Bacchus, as described by Hero of Alexandria. Charles B. Bunnell, "Hero's Automotor Dating 200 BC," *Science and Invention* 12 (1924): 662. *From the author's collection*

milk spurted from his staff and wine poured from his vessel, which ignited a second altar, all accompanied by music and dance.

The whole shrine was propelled by a mechanism hidden inside the large base that moved on three wheels connected by an axle around which a rope was wound that passed upward and over pulleys. The rope was fastened to a ponderous lead weight that rested on sand in a box with a hole that allowed the sand to slowly escape. The lowering weight moved the wheels' axle and thus the whole shrine.

The altar fires were lit when a metal plate was drawn aside, again by a weight, exposing combustible material to a candle burning below. The milk and wine came from two separate reservoirs hidden inside the roof of the temple. Tubes transported the liquids down the temple columns, under the floor, and up the statue to the hands of Bacchus. The sounds emanated from falling granules of lead that bounced from an inclined tambourine onto cymbals.[19] The garlands, guided by fine threads, dropped on the nymphs from a space hidden above in the temple soffits.

Once Hero perfected a specific working mechanism, he devised alternative uses in order to create other miracles. For example, the turning axle that moved the Shrine of Bacchus was redesigned to open the doors of the Temple of Isis. In this case, the shrine's double doors opened when an altar fire expanded air in a closed space. The air pressure pushed water through a tube into a suspended bucket. When the bucket became sufficiently heavy, it pulled cords wrapped around two spindles that functioned as the temple's door hinges.

A second book by Hero, *Spiritalia*, describes many ordinary devices like fire extinguishers and lamps. However, of the seventy-six "problems" considered in the book, twelve relate to the workings of magical altars and at least forty concern magical cups, pitchers, and sacrificial vases. All of his vessels contained concealed cavities that dispensed or retained fluid depending on the placement of the thumb over small openings. Think of placing your finger over the end of a drinking straw in a glass of water. Air pressure holds the liquid inside when you raise the straw, but it quickly runs out when you lift your finger. Hero used that same principle, but the critical tube was hidden inside the vessel, curving around to exit where the thumb would naturally rest on the handle. Thus, a priest could pour water, wine, or oil from one container on demand. Magic historian H. Adrian Smith said that Hero's vases are the most ingenious devices he has ever seen,[20] and a complete issue of the *Journal of Magic Research* was dedicated to calculating their efficiency.[21] We can only imagine the impact they produced at oracle sites where they were commonly used.

ORACLES ON FIRE

Many ancient oracles sites featured fire and flames among their attractions. For example, one priest devised a statue of his god that could mysteriously extinguish the flames of a blazing fire.[22] The secret, like many deceptions, was simple once known. The construction of the idol was ingenious. The clay statue was filled with water held in place with a large wax plug. Once the wax melted in the fire, the emerging water quickly doused the flames.

Saint Augustine described a lamp in the temple of Venus that burned continually through all seasons. He presumed the wick was asbestos, but he concluded the whole effect was probably the work of demons. Similarly, in ancient Greece and Italy, there are numerous descriptions of perpetually burning lamps (including Plutarch's mention of one at the shrine of Ammon[23]), but French mathematician and historian Jacques Ozanam, writing in the seventeenth century, believed these accounts were misunderstandings. The secret, Ozanam believed, was that crafty priests surreptitiously replenished the lamp with oil from secret reservoirs using hydraulic principles.[24]

Horace described a sanctuary where incense kindled itself in honor of the gods, and another in which an altar spontaneously ignited as a sacrifice to Jupiter.[25] These "divine fires" were believed to exclude the possibility of human intervention, but we now know that early priests had access to easily flammable materials, including naphtha. This highly combustible mineral oil was used at altars in areas where it naturally occurs. One naphtha flame is believed to have burned continuously for a thousand years,[26] and an everlasting fire near the Caspian Sea is said to have been burning since the time of Noah's flood. Several temples nearby are all built within a two-mile area where clay soil overlies naturally occurring naphtha.[27] Science scholar John Phin suggested that perpetual lamps are the counterpart of perpetual motion, both being impossible, yet the fire-giving wells of the Caspian Sea can be considered the counterpart of electric lighting from power generated by a mountain stream.[28]

Remember those lightning bolts said to have been thrown by Apollo at Delphi? These were likely phosphorus compounds.[29] Formulas for their production were disclosed only from father to son, and any inquirer was told that the secret could be obtained only by direct revelation from angels.[30] This was useful stuff. An army attacking Delphi would have traversed some narrow ravines where the crags and precipices would have been easy hiding places for a small squadron of soldiers to throw burning materials down on advancing enemy troops. Moreover, those same canyons had rocks that could have been detached to crush those below.[31] On the other hand, a lightning storm with flash flooding, common in that area, would have been sufficient to discourage any army into retreat. By whatever means it happened, the lightning-toss and boulder-drop events were subsequently woven into stories of miracles.

Since the earliest of times, lightning has been mimicked by the flash of lycopodium powder, the spores of certain ferns.[32] A similar effect can be obtained by throwing rosin at a flame[33] or by exploiting the flammability of various chemicals.[34] Thunder was produced by rolling iron balls down metal rain gutters,[35] pushing large stones down wooded planks, or shaking sheets of brass or iron.[36] Chemical detonations also imitated thunder;[37] a formula of saltpeter, salt of tartar, and sulfur makes "a fearful noise, like thunder, as loud as a cannon."[38] Hippolytus informs us that such explosions had an unnerving effect in caves and grottos that housed oracles. I believe him.

WALKING ON FIRE

Walking on burning coals has been a compelling sight since the time of ancient Greece. Devotees of Diana in Cappadocia received much veneration because of their ability to walk barefooted over burning coals, although one cynic said that he believed more in the preparation of their feet than the sanctity of their souls. Members of a certain family in central Italy performed this miracle annually in a local temple of Apollo, an ability that secured their exemption from military service and brought considerable tourist money to their city. Early fire walkers protected themselves against the pain by creating calluses, using chemical preparations with low heat conductivity, applying asbestos and soot, and relying on chemical unguents.[39]

The first scientific investigation of fire walking occurred in 1934 when Harry Price placed an advertisement in the *London Times* asking for volunteers.[40] He received hundreds of letters from people who said they had witnessed fire walking, but no one was prepared to step out and do it. Then in 1935, Price met Kuda Bux, a young Kashmir magician with a sensational blindfold act, which is discussed later. Bux stated that he had walked on fiery pits during religious ceremonies in India, and he agreed to submit to any test Price could devise. On Monday, September 9, 1935, three tons of wood blazed for hours before Bux declared conditions proper. His feet were swabbed for residue of any hidden substance and then washed. Bux said a prayer from the Koran, stepped onto the embers, and made four deliberate steps across the trench. At 430° Celsius (806° Fahrenheit), the pit was so hot that those who raked the fire were forced to use wooden shields for protection. Bux made three more trips across. His feet showed no sign of damage or ill effects.[41]

At this point, Digby Moynagh, editor of *St. Bartholomew's Hospital Journal*, expressed an irresistible desire to try the walk himself. Against everyone's advice but encouraged by Bux, Digby strode out into the inferno in his tweed slacks and sweater. He admitted it was extremely hot, but his feet were spared except for blisters that formed a half-hour later.

Fire walking can be safe if each foot is in contact with the embers for less than one second with a limit of about four steps for the total walk.[42] (Individuals who seldom wear shoes are better protected during fire walks because of calluses.) The temperature of the fire is not the critical issue; the key is an understanding of the transfer of heat. Burned wood produces carbon, which is a poor conductor of heat. Don't even think about walking on copper or aluminum, which are thousands of times more conductive.[43]

Because the underlying principles are not intuitively apparent, walking on fire has been exploited since the time of the Greeks to the present. Tony Robbins, late-night television salesman and self-appointed motivational guru, implies that he has special knowledge and motivational expertise for teaching his seminar participants how to walk on fire.[44] Like the priests of ancient Greece, Robbins prefers that the public buy (literally and figuratively) his belief system and ignore the scientific explanation.[45] Robbins knows exactly how to prepare the embers and when it is safe to walk, although these factors apparently were neglected when twelve lanes of fire were constructed for his walk at a San Jose workshop.[46] On that day, at least twenty-one people were treated for burns, including three who were taken to a local hospital. Again in 2016, about thirty people were hospitalized after a fire walk in Dallas.[47]

HOLY SMOKE AND MIRRORS

Mirrors are inherently fascinating, and scoundrels have devised many ways to exploit their use. Obsidian mirrors were available in Turkey in the sixth millennium BCE.[48] Polished bronze and copper mirrors were created in Iran about 4000 BCE and were later available in Egypt. Silver and gold gilding created mirrors for the ancient Greeks. Early Roman writers ridiculed the extravagance and vanity of the age, decrying every young woman's demand for a silver mirror.[49] Polished metals were sometimes formed into washbasins for purposes of grooming. These basins were used by wizards for scrying, which means looking into the past or seeing at a distance. Glass mirrors with a silvered back as we know them today probably originated in Phoenicia (Lebanon) but were available in Rome around the third century CE. Beckmann says the question is not when glass mirrors were first available but when they were first better than metallic reflection. His surprising conclusion is that efficient glass mirrors were not available until the seventeenth century.[50]

Hero understood that the mirrors he had available could be used in temples to create startling illusions.[51] For example, he instructed priests "to place a mirror so that one approaching it sees neither his own image nor that of another but only the image which we select." Pliny[52] (23–79 BCE) said that mirrors at the temple of Hercules projected images of gods into the sky, an optical illusion created by the concave mirror.[53] The concave mirror was also used to project startling images into a jar filled with water or into a cauldron.[54] This illusion is so convincing that a San Francisco nightclub used the principle to feature a nude girl in a tank of goldfish.[55]

Aesculapius was exhibited to his worshipers in his temple at Tarsus, through what was believed to be a projection lens (a primitive magic lantern) or a reflecting mirror. Vapors and incense smoke apparently create a satisfactory screen for a projected image,[56] but smoke can also provide cover for an actor representing a deity to mysteriously appear. We still commonly say that a puzzling occurrence has been accomplished by "smoke and mirrors."

Priests created images of the moon and stars on temple or cave ceilings through the use of a large dish of water with a mirror in the bottom. By lighting a candle above the dish, the mirror reflected light on the ceiling. Similar effects were accomplished by draping a dark cloth over a drum attached to the ceiling. A candle was placed behind the translucent head of the drum so that the priest could call down the moon by whisking the cloth away.[57] For stars, the scales of fish were stuck to the ceiling after being soaked in water and gum. The scales reflected light in a way that mimicked the night sky.

Hippolytus said priests attached a ball of pitch on a string to the leg of a hawk. On command for the spirit to appear, an accomplice ignited the pitch and released the frightened bird, which flew upward, whirling flames across the dark night sky. Hippolytus assured readers that dupes were horrified by the demon.

Hero and Hippolytus confirm Fontenelle's belief that priestly cheats were sufficient to explain the miracles at oracles. But did Fontenelle convince his readers?

THE SUCCESS OF FONTENELLE

Fontenelle wanted his book to cast doubt on demonology, which was so prevalent in his time, and he succeeded. He lived to see French courts reject the ravings of demonologists and stop consulting witch-finding manuals. Judges asked for evidence, and executions ended in 1745.[58] Many historians believe that cessation of such pervasive belief in witchcraft, magic, and demonology represents one of the more significant intellectual changes in human history.[59]

Fontenelle's existential funk ended after he wrote *The History of Oracles* and another book discussed in a later chapter. He lived an esteemed life, becoming secretary of the French Academy of Sciences.[60] The whole city of Paris celebrated his contributions to the intellectual culture of France on his one-hundredth birthday. He missed the festivities by one month.

Fontenelle's book on oracles became a modern narrative about emancipation from civil and religious tyranny, and he inspired others to fight against superstition and trickery. In 1828, another Frenchman, Eusebe Salverte, wrote *The*

Philosophy of Magic, in which he discussed how Christianity promoted prodigies and miracles to convince believers. Salverte used Fontenelle's strategy of providing natural explanations for "mysteries" that the church had exploited.

After Salverte wrote his treatise, Englishman Sir David Brewster published a similar book about the ways in which scientific discoveries were hidden from common people for purposes of maintaining power. "The prince, the priest, and the sage were leagued in a dark conspiracy to deceive and enslave their species."[61] Fontenelle, Salverte, and Brewster all came to the same conclusions: there are always unscrupulous people ready to exploit a mystery of nature, and they find uninformed and gullible followers.

We turn now to Lorenzo Valla, who confronted the question of what to do when handed a forgery that gave extensive authority to the most powerful living person.

II

REVISING AND REVISITING THE PAST

5

UNCOVERING
A FORGERY

*A great deal of falsehood appears to have got mixed up with a
very small amount of physical truth.*

—Brewster, *Letters on Natural Magic*, 1839

In the early part of the Italian Renaissance, Alphonse of Aragon proclaimed
himself king of Naples, but Pope Eugenius IV was not about to let this
Spaniard get his foreign foot in the Italian door. Eugenius had experienced all
the challenges to his papal authority he could tolerate, and he was determined
to keep this unwelcome immigrant out.[1] Alphonse's claim was based on a
dubious entitlement, and the pope held all the power because of a document
known as the Donation of Constantine. It gave vast temporal and spiritual
powers to successive popes. Alphonse was in need of some serious help, so
he reached out to Lorenzo Valla, someone who was also facing conflicts with
the church.

THE PENETRATING EYES OF A LYNX[2]

Lorenzo Valla (1407–1457) was a brilliant young humanist, a product of the
emerging Italian Renaissance. His personal goal was to become a papal secretary, but he was so young that his impressive university qualifications were
insufficient to swing the appointment. Three years later Valla's application to
work at the Vatican was again rejected. In response to this age discrimination,
Valla threw himself into the study of Greek and Roman manuscripts that were

being retrieved from monasteries and neglected libraries, often in places where monks had no appreciation for the treasures they were guarding. Valla became a popular university lecturer, but, typical of many humanists, he was usually at odds with the politics and course content of the schools he visited.

Valla gained a reputation for openly discussing controversial topics, which made him a person of interest to inquisitors. Their attention changed to rage when Valla reported his discovery that the Apostles' Creed was not written by the Apostles; it was constructed in the sixth century during the Second Ecumenical Council at Constantinople.[3] Valla was preparing for his confrontation with inquisitors when Alphonse, king of Naples, spirited him away to be his personal secretary, a position something like chief of staff. As indicated earlier, Alphonse had good reason to bring Valla to Naples.

Alphonse was in deep ecclesiastical soup. He told Valla to forget about the Apostles' Creed and find a way to block the pope's authority as the arbitrator of his entitlement to the throne in Naples. This was no small task. When Henry VIII wanted the pope to grant him a divorce (many decades later), his cadre of scholars spent months scouring the libraries of England to find documents that might support his questionable demand.[4] But Alphonse had recruited a first-round draft choice in Valla. He was a motivated player with a score to settle with Rome. Valla's skills set him apart from others. When Alphonse handed him the Donation of Constantine, Valla immediately knew that the document was a forgery. Valla understood the enormity of his conclusion. Exposing the Donation as a fraud would make his troubles with the Apostles' Creed look like a playground squabble. This was a game changer.

The Donation was purportedly a document that Roman emperor Constantine (ca. 288–377) gave to the church because he had been cured of leprosy by Sylvester, bishop of Rome. In grateful response, Constantine gave him the Lateran Palace, the city of Rome, Italian territories, western provinces, and other lands and responsibilities. Sylvester got Constantine's crown of pure gold with precious gems, as well as his shoes, felt socks, and other accoutrements of status and power. In sum, the document conferred vast temporal and spiritual authority to the Roman Church.

But, alas, the leprosy cure was a legend, and the Donation was a later revision of history by the church, just as the story of Croesus was a rewrite by Delphic priests. Others had suspected something was amiss with the Donation, but no one had the evidence or courage to act. Valla had both.

Valla knew he had better be right because the church had the power to make him regret his conclusion, even if he was correct. To appreciate the genius of Valla's response, we must first review how he accumulated the knowledge to

A detail from the Donation of Constantine painted in Raphael's workshop about one hundred years after Valla proved it never happened. Fake news from the 1520s. *Getty Images*

confirm the document as a forgery. We begin with a review of some history associated with the fall of Rome.

THE FALL OF ROME

Constantine, you may remember, was the first emperor of Rome to tolerate and even embrace Christianity. After the reign of Constantine, Roman civilization began to collapse in the turmoil of political, economic, and military changes that have been interminably discussed and debated by historians. Repeated invasions from northern tribes resulted in continual destruction of the city with terrible consequences for citizens.[5] In most raids, marauders grabbed horses and women while ignoring books and cultural artifacts. But those treasures not inadvertently destroyed were lost or disregarded as citizens struggled to survive. Although the term is controversial today, the next five hundred years became known as the Dark Ages—and for obvious reasons.[6]

Edward Gibbon suggested that the fall of Rome should be marked at 476 CE because that was the first time a foreign chieftain stayed to rule the city. In agreement, Michael Malone said that the country's rich cultural tradition simply disintegrated when Odoacer, that foreign chieftain, became emperor.[7] With the culture destroyed, scholars focused their attention on the Bible as the centerpiece of all knowledge. It was commonly believed that God himself had directly dictated the words; therefore, the scriptures were considered inerrant and self-authenticating. Other books were unnecessary, and reading those old Roman poets might even endanger one's soul.[8]

At the beginning of the Italian Renaissance, the old books of the Roman Empire were retrieved from forgotten places and reexamined. The old guard scholars (known as scholastics or schoolmen) were in conflict with the upstarts (humanists) who eagerly devoured this unsavory stuff. Because Valla had studied those old books, he immediately identified problems with the Donation that others dismissed. But before specifying what he observed, it is helpful to understand why some dismissed the treasures being recovered.

CURIOSITY ABOUT SECRETS

Augustine said that probing the nature of things, as the Greeks had, was not necessary because Christians only needed to know that God created all things. Investigating things that God had not revealed in the Bible was presumptuous.[9]

Curiosity involved prying into his secrets. It was meddlesome; it was "lust of the eyes." Augustine said that astronomers who predicted eclipses, for example, lost their sense of wonder because they diverted attention from God to themselves.[10]

Similarly, Thomas Aquinas struggled to determine what things were proper to be curious about and how much God would allow to be known.[11] Seeking knowledge about the hidden mysteries of nature might convey one into dangerous territory. Historian Stuart Clark[12] suggested that our modern mind-set must be completely set aside in order to grasp the significance of these basic beliefs. Curiosity was sinful because it was the desire to know that which was not proper to know. Anyone who ventured into this business of finding things out would most likely be tempted by the devil, who offered information for a price—one's soul. Perhaps angels could impart knowledge about the natural world, and imploring their assistance was the essence of "higher magic," later called "white magic."[13] But inquisitors did not approve any rituals that implored angels for knowledge because of the possibility that demons might show up pretending to be the good guys.[14]

In this bleak intellectual environment, lost manuscripts started to emerge from neglected libraries, and those old Roman documents revealed the history of a forgotten past.[15] Curious people like Valla were eager to learn what others rejected and even feared. The new humanists dismissed the idea that curiosity was sinful. They said the punishment of Adam and Eve was for curiosity about circumventing moral responsibility, not curiosity about natural knowledge.[16] Humanists were not willing to view history through the peephole constricted by scholastics and theologians. Scholastics pointed to the ancient ruins of Rome and proclaimed that all was vanity. Humanists observed the ruins of Rome and longed for the splendor of the old days that might enrich their lives, but their eagerness for freedom got them in trouble.[17] Scholastics wondered if that curiosity was really an interest in those old mythical religions. Maybe they had discarded Christianity. Suspicions started boiling when humanists discovered discrepancies in biblical translations, but inquisitors blew a gasket when Valla claimed the apostles did not construct the Apostles' Creed.

THE BATTLES BEGIN NOW

Renaissance humanism, unlike modern humanism, was not a rejection of religion but the rejection of clerical traditions, constrictions, and superstition. The humanists also dismissed the earlier belief that the body was the source of all evil and that the lust of the flesh was the worst of all sins. Instead of considering the body contemptable, they viewed humans as the pinnacle of God's creation, something marvelous and divine. And if the human body was the work of God, then so was beauty, desire, and sexuality. Valla brought this forbidden topic into the open with his *De Voluptate* (*On Pleasure*), in which he openly discussed free love. Valla wrote this essay about sex and alcohol as if he were arguing the current Christian point of view, but some of his contemporaries, and modern scholars as well, believe he was secretly embracing Epicureanism, which placed him more in line with St. Patrick than St. Augustine.[18] Once the body was viewed without shame, Renaissance artists, artisans, and architects could depict splendor never previously experienced.[19] Human passion and pleasure was acknowledged and accepted by some but feared and repressed by others.

Modern historian Erika Rummel reviewed the bitter exchanges between Renaissance scholastics and humanists, which are thigh-slapping funny from today's perspective.[20] They spread gossip and called each other names that would embarrass pirates. Scholastics were frustrated with humanists, who debated with metaphors and historical examples, whereas scholastics were determined to avoid fallacy by employing reason as outlined by Aristotle. Scholastics could endanger humanists, however, by labeling them as sophists, which easily became inflated to heretic. Indeed, some humanists working for the church secretly published anticlerical writings. In an attempt to stop their clandestine activity, the Vatican introduced *The Index of Prohibited Books*.[21]

Although scholastics viewed humanists as their worst enemies, scholastics themselves often disagreed with each other and frequently fought among themselves.[22] Similarly, humanists tangled with each other over inconsequential academic issues and trivial personal matters. Scholastics and humanists of every variation were employed in the papal court at the time of Valla. Modern historian Stephen Greenblatt provides exquisite details about how these men competed against each other for advancement, power, and the considerable profit available in papal appointments. Hundreds of notaries, theologians, lawyers, and secretaries arbitrated two thousand cases a week from all over Christendom. Serious and absurd conflicts were sent to Rome for settlement, sometimes while still being adjudicated in secular court. Additionally, the pope had a large entourage of courtiers, advisers, and high-ranking servants for his political and ceremonial assistance. But sadly, the Vatican was in a state of moral bankruptcy "in which crime, moral outrage, fraud, and deceit take the name of virtue and are held in high esteem."[23]

Poggio Bracciolini, personal secretary of the pope, unabashedly called this institutional morass "the lie factory."[24] No one was spared and no topic was considered off-limits for accusation or intimidation. Valla and Bracciolini became bitter disputants, and the older and more politically powerful Bracciolini might have ruined Valla if he had started his career at the Vatican as he originally intended. Their fight was often over matters of writing style, a topic that seems trivial today. However, Valla considered himself the authority on this topic, outlining the rules for Latin grammar in his *De Elegantiis Latinae Linguae*.[25] John Symonds says that Valla and Bracciolini spared no vituperation in their pamphlets and letters to each other, vomiting forth obscenities and libels never before seen in the light of day.[26]

It is easy to view the humanists as the heroes of this period. They brought learning to Italy, creating a home for knowledge and culture. In a world stifled by fear, they created excitement by uncovering the forgotten secrets of ages past. They revived classical Latin, in which learned individuals of all nations could mutually interact. Their conflicts with Aristotle opened new ways of viewing the world.

But Symonds reminds us that the humanists, for all their importance in the advancement of knowledge, were also denizens of no country but the fantasy land of antiquity. Although they opened the past, they were often ready to believe the most incredible nonsense. They were skeptical about obvious truths and devoid of reasonable circumspection. Many abandoned the moorings of the common Christian faith and wrapped everything in Greek and Roman terms, some even changing their names to sound ancient. Their goal was to persuade by eloquence and dazzle by rhetoric, but they often resorted to attack with vicious personal ridicule. They helped create a society that disconnected philosophy and moral conduct. Platonists and Aristotelians sometimes settled disputes with physical attacks. Rome became known as a sink of all things shameful and abominable. The city was filled with prostitutes and priests with syphilis, a situation not lost on Luther when he visited. Bracciolini himself fathered twenty children (that he admitted) while working as a papal official, some of them with his wife.[27]

Little wonder that inquisitors attended to theological disputations while ignoring behavioral improprieties: no one could cast the first stone.[28] Historical tradition also bent inquisitors toward censorship of intellectual activity. The early church fathers wanted the respect of contemporary philosophers, so they struggled to gain intellectual acceptance by engaging in disputations and scholarly trivia.[29] Further, the various sects of early Christianity fought each other about belief while the moral behavior encouraged by Jesus was mostly ignored.[30] Consider, for example, the consequences of how all this worked out in an event of 1476.

Galeazzo Sforza, fifth duke of Milan, was attending high mass for the feast of St. Stephen when he was assassinated.[31] His sudden death created a theological problem because he had not had the opportunity to confess his sins. To save his soul, his wife petitioned Pope Sixtus IV to grant a posthumous absolution. She gave the pope a list of his sins that need a little attention: acting like a tyrant, making war unjustly, sacking cities without mercy, robbery, extortion, negligence of justice, illegal enforcement of taxes from clerics, rape of virgins and wives, whoring, simony, and on went the list of his indelicacies. The pope had no reluctance in complying with her wishes, perhaps because Sixtus was known for his own peccadillos, such as selling indulgences on a previously unparalleled scale to support questionable activities. He was also responsible for the 1478 bull that started the notorious Spanish Inquisition associated with the brutality of Torquemada.

And speaking of the Inquisition, a death penalty was always a possibility if the church handed over an adjudged heretic to secular authorities, although the consequences of an inquisition were usually more limited to social and economic sanctions.[32] Nevertheless, excommunication was more than being kicked out of one's church. It could result in loss of land and property, prevent inheritance, invalidate a will, deny commercial and judicial services, and cause the loss of employment by decree or social ostracism.[33] Valla understood that picking a fight with the church was not a career-enhancing activity.

DAVID CHALLENGES GOLIATH'S ENTITLEMENTS

Valla had spent his career reading old Roman documents, and he knew that language and writing had evolved over the centuries. Thus, he immediately recognized that the Donation of Constantine was written in medieval Latin when the text should have been entirely in classical Latin. At least ten popes had cited the Donation of Constantine for the maintenance of their own power,[34] even though some had privately understood it to be a forgery.[35] The document had been challenged before the time of Valla, but something different happened when it reached his hands in the fifteenth century. Valla moved beyond complaining into exposing the Donation as forgery. Once he read the document, he realized his task was not merely to find loopholes for Alphonse but to show everyone that the document itself was not what it claimed to be.

Valla laid out his conclusions in a celebrated treatise that acknowledged that he was not merely writing against authorities of the past, as he had in his other scholarly work. His conclusions portended significant consequences for the pope who had immediate power over him. Knowing that he was at risk for excommunication or other untoward consequences, Valla began by citing church regulations that asserted the pope "may not bind nor loose anyone [in a way] contrary to law and justice."[36] Valla understood this legal constraint of the pope was a meaningless protection; nevertheless, he let it be known that his treatise was not merely an opinion but truth based on evidence. Before Valla, the assertions of the church were accepted as truth. Valla laid out another way of establishing truth: historical facts.

Valla reviewed the historical reality in which Constantine and Sylvester lived. He argued that Constantine did not have any authority to make a transfer of political power, and the church at that time was not capable of managing the governmental complexities of Rome, let alone all the other cities and provinces. Such a transfer could not have happened because the sociopolitical system would have been thrown into chaos. Constantine's sons, family, and all the citizens of Rome would have revolted. Finally, the church would not have accepted this gift because its mission was strictly spiritual.

Most scholars now agree that Constantine did little during his reign to support the church because his Christianity was more political expediency than personal conviction.[37] For example, his goal in calling the Council of Nicaea was to obtain unity among Christians, not to divine the correct doctrine to combat schism. He was never devout or penitent, and he ordered his own son's execution in a fit of concern about loyalty. Constantine was not baptized until the end of his life, when he had no more strength to sin or to kill anyone else, and the rite was performed by a heretical Arian bishop in distant Helenopolis.[38]

When Constantine died, Roman territory was divided among his remaining three sons and two nephews. His biggest gift to the church may have been his transfer of the Roman capital to Constantinople, which then enabled the church to fill the vacuum of power in Rome. This move allowed later Christians to promote the belief that Constantine had abandoned Rome, leaving the city in the hands of Sylvester.

Valla pointed out, contrary to the claim of the Donation, that Constantine did not have a bejeweled golden crown; he had a diadem made of cloth or silk. Valla then ridiculed clerics of his day who were wearing vestments that reflected power, wealth, and rank, when Jesus had presented himself humbly on the colt of an ass. The church, Valla insisted, should follow the message of Jesus to feed widows instead of squeezing taxes out of people in order to hire soldiers to wage war and to stop endorsing superstitions and creating fables. He said the church should denounce cheating like those speaking tubes hidden in statues that frighten the ignorant.[39]

Historians are now in agreement with Valla that the Donation of Constantine was created in the eighth century, when times were desperate for both the church and Rome itself.[40] Pope Steven II found himself without protection from the military attacks of the marauding Lombards. Rome, and all of Italy for that matter, was so unstable that Steven decided to seize political control. To explain his grab of temporal power, he directed some monks to construct a justification for his bold maneuver. Their solution was to forge a deed from Constantine that transferred political authority to the church.[41]

Valla knew that language changed dramatically after Constantine and the decline of Rome in the fifth century. However, those eighth-century monks who wrote the forgery were not fluent with the language used centuries earlier when Constantine was in power. The difference would be similar to comparing a contemporary document with one written by Shakespeare.[42] Valla was one of the few scholars who could easily distinguish writings of the fourth century from those of the eighth century. Thus, he immediately recognized that the Donation of Constantine was full of bloopers: "contradictions, impossibilities, stupidities, barbarisms, and absurdities." After reviewing one particularly fractured passage, he blurted out in disgust, "O sancte Jesu." (Oh, holy Jesus!)

You can hear Valla howling with scornful laughter down there in his cubicle. The language and grammar of the forgery violated the style associated with Constantine's era. He identified many anachronisms, words that could not have appeared in Constantine's era. For example, Constantine called Sylvester "pope" when that title had not yet been applied to the bishops of Rome.[43] References to the New Testament were from the Vulgate Bible, which was compiled by St. Jerome, who was born some twenty-six years after Constantine supposedly signed the document.[44] Valla piled on example after painful example. His work remains the seminal literary analysis of forged documents, a coherent body of knowledge about how to evaluate questioned writing.

One twentieth-century scholar concluded that the Donation of Constantine "has probably had more influence on the course of human history than anything else of human invention."[45] Given the subsequent development of western civilization, "one is disposed to wonder whether falsehood rather than truth has not had more permanent effect on the destinies of mankind."

CURIOSITY REVISITED

We noted that Valla was summoned by the Inquisition for his correct claim that the Apostles' Creed had not been constructed by the apostles. In fact, everywhere Valla looked, he saw mistakes, misunderstandings, mistranslations, and lies that no one else had noticed—or had discovered but had been ignored.

As an example, Valla wrote extensive notes as he examined old Greek texts of the Bible and fragments of scripture that Jerome had translated for his creation of the revered Latin Vulgate Bible. Valla discovered contortions of syntax and flawed interpretation so serious that he labeled them "barbarous."[46]

One of the mistakes Valla noticed in the Vulgate concerned the meaning of curiosity.[47] St. Paul told the Romans who converted to Christianity not to be proud or consider themselves better than the Jews. Linguistic confusion resulted in translating this passage of scripture (Romans 11:20) as a rebuke of intellectual curiosity instead of a condemnation of moral pride as Paul intended. Century after century, this section of Paul's letter was interpreted as a warning not to seek illicit knowledge of "higher things." Modern historian Carlo Ginzberg shows how this misrepresentation was incorporated into literature and art of the time. The restraint of curiosity became a medieval theme applied to religion, politics, nature, and the cosmos. Thus, as we see later, theologians feared that Galileo's search into things hidden in the heavens was a dangerous activity. Similarly, Lutheran Philipp Melanchthon accused Galileo of being motivated by the "love of novelty" instead of accepting the truth as revealed by God.[48] Such restrictive beliefs permeated the fabric of society.

NO TRESPASSING

Valla's comments critical of religion, especially the scriptures, were not appreciated by scholastics and theologians, a problem encountered again later with Galileo. Valla was informed that he had put his "sickle into this crop that was not his." His detractors said he was not a trained theologian, and he had no right to interpret the meaning of scriptures. Valla was told in no uncertain terms that these tasks must be left to theologians. Valla replied that their knowledge of Aristotle did not give them any special insight into the meaning of early Christian texts.

And speaking of Aristotle, Valla pointed out that scholastics were reading Aristotle's works that had been translated from the Greek into Arabic and then into Latin. He recommended they read the original Greek instead of their corrupted translations. They might then understand that one of the more popular and widely known works attributed to Aristotle, *The Book of the Secret of Secrets,* was a pseudo-Aristotelian book—a forgery.[49] This text purported to be the philosopher's teachings—his secrets—which he reserved for a few of his intimate disciples. The secretive style implied that those able to penetrate the enigmatic parables and riddles would have knowledge unavailable to common people, and the uninformed scholar would be kept busy chasing nonsense.[50]

Valla concluded that Aristotle was not worthy of the hero worship he was given by scholastics. Valla tweaked their noses by suggesting Aristotle had not accomplished that much for the benefit of mankind, such as giving public counsel, administrating provinces, developing therapeutic treatments, governing with justice, issuing decrees, writing history, or composing poems. He said Aristotle's followers were so devoted to him that they refused to acknowledge when he was wrong and thereby had closed their minds to the truth. He sarcastically mused that they were zealous about Aristotle because he was the only philosopher they knew. To add further insult, Valla suggested corrections and simplifications to some of Aristotle's theories. Others had enumerated the costs of spending one's life focused exclusively on Aristotle, but Valla pointed out that zeal for Aristotle had diverted focus away from exploring new intellectual territory.

VALLA VS. THE WORLD

It is perhaps no surprise that Valla was brought before another Inquisition in 1444.[51] However, his accusers never mentioned his condemnation of the Donation or his earlier problems with the Apostles' Creed; instead, they charged him with criticizing Aristotle and Boethius. Valla had enemies everywhere. Some said he was a dissolute satirist to whom nothing was sacred. Others said he had repudiated the faith of his fathers. The papal secretary, Poggio Bracciolini, denounced Valla as a coward, a notoriously immoral writer, and a heretic.[52] Nor did Valla gain friends in Italy when Martin Luther praised him and incorporated his criticism of the church into his Protestant cause.[53]

Looking at his life, most modern scholars have concluded that Valla challenged the idea that the classics and Christianity could be harmonized, and he did not claim Christianity was false. He believed he was helping Christianity by attempting to detach theology from Aristotle.[54] Unfortunately, his critics could not make these subtle distinctions,[55] nor apparently can the *Catholic Encyclopedia*, which says dismissive and degrading things about his scholarship and, ironically, his morality.[56]

Valla expressed serious disagreements with the way Christianity was practiced by the church. He will be remembered, however, for changing the course of history with his remarkable work on the Donation.

VALLA DIES AND GOES TO HEAVEN

In reviewing Valla's life, we have covered four controversial topics that he brought out of the shadows: (1) authorship of the Apostles' Creed, (2) exposition of the Donation of Constantine forgery, (3) identification of translation problems in the Latin Vulgate Bible, and (4) criticism of Aristotle and his followers. Not many historical individuals have addressed such a range of sensitive topics in which the stakes were so high. Valla asserted specific facts of history and stood by them despite the consequences.

A current eulogy for Valla would note that a few years after his death, scholars uncovered even more evidence that the Donation was counterfeit. And the problem of forgery went even deeper. Early church historians regularly replaced original documents with forgeries in an attempt to portray a more venerable history.[57] As a result, the Christian tradition is filled with confusion, starting with the gospels and epistles. Early manuscripts were falsely attributed to Jesus, Mary, and church patriarchs.[58] The scope of Valla's scholarship has, of necessity, continued to expand.[59]

Valla also set a new pattern for scholarly debate and historical reporting. He did not blindly cite authorities, assert meaningless rhetorical questions, or rely on formal reasoning. Instead, Valla spoke from his experience and knowledge of history, presenting evidence in ways that his detractors had not previously experienced. His learning was driven by curiosity untethered from theological

Valla spends his last days in the Vatican library. *Getty Images*

constraints. He did not explain away or excuse contradictions or problems in scriptures but stated what he observed, whether popular or not. When he asserted that the pope had to treat him within the law, he planted his flag in the ground of historical facts, ignoring whether it was in or out of current orthodox boundaries.

Valla was once chastised for mentioning that a king had fallen asleep and snored during the oration of an ambassador. A critic complained that Valla was indecorous to describe this royal indignity, which was improbable and damaged citizens' confidence in the king. Valla championed a new standard of historical reality. Whereas Herodotus told moral lessons of history, Valla focused on the reality of events. He declared: "I have not recorded what people ought to think but what they do think."[60]

After Pope Eugenius died in 1447, Valla was invited back to Rome by Pope Nicholas V, the first humanist pope and the official founder of the Vatican library. Nicholas said the only things worthy of spending money on were buildings and books. He sent his agents all over Europe in search of unknown manuscripts, Christian and pagan, and he hired forty-five copyists who created the nucleus of today's Vatican Library. Valla was hired as the Latin expert and entrusted with the task of translating Herodotus and Thucydides.[61] Not surprisingly, he also identified forgeries and discovered bogus texts from Alexandria. Valla's *On the Donation of Constantine* was placed on the list of prohibited books in the mid-sixteenth century because Luther continually cited it. But Valla would have laughed knowing that Poggio Bracciolini had written a salacious book that suffered the same fate.[62]

As Valla worked amid the splendor of the Vatican Library, he must have believed he had died and gone to heaven. But he was, in fact, buried in 1457.[63]

6

ASTROLOGY
AND PROBABILITY

*It is my aim, with God's help, to show from experience, from hu-
man reason, and from authority, that it is foolish, wicked, and
dangerous even in this life, to set one's mind to know or search
out hidden matters or the hazards and fortunes of the future.*

—Nicole Oresme, *Book of Divination*, 1350

66 **I**ntroducing in this corner, Gerolamo Cardano, mathematician, natural
philosopher, court physician, and visiting professor of medicine at Pavia,
Padua, and Bologna." *Drunken cheers erupt throughout the plaza.*

From afar you might think this was a boxing match, but Renaissance professors often challenged each other to solve mathematical problems in public
squares.[1] Cardano (1501–1576) was one of the best, and the competition
provided him a little extra money when he was travelling. His math skills were
decades ahead of everyone else's, so he kept some of his knowledge secret.
Cardano was confident in debating nearly any topic of the day because he possessed encyclopedic knowledge that he demonstrated with snappy insults and
rapier wit.

What better way to spend time on a Saturday afternoon before soccer
rivalries? In truth, these contests might not have been much different because
spectators often placed bets on the outcome and disputes sometimes ended in
brawls fueled by excessive alcohol.

When he was young, no one anticipated the amazing achievements of Cardano. He was the bastard child of a twenty-year-old mother and a fifty-year-old
father. His mother tried in vain to abort him with various medicines.[2] He was

physically and emotionally abused by both parents, and he was nearly given up for dead several times because of illnesses. His wet nurse died of plague, as did all of his siblings, aunts, uncles, and cousins.[3] His life was interrupted by war on at least three occasions.

Cardano's wife died, leaving him with three children under the age of twelve. One of his two sons was later tortured and executed for poisoning his wife, and the other was a petty thief addicted to gambling. Cardano lived in poverty and often in hunger until around age thirty-seven, but even afterward he was besieged with real and imaginary diseases, superstitions, irrational fears, nightmares, and dread of dying. Near the end of his seventy-five-year life he was imprisoned by the Inquisition on charges of heresy because he cast the horoscope of Jesus, a project that got mathematician and poet Cecco d'Ascoli burned at the stake in 1327.[4] It seems appropriate, therefore, that when Hamlet first appeared at the London Globe Theatre to deliver his famous soliloquy, "To be or not to be," he was holding the 1573 English edition of Cardano's *Consolation: A Dissertation on Grief.*

We concentrate on Cardano in this chapter because he promoted two different ways of finding knowledge about things hidden. He sought understanding through the casting of horoscopes, and, more importantly, Cardano showed that some future events could be determined in a way never previously considered—through mathematical probability.

INDEFATIGABLE LABOR

After relentless harassment by his mother, Cardano's father reluctantly agreed to pay his tuition for college. At the University of Padua, Cardano consistently embarrassed his professors so soundly in disputation that someone said he must have enlisted the assistance of the devil. Accusations like that continued to smolder around him for much of his life, and he did not help himself by mentioning that he sometimes communed with a genie in his dreams. However, he denied ever consulting the devil, and seventeenth-century French librarian Gabriel Naudé defended Cardano by asserting that demons were not necessary to explain his keen perception and vast learning. He acquired his knowledge and skill the old-fashioned way, through "indefatigable labor."[5]

Cardano invented a device for holding a compass steady on a ship; the suspension compass is still used as the universal joint in automobiles. He suggested improvements for practical use in lamps, furnaces, chimneys, irrigation pumps, and clocks. He created a cipher mechanism with cogs to code secret messages.

Cardano obtained a medical degree but was not admitted to the College of Physicians in Milan because of his illegitimate birth. As a result, he practiced medicine in a rural area outside of Padua, where his luck turned around after a friend inherited a printing press. Cardano now had an outlet for disseminating his vast knowledge, and his first publication detailed the errors of physicians.[6] This work rejected much of what he was taught and avenged his exclusion from his professional organization. Unfortunately, his book was peppered with grammatical errors, which provided an opportunity for his infuriated colleagues to humiliate him while ignoring the substance of his arguments. However, the book sold well because he revealed the truth about the dangers of medical practice at that time.[7]

In Cardano's day, physicians were trained in philosophy. They were expected to provide expositions of authoritative texts, and Cardano produced more than twelve hundred folio-sized pages of commentary on Hippocrates.[8] In contrast to others, however, he was critical of how his predecessors approached their responsibilities. His ideas became so popular that a senator accused the College of Physicians of excluding Cardano because of his superior understanding of the responsibilities of his profession. Despite Cardano's illegitimate birth, the senator forced the college to accept him into its organization, and his popularity spread throughout Europe.[9]

Physicians of Cardano's day reasoned out their treatments based on medical theories and expressed disdain for the "empiric" who was untrained in reason and thus relied on testing various remedies to see what produced the best outcome.[10] Cardano focused on outcome more than theory, learning from the successes and mistakes of his practice. He could see that doctors were often as dangerous to patients as the diseases they attempted to treat. Thus, he opposed bleeding because patients often died following this treatment. Physicians commonly boasted of their successes, especially with patients declared incurable by others. Cardano paraded his successes, but he also admitted that his cures for tuberculosis were wrong after he conducted long-term follow-up. Cardano believed the obligation of the physician was to determine a diagnosis, provide treatment, and suggest a prognosis. Philosophical discussions were irrelevant and should be avoided.

UNCOVERING THE HIDDEN MYSTERIES OF LIFE

Cardano revised medical care, and he also attempted to understand other areas of knowledge. A new conception of reality was needed, and Cardano headed out to find it, pushing forward against the heavy restraints of tradition. Before the scientific revolution, medieval and Renaissance intellectuals who had moved past fear of curiosity believed that everything they needed to know was available, waiting to be discovered.[11] Knowledge existed somewhere out there, probably complete and organized. Maybe angels or demons could be conjured into revealing their secrets, but to be worthy of secret information required personal preparation like fasting, prayer, celibacy, or living a holy life. In that process, some individuals endured intense rituals, creating "beatific visions" that were likely related to hypnotic states or even delirium. The ceremonies described in early magic texts suggest powerful emotional experiences associated with feelings of oneness with God and nature.[12]

But there was no consensus about where this knowledge was hidden. Astrologers studied the connection between heaven and human affairs, certain that everything could be understood by exploring messages in the sky. Hermetic sages looked to the macrocosm to learn about the microcosm. Humanists thought lost knowledge had been saved in ancient pagan documents and, along with alchemists, searched for authors like Hermes Trismegistus.[13] Theologians believed knowledge was lost when Adam was evicted from paradise or when nations were dispersed with different languages at the Tower of Babel.[14] Or maybe philosophy and science were intentionally destroyed in Noah's flood, and now everything needed was provided in the Bible.[15] Even the new Protestants were enchanted with the past. Luther, Zwingli, and Calvin had rediscovered the "real" meaning of the gospels.[16] Jewish kabbalists searched their sacred texts for the hidden path to knowledge, which gave Christians the idea that they could similarly find secret messages within their own scriptures.[17] Some held a contrary view, believing that Noah's three sons each went a different direction after the flood, dispensing wisdom to the known world. Thus, scholars concluded that art, science, and religion were historically linked, and the truth was dispersed in various cultures.[18]

It is difficult for us today to appreciate how these belief systems limited the understanding that opinion needs to be confirmed by reality. For example, early physicians were as confident in the early works of Galen as scholars were of Aristotle. In fact, dissectionists were so confident in Galen's texts that they ignored differences between what he said and what they observed inside human bodies.[19] When Vesalius (1514–1564) attempted to point out these discrepancies, he was labeled arrogant, ungrateful, and a monster of ignorance.[20] Rather than admit Galen was wrong, critics of Vesalius declared that men were physically different during Galen's times. We see here otherwise intelligent individuals clutching their belief systems and talking nonsense rather than confronting and learning from the reality facing them.

No medieval philosopher had a goal of making progress through study of the natural world.[21] Mathematicians were stymied because few ancient texts were available, and formulas in cuneiform text were difficult to decipher. Comprehension was best when concepts were transmitted directly from teacher to student, and those relationships had broken down.[22] Astronomers believed Ptolemy's great work was unlikely to ever be surpassed. Renaissance artists made progress in their craft, but their attention was directed back to the more perfect past, which they depicted in the idealistic beauty of ancient mythology and Greek landscapes. Humanists reviving the Greek and Roman writers sparked intellectual activity, but their vision was in praise of antiquity.[23]

This attachment to the past and reliance on authority was challenged when early explorers returned with artifacts and observations of new lands.[24] Uneducated seamen brought back plants, animals, and cultural artifacts—mysteries of nature—that were unknown to those revered ancient authorities. Compelling eyewitness reports changed the understanding of "truth." Over time, experience came to be valued more than opinion. Verification of an idea evolved from citing ancient authorities to relying on dependable observation. A "fact" now implied a statement confirmed as trustworthy. For example, habitation at the equator was established as a fact by Portuguese adventurers. Aristotle was wrong when he declared that it was too hot to live in the tropics. Philosophers had reasoned that people could not exist on the other side of the Earth, because things would be upside down and fall away from the Earth.[25] Similarly, Augustine declared that the lands across the sea could not be populated because the descendants of Adam could not have travelled there. Theologians insisted inhabited landmasses must be connected because the disciples of Jesus preached to the people of every nation.[26] Cartographers constructed maps accordingly,[27] but sailors rejected maps based on prevailing opinion; they demanded guidance based on reliable information. The voyages of Columbus and Magellan created subtle but significant changes in thinking by undermining the opinions of theologians, philosophers, and Ptolemy's geography. At least for some people of that day, experience became accepted as the ultimate authority.

Cardano lived during this time of transition. For the first time, ships sailed away from coastlines and into unknown climates; the size of the Earth doubled.[28] As he explained, "I was born in a rare century, which has come to know the whole world."[29] He delighted in describing new objects without an emphasis on the glory of God. Instead, he was determined to show that surprising things and events have natural explanations. He said that variations in climate and soil could produce plants and animals in Africa that were strange to European eyes.

An unusual object no longer needed to be considered a prodigy; there was no need to speculate about its theological meaning.

Cardano urged intellectuals to turn their attention away from ancient authorities so they could attend to what their senses perceived. Important secrets were hidden in every conceivable subject. All branches of knowledge had territories yet unexplored, much like the globe itself.

Despite his vast learning and forward thinking, Cardano was still a man trapped in a time of uncertainty. With no clear path to follow when looking outward for knowledge, he resorted to searching for secrets in the writings of ancient authorities. His books included dissertations on topics associated with superstition and fanciful thinking: chiromancy (palm reading), geomancy (divination through patterns), cabalistic symbolism, physiognomy, and the interpretation of dreams, which he believed were coded messages that could reveal secrets.[30] And astrology.

ASTROLOGY

The early church fathers, especially Augustine,[31] universally condemned astrology. Hippolytus, whom we met previously, described the physical and psychological characteristics believed present in persons born under each of the twelve signs of the zodiac. His astrological profiles are so vivid that they would appear contemporary if printed in tomorrow's newspaper. However, Hippolytus ridiculed those who presumed that heavenly bodies could influence one's appearance or conduct. Why, he asked, should we believe a constellation has any relationship to an individual? It would be foolish to believe that all people born under Leo embodied the characteristics associated with lions.

Astrology returned to the foreground of Italian life at the beginning of the thirteenth century and exerted an unrelenting grip on the most learned intellects.[32] Thomas Aquinas (1225–1274) promoted astrology for purposes of meteorology, medicine, and some physical processes. Subsequent astrologers concurred that God worked according to a plan, which was apparent from their observations of the heavens, so they searched this "natural theology."[33] Astrology was the science that complemented the revelations in scripture. Both illuminated the hidden path.

Pope Paul III, who unsuccessfully solicited Cardano to be his personal physician, was so concerned about the influence of the stars that he entered into no activity without proper astrological assurances of safety.[34] Pope Leo X believed his support of astrology was a credit to his pontificate.[35] Nevertheless, the whole issue of prognostication was problematic for the church. King Solomon and the prophets were diviners of the future, and Joseph (with his coat of many colors) saved the Egyptians from starvation when he predicted a famine.[36] The stars and planets announced the birth of Christ, but foreknowledge was the prerogative of God alone.[37]

In Cardano's time, those trained in astronomy and math were expected to construct horoscopes, and he made considerable money by telling the wealthy what they wanted to know. Moreover, Cardano pulled another clever publicity stunt by writing a book about the proper method for interpreting the horoscope. This positioned him as an expert, just as his book on medicine expanded his reputation.

In one section of his treatise, Cardano wrote something like *The Secrets of Successful Astrologers*. He laid out strategies that astrologers could use to convince people to put cash on the line for a reading. For example, he advised the astrologer not to tell his client that he was in danger. Instead, the wording should imply a *risk* of danger or give a *warning* of danger.[38] Risks were everywhere during the Renaissance, so almost any negative occurrence could be construed as a fulfillment of the forecast. Priests at Delphi would have applauded Cardano's ability to throw the ambiguous curveball.

Cardano also recommended that the clothing and manners of the astrologer emphasize dignity because people judge the credibility of a horoscope by their impression of the astrologer as much as from the reading itself. He recommended against giving readings to skeptics or those who would put the astrologer to the test. He also advised against a public practice in which one's work would be made widely known. A whole city might laugh at the astrologer's mistakes.

These admissions of Cardano suggest that he knew exactly what he was doing when he prepared a horoscope, namely capitalizing on clever wording and personal appeal to exploit customer gullibility. However, the truth is more complex; bright people like Cardano can hold contradictory beliefs. Cardano appeared to have unbounded faith in astrology until the end of his life. He failed to understand that astrology was merely the context in which he could expound his knowledge and present his recommendations to help people manage their future.

THE DECLINE OF ASTROLOGY—AND CARDANO

It is tempting to suggest that Cardano was responsible for the decline of the astrology he promoted. One could make the case that marine insurance—based on probability, which Cardano helped establish—provided better protection than astrological reassurance. Moreover, the probability of a safe return increased with better ships and navigational systems, like Cardano's compass that remained level on choppy seas. However, marine insurance was available before Cardano's mathematics of probability,[39] and people consulted astrologers about whether to buy marine insurance. Astrologer William Lilly's casebook contains several entries relating to clients who asked him whether or not to buy insurance, and he advertised his success in this endeavor.[40] Consider the anxiety each time a ship left an Italian port in quest of spices and exotic goods. The ship's owner worried about his extensive investment, and families realized the possibility of never seeing their sailors again. The *assurance* of an astrologer was likely more comforting than *insurance* on parchment.

What, then, caused astrology's fall from favor? Modern historian Keith Thomas suggests that astrology became less convincing as the geocentric view of the universe broke down, which did not happen until after Cardano. The Earth, as envisioned by Copernicus, was reduced to the embarrassing status of a planet. Astrologers scrambled to revise their theories based on the findings of astronomers, but the two worldviews were light-years apart. Philosophers had ridiculed astrologers since the earliest times, but now the penetrating arrows of satire hit the charlatan with deadly blows. The "doctor of the seven lazy arts" predicted with cold sarcasm that "more ice will appear in winter than in August."[41] In 1620, John Melton defined astrology as

> An art, whereby cunning knaves cheat plain honest Men, that teach both the theory and practice of close cousenage [deception], a science instructing all the students of it to lie as often as they speak, and to be believed no oftener than they hold their tongues; that tells truth as often as bawds go to church, and witches or whores say their prayers.[42]

A stargazer of 1680 said:

> Some astrologers will be so bold as to expose their ridiculous conceits in the eyes of the world, and before persons of judgment and authority; which may prove their ruin beyond hopes of recovery.[43]

The granddaddy of all astrologer parodies was concocted as an April Fools' joke by Jonathan Swift.[44] Swift (1667–1745) was disgusted with astrologer John Partridge for his talk-show spewing of violent predictions. Swift placed Partridge on the receiving end of a prediction.[45] Writing under the pseudonym of Isaac Bickerstaff, Swift published a bogus astrological horoscope of the death of Partridge. The day after his supposed demise, Swift published a forged pamphlet in which Partridge admitted that his astrological writings were all an imposition to make money. Partridge naturally denied everything, which gave Swift an opportunity to question why anyone would want to claim to be that dead charlatan.

Astrology's scientific pretenses were gone.[46] Knowledge came from telescopes, not from astrologers consulting horoscopes. Universities banished astrologers, and medical astrologers quietly disappeared. Cardano's astrology and occult practices made him an easy target of derision as time passed. Despite his considerable achievements, he almost disappeared from public awareness. However, his name is known in a couple of minor historical niches: gambling and statistical probability.

GAMBLING AND SUPERSTITION

Cardano supplemented his income with gambling, and he probably made more money betting than he did from his first job as a physician near Milan.[47] He acknowledged that he became addicted to gambling while in college, which caused him to neglect learning and responsibilities.[48] He gambled with kings and paupers. He knew the lowlifes who constructed crooked dice and the card sharps who cheated the unwary.[49] He learned the card manipulation tricks of gamblers as well as their psychological ploys. He described how cheats marked the sides of cards, soaped the surface of critical cards so that they could be identified by their glide when dealt, and identified cards through the use of a hidden mirror. In the middle of the sixteenth century, he exposed card tricks that still appear modern today.[50]

Cardano also understood a deeper truth about gambling. He was apparently the first to say that gambling can become an addiction or a chronic disorder. He provided a medical conceptualization for what everyone else labeled as a moral deficiency.

> Even if gambling were altogether an evil, still, on account of the very large number of people who play, it would seem to be a natural evil. For that very reason it ought to be discussed by a medical doctor like one of the incurable diseases.

Despite his confidence in astrology, Cardano admitted that he had never known an astrologer who was lucky at gambling or, for that matter, anyone who was advised by an astrologer faring well at gambling. The astrologer might occasionally make the right forecast, he said, but after a correct guess the gambler would immediately thereafter go wrong because success makes people more venturesome whereas the path to subsequent error is always steep.

Instead of consulting constellations, Cardano endorsed pearls and onyx as lucky charms, and he recommended taking a position at the table that faced a rising moon. Cardano was certain he had remarkable luck due to supernatural influence, but his discussion of luck is not convincing. Gamblers have difficulty avoiding the lure of superstition, and Cardano was no exception. He remained severely superstitious in many different areas his whole life. However, he had good reason for some of his success. Cardano understood probability as no one ever had before.

LUCK AND PROBABILITY

Cardano wrote his first book on mathematics at the age of nineteen. He loaned the manuscript to a friend who promptly died of the plague and Cardano never got it back. His greatest mathematical accomplishments were presented in 1545, *The Book of the Great Art*. Biographer Henry Morley said no algebra book of more importance ever appeared during Cardano's lifetime, and it was only one of twelve books he published that year.[51] Cardano introduced cubic equations, rules for biquadratics, and algebra for the resolution of geometric problems. Most of his mathematics explored practical matters including the first explanation of double-entry bookkeeping,[52] and he was the first to introduce the word "million" in print. His claim to mathematical fame, however, rests on his study of statistical probability.

The Greeks associated chance and inexplicable events with the whims of the goddess Tyche, and the Romans accepted her as Fortuna, similarly intervening capriciously and indifferently in the affairs of mortals. Fortuna remains with us today, and Frank Sinatra sang his paean to her in "Luck Be a Lady Tonight." John Calvin thought Fortuna (and chance) were a myth. But Cardano showed that Fortuna intrudes her whims according to rules of statistical probability. Knowledge of probability created an opportunity to explore the mysteries of nature in a profoundly different way.

Cardano's pamphlet *The Book on Games of Chance* had little impact on mathematicians because they considered it to be hardly more than the hodgepodge of a gambler's manual. We now know, thanks to mathematician and Cardano biographer Oystein Ore, that Cardano's accomplishments went beyond the first simple steps of probability.[53] Cardano understood the fundamental principles of large numbers, repetition of events, and complex issues relating to chance with cards. Ore concluded that despite some shortcomings and mistakes, Cardano's erudition was remarkably successful and a great step forward, particularly when he lacked the symbols that Descartes would later introduce.[54] Previous attempts to explain games of chance were erroneous and showed no grasp of the basic ideas of probability theory. No further advance in probability would occur until the notable work of Pascal a full century later.

Cardano captured some observations that now seem perfectly obvious. For example, when throwing a single die, each of the six sides has an equal chance of appearing. When two dice are thrown, there are thirty-six possible outcomes (6 × 6). The face sum of two can occur only if each die falls as one. However, a sum of three can result in two ways: the dice may fall as one and two or as two and one. A total of seven can occur in six ways. Therefore, the odds of throwing two are not equal to the odds of throwing seven.

A gambler might have sensed that betting on a throw of seven was more advantageous than betting on a two, but the knowledge of mathematical expectation did not exist until Cardano.[55] Even so, Cardano might have conceived the results as a propensity rather than a fixed natural law. Nevertheless, his mathematical knowledge and intuition undoubtedly gave him an enormous advantage in placing bets.

Cardano also worked out the probabilities when throwing three dice, which was common in games at that time. His comprehension was not as extensive when he turned his attention to cards. However, he recognized an important point that card counters exploit: as more cards are drawn, the probability of a correct guess increases.[56]

RULES FOR FOOLS

The inevitability of statistical probability quickly moves beyond the obvious into situations that are counterintuitive. Once that boundary is crossed, a whole new territory is open for exploitation. Gamblers, magicians, and con artists have devised wagers that appear fair on the surface but always result in loss in the long run.[57] We are easily lured by the possibility that we might be able to exploit luck to make an exchange to our own advantage. We are defeated by our own ignorance in the heat of gambling, like the fool advised by the astrologer. We believe in the simplicity of "runs" and that some people attract luck better than others.

More dangerously, we inject personal belief into situations that are indifferent to our expectations. In gambling, we believe that success after a run of failures means we will now be rewarded for staying with the game. This ability to create meaning and expectation is what makes us human.[58] Our brains are overly efficient in helping us make sense of the world. Problems arise, however, when we find meaning where none exists. Herein lies the most common cause for belief in the paranormal. We are biologically programmed to create meaning, especially in an emotional context; inanimate objects and random events evoke explanations based on presumed forces.[59] When something unusual happens, we ignore the fact that things just happen—good and bad. We did not evolve to accept mathematical odds, which is why someone of Cardano's genius was necessary to figure out how Fortuna was playing her game.

Numbers and mathematics emerged early in human history and developed over the centuries, but probability theory was not meaningfully discussed until Cardano. Pascal, Huygens, and other European savants following Cardano maintained an extensive correspondence about how small changes in wording and assumptions require different statistical procedures. Probability was not incorporated into research until the nineteenth century, and a book on proper procedures for designing an experiment did not appear until 1935.[60] Scientists and mathematicians continue to explore statistical predictions, and the context of gambling remains a helpful paradigm.[61]

Individuals can live meaningful lives without understanding probability theory. We don't need to be right all the time, just enough to survive. But beware. If a sucker is born every minute, then someone exploiting probability is born every hour to fleece those sixty suckers. Only an epidemiologist can estimate how many statisticians must be born to count the swindlers and educate the suckers.

7

NATURAL MAGICK

The aim of magic is to recall natural philosophy from the vanity of speculations to the importance of experiments. . . . Although things lie buried deep beneath a mass of falsehood and fable, yet they should be looked into.

—Roger Bacon, *Novum Organum*, 1628

Giovanni Battista Della Porta (1535–1615) was raised in a wealthy Naples family that could afford good homeschool tutors. He was writing essays in Latin at the age of ten,[1] and in his early twenties he wrote his first book with the provocative title of *Natural Magick.* Thirty years later (1589) he published a major update, which now becomes the focus of our attention. His stated goal was to reveal the mysteries of nature and the secrets of men.

Porta acknowledged the assistance of colleagues in preparing his second edition. These were so-called men of leisure who joined his Academy of the Secrets of Nature. But in the sixteenth century, inquisitors had their eye on anyone meddling in "magick" and secrets. Why, then, did Porta include "magick" in the title of his book and reveal that his academy was interested in secrets? It was like waving a red flag in front of the inquisitional bull. The attack was inevitable.

To answer these questions, we begin by considering the role of Italian Renaissance academies, which we encounter again in later chapters.

ITALIAN ACADEMIES

In sixteenth-century Italy, intellectual life was still mostly focused on the past. University students debated questions that had little relationship to reality in a style formulated by Aristotle. Philosophers looked for answers in the writings of the ancients; theologians believed everything one needed to know had already been revealed. Humanists focused their attention on the newly discovered texts of the ancient Greeks and Romans. Even the spectacular advances in art were expressed in paintings of the idyllic past.

Yet Italian cities began to flourish with advances in architecture, crafts, trades, and commerce. The wealthy collected treasures brought back from explorations of the new world. With better economic conditions, people were less dependent on the church, paying more attention to this world than the next. Nevertheless, the church and state exerted considerable control over the personal lives and thoughts of citizens. Porta fought for the right to search out his own truth without restraint by religious authorities or regard for the opinions of ancient authorities. He suggested that curiosity was not something to be feared; curiosity inspired satisfaction and pleasure by revealing knowledge. These basic rights that we take for granted got Porta into trouble—and even tortured.

As a way of discussing sensitive topics in Porta's time, like-minded individuals throughout Italy gathered in informal academies to discuss the changes around them and newly emerging possibilities. They created a milieu of openness, tolerance, and intellectual stimulation, which is necessary for the development of science.[2] This social environment was a potential context for religious doubt and disbelief. Modern historian David Wootton suggests new ideas and cultural diversity create questions about which belief, if any, is true.[3] For this reason, secular and religious authorities were concerned that academies might be discussing the emerging Protestant Reformation, those old pagan gods, or the failed theocracy of Savonarola, a charismatic Dominican friar who ruled Florence in the 1490s. These private enclaves were viewed as a threat to social, political, and religious stability.[4] Therefore, most academies disguised their purpose by identifying themselves as reading societies, which discussed innocuous topics like art or travel. They gave themselves whimsical names to divert the attention of authorities while secretly considering topics not commonly open to public discussion.

At first glance, Porta's academy would not appear to be a threat because of its focus on exploring nature. He was, in our view today, simply a moderately wealthy guy who felt the thrill of discovering the hidden mysteries of the natural world. To join the Academy of the Secrets of Nature, he required prospective members to present a new discovery, something not previously known or considered. But even the most innocent-sounding topics presented threats at that time. Curiosity about the natural world was considered meddling into knowledge that God had not revealed. A search for the secrets of nature suggested activities of magicians who might be tempted to enlist the assistance of devils.

NATURAL MAGIC

Porta said there were two kinds of magic: (1) sorcery, which has to do with foul spirits and consists of "enchantments and wicked curiosity," and (2) the hidden secrets of nature. He wanted nothing to do with spirits. Instead, Porta noted that when the ancients discovered something from dark and hidden places, they experienced their finding with so much pleasure that they called it magic. It was magical.[5] Magic and curiosity, he asserted, could have a wicked side, but he rejected the idea that men should avoid new knowledge and shun curiosity. He dismissed the old magic of enchantments and celebrated the new magician, someone who would legitimately be curious about:

> The effects of fire, earth, air, and water, the principal matter of the heavens; and what is the cause of the flowing of the sea, and of the divers[e]-colored rainbow; and of the loud thunder, and of comets, and fiery lights that appear by night, and of earth-quakes; and what are the beginnings of gold and of iron; and what is the whole witty force of hidden nature.[6]

Fire, earth, air, and water were the four elements that Aristotle used to explain "the principle matter of the heavens" in his *Meteorologica*.[7] Renaissance readers would have been acquainted with that book, but people were no longer satisfied with Aristotle's vapors, exhalations, and reflections to explain natural phenomena. Rainbows and thunder were given meaning in the scriptures, but their theological interpretation did not include scientific explanations: refraction in rain droplets resulting in a spectrum of light or expansion and contraction of air causing thunder. Comets and earthquakes provided opportunity for astrologers to blather about signs of doom, but that avoided their natural explanations. Alchemists raved about gold and iron, each with their own meaningless metaphors. Natural magic was the collection of scientific marvels out there waiting to be discovered. Porta was ready to search the "whole witty [rational] force of hidden nature."

In Porta's time, scholastics and theologians were satisfied to say that God—or the devil—was the responsible agent for something, thus ending all discussion. Nothing else was needed because that was a satisfactory explanation for them. They believed natural explanations missed the point, casting doubt about God's plan and purpose. For them, natural explanations failed to explain the ultimate reality or provide any transcendent understanding. Even further, they viewed natural explanations as a rejection of God. They were suspicious about scientific investigation because they conflated divination with exploration of nature's hidden secrets. Nicole Oresme said that alchemists exploring the secrets of nature are like the astrologists searching the secrets of fortune, and to make such inquiries they encounter the deceits of the devil.[8] Both activities were idolatrous—seeking knowledge instead of God.

Historian Basil Willey noted that individuals react to explanations according to their belief systems.[9] An explanation is satisfactory if it gives the experience of closure such that further questions are unnecessary. A good explanation creates a sense of understanding that was not previously evident, and confusion is no longer present. The Renaissance era was the first time that two different interpretations of an event or object were available: natural and religious. Both have their own appeal, and both can provide satisfaction. Porta was ready for the new magician to provide natural explanations.

Before Porta, religious and philosophical scholars kept ideas hidden in order to keep knowledge from those considered unworthy. When learned individuals captured some knowledge, they hid that information in metaphorical language or Latin to separate themselves from common people. In another context, one can see that with the rise of trades and craftsmen, a different kind of secret was hidden for a different reason. Guilds kept secrets in order to maintain their advantage in trade. Their secrets were hidden from potential competitors to ensure economic success.

Porta wanted to share all the secrets he could uncover. Books of secrets were common but useless. Most were filled with bogus medical advice or half-baked suggestions regarding health and beauty.[10] They were generally compiled by charlatans who established their credibility by flaunting their travels and the travails they endured while obtaining the offered recommendations. But the substance of their remedies were trivia if not pure nonsense.[11] Porta said he would share only things that really worked no matter where the information came from: "by most earnest study, and constant experience, I did both night and day endeavour to know whether what I heard or read, was true or false."

All the quacks, empirics, charlatans, alchemists, astrologers, and even the educated doctors and philosophers said "trust me." Porta said "show me." This difference that Porta emphasized is worthy of a brief illustration.

GET ME A BOTTLE OF THAT STUFF

Pope Felix V, who reigned from 1566 to 1572, was groaning in pain and partially paralyzed when he asked his physician, "What do you have to say about that medicine of the philosophers that they call the elixir?"[12] Felix wondered whether an alchemist might have a potion to ease his suffering. His physician, Guglielmo Fabri, did not respond to this question by locating someone with knowledge about this rumored universal medicine nor did he retreat to beakers and furnace. Fabri's education taught him to answer this question through disputation of theories, which had no underlying association to evidence. Thus, Fabri wrote out his response in typical form of the day: by reviewing the opinions of authorities. He considered the possibility of transmutation, the therapeutic efficacy of potable gold based on its "virtue," and the interpretation of occult terms associated with alchemy. Fabri's answer was designed to show his acquaintance with the topic, his ability to weave together interesting themes, and his knowledge of symbols. Transmutation, he said, was the purification of metal, the removal of "corruptible superfluities." Whoever possessed the elixir obtained infinite riches and could overcome all diseases. Fabri responded to the pope's question, but his answers were all metaphorical and allegorical. A satisfactory answer for Fabri did not relieve the agony of Felix.

Alchemists chased this folly because they believed that metaphors expressed the reality of nature. Unfortunately, these metaphors led to belief in imagined powers.[13] Medicine was imprisoned by a belief that all things had purpose and were endowed with virtues bestowed on them for the benefit of humanity.[14] The result was misleading conclusions, needless passions, unwarranted beliefs, and, as Felix would affirm, unremitting misery. Objects were given attributes from which cures were presumed through deductive reasoning.[15] None of this led to anything helpful except by chance.

Porta's book promised natural explanations with evidence-based medicine, not the spinning of metaphors.

WHERE'S THE BEEF

The first expensive book I ever purchased was Porta's first English edition of *Natural Magick* (1658). When I first started reading, I failed to understand the steps Porta was taking out of intellectual darkness in advance of his contemporaries.[16] Instead, his shortcomings jumped out at me. His testing of medical recommendations was particularly disappointing. He endorsed procedures if they were approved by authorities he trusted. I was disappointed that Porta passed on some real whoppers. After ridiculing the foolishness of ancient authorities, he reported with tabloid soberness the old legend about two mountains so rich in lodestone that they drew the nails out of ships that passed between them. And there were lots of others.

Despite my disappointment, I immediately appreciated the camaraderie of his academy as the members identified charlatans, impostors, and counterfeiters. For example, Porta exposed cheaters who adulterated food and common products like soap, oils, and musk. He revealed how wax was diluted with bean meal in order to make cheap candles and how to recognize brass forged to look like silver. It was like informing readers how to spot a fake Rolex. His colleagues experimented with invisible ink and formulas for the beautification of women; they staged magical illusions and devised strategies for altering one's appearance for disguise. They improved the camera obscura, warped mirrors for distorting images, and devised instruments for copying pictures and projecting apparitions. And, of course, Porta described how statues could be made to blow trumpets, eavesdrop on conversations, and scare the bejesus out of "ignorant men and simple women" through the use of hidden speaking tubes.

If nothing else, members of his Academy of Secrets must have had a grand time laughing at all of these secrets and playing practical jokes. Porta must have spent a fortune on the wine and pizza they consumed. And speaking of wine, Porta designed a goblet that turned wine into water when given to a buddy, all thanks to the action of specific gravity.[17] For a laugh a minute, he devised a vessel that he could drink from but that no one else could get a drop without knowing the secret. Even wackier, he constructed a glass that violently spewed wine in your face after a certain portion had been consumed.

It was easy to understand why Porta's book was so popular and important. Porta was not writing to the academic elite but for common people. Despite my initial disappointment, Porta really did primitive testing of some things, if only by trial and error. Nowhere is this more evident than with his study of lodestones; he claimed to have completed more than two hundred experiments.

THOSE TRICKY MAGNETS

Plutarch said that garlic was at great enmity with lodestone, and Ptolemy said sailors avoided onions and garlic on their voyages so that their navigational system remained accurate. Porta stepped right up, just as he promised in his introduction, with a test to determine if these writers were right or wrong. After eating garlic, Porta breathed and belched on his lodestones, but they still worked fine. Then he rubbed his magnets with garlic, still without effect. He wondered if he was missing something. Sailors laughed when he asked them, dismissing the idea as an old wives' tale, or in this case, a stupid philosophers' tale.[18]

Porta also tested the belief that diamonds and goat's blood created discord with the lodestone. He chided those who propagated such lies, suggesting that "Truth must be searched, loved and professed by all men; nor must any men's authority, old or new, hold us from it."

Porta presented some good examples of the characteristics of magnets, and he was the first to use a scale to measure the force of magnetic attraction.[19] The most memorable part of his chapter, however, concerns his creative use of the magnet to amaze and deceive. For example, Porta reminded readers that a magnet could pass magnetic attraction through one iron ring into another that was not directly touching the magnet. Thus, an iron ring hanging below a lodestone could pick up another ring, and so on, such that a chain of rings would connect together on the *outside* surface. Porta did not indicate where he learned this trick, but centuries earlier Augustine said he was "thunderstruck" when he first saw this wonder.[20] Some early magician (and Porta) understood that rings connected on their outer edges made a stronger visual statement than picking up one nail with another. The trick is even more compelling when the magician secretly conceals the magnet in his hand. Pressing the magnet against the top iron ring keeps the lower ones together, leaving the observer confused if not thunderstruck. Kircher's publications on magnets (four decades later) have pictures of this external concatenation of rings on the title page and frontispiece of his books.[21]

Porta created other illusions using hidden magnets. For example, he prepared for one demonstration by crushing lodestone to the consistency of sand, which he mixed with a differently colored nonmagnetic sand. He could mysteriously separate the two colors by attaching a magnet to his knee, which he surreptitiously moved without detection. Porta was really proud of this trick: "Of so many strange miracles in nature, there is none more wonderful than this."[22]

Like all good creators of magic, Porta designed another showstopper using the same method to create a different effect. For this new presentation, Porta placed two different colors of lodestone sand on opposite sides of the table,

Frontispiece of *Magneticum Naturae Reghum* (1667) by Athanasius Kircher showing magnetic rings held together on the outer surface. Kircher divided the magnetic realm into inanimate, animate, and sensitive: the inanimate magnetic needle, the animate cock and stag, and the sensitive palm tree. The two snakes at the bottom center relate to the "snake stone" that magnetically cured snakebites. *From the Division of Rare and Manuscript Collections, Carl A. Kroch Library, Cornell University*

each representing an army. By manipulating two magnets, one on each knee, the shards of sand began to tremble and then advance toward each other. In Porta's own words, the drama unfolded as follows:

> And so the troops came on, and showed the form of a Battle; and you might see them sometimes retreat, sometimes march forward; sometimes to conquer, and sometimes to be conquered; sometimes to lift up their Spears, and lay them down again.

This reenactment was so convincing that Porta proudly declared that "many ignorant of the business thought it was done by the help of the devil."[23] And speaking of the devil, Porta created another stunner that convinced some he was in league with Satan himself. Porta placed a single sheet of paper on a wall, where it mysteriously remained without falling, then the paper moved in response to his commands. Spectators did not know that Porta had attached a thin piece of iron, maybe a needle, to the back of the paper. A confederate behind the wall could hear his voice and manipulate a magnet.[24]

Of all the tricks that Porta described, one in particular remained popular with professional conjurors for another three hundred years. The prop was a small boat with a wooden oarsman who had hidden magnets fixed to his feet and the oars. With the boat in a basin of water, Porta could direct the oarsman to row the boat in any direction.

The secret movement of a small boat in water was previously described in 1550 by Cardano, our gambling mathematician.[25] Porta undoubtedly knew of Cardano's work (although he never gave him credit), and he changed Cardano's little effect into an oracle. He placed a magnet inside a small wooden swan that floated in a basin. Then he placed letters around the edges so the swan could move about to spell words. This trick became known as the magnetic swan, the educated swan, the wonderful swan, the sagacious swan, and the learned swan,[26] although other aquatic creatures were also constructed.

Just as Porta improved what he learned from Cardano, so magicians modified Porta's presentation into other variations. Mechanisms to move the swan were concealed inside tables. Automata with magnets were so highly developed by the eighteenth century that they remain marvels even today.[27] In one variation, the hours of the day were painted on the rim of a basin of water. A floating turtle's ever-changing position indicated the correct time, a result of a clock with a magnet attached to the hour hand hidden beneath. At the end of the eighteenth century, Englishman William Hooper filled a whole chapter describing sophisticated magical apparatus that relied on magnets as the underlying principle.

Some enterprising craftsman could bring these marvelous tricks and mysterious novelties up to date as modern wonders.[28]

THE DEVIL IS IN THE DETAILS

Did Porta get in trouble with the Inquisition because someone believed he solicited satanic powers to make sand jump around on a table? Any investigating priest would be embarrassed, one would think, once Porta revealed his magnet. Two magic historians recently offered the opinion that, contrary to popular belief, sixteenth- and seventeenth-century people "understood that magic tricks were not the work of the devil."[29] They go on to say that "jugglers performed magic tricks without persecution, because their audiences understood that they were watching tricks." They say those spectators were not as gullible as we have portrayed them. Porta's experience clearly contradicts this idea. Some background information will help us understand this pervasive problem.

Italian society was highly regulated by civil and ecclesiastical authorities. Anyone involved in any form of resistance, change, innovation, or suspected demonic activity could be arrested. For example, Balsamo, better known as the Count of Cagliostro, died in an inquisitional prison in 1795 for his promotion of masonic rituals. The level of social control is difficult to imagine for those of us who have not lived under a totalitarian government.[30]

Licenses and permissions were required for activities at all levels of life in Renaissance Italy; the church and the state each had a hand in the process—a hand that was sometimes open for a fee or bribe. Beggars needed a license as well as physicians. The Inquisition monitored beggars for misrepresentations and physicians for their compliance with an edict from the Council of Trent. Under pain of exclusion from the church, patients were required to consult a priest before receiving treatment or prescription. Any physician who believed that illness had natural causes was at risk of being labeled an atheist.[31] Indeed, English physician and scholar Sir Thomas Browne said that those who deny witches and spirits are not merely infidels but atheists.[32] Rigid belief was not just a residue of the uneducated and superstitious.

On the social ladder below physicians and apothecaries were travelling folk healers and nostrum sellers who needed approval from physicians to sell their remedies and services. These common peddlers were known as charlatans, a word that has come to denote deception and quackery. Charlatans used a spectrum of entertainments to attract customers, including conjuring tricks.

However, their handbills purposely avoided mentioning magic because that attracted the attention of inquisitors.

When nostrum peddler Nicolo Barbieri applied for his license in one Italian town, the bishop refused because he had prior knowledge of Barbieri's pitch. Barbieri's jester entertained audiences with conjuring tricks that the bishop mistook for demonic magic.[33] In another instance, a female who performed mental tricks was accused of "doing it by the Devil's means."[34] A couple of charlatans were summarily denied a license to perform their magic after a priest threatened them with imprisonment.[35] He said he would not allow necromancy in his diocese. He had seen their performance in other cities and deemed it devilish.

Thomas Betson, an English monk with a reputation as a prankster, exposed in his notebooks the secrets of magic that he used as entertainment. In one of his tricks, Betson attached a hair to a coin that allowed him to surreptitiously pull it across a table. He said that "many people will think it is done by magical art."[36]

Witchfinder Martin Del Rio, a well-traveled Jesuit, said that "artificial magic" (conjuring tricks and illusions) might be nothing more than entertainment.[37] Nevertheless, it meddled with people's perception. He mentioned a young female who provided after-dinner amusement for a gathering of nobles. Among other tricks, she performed what would be called the torn-and-restored napkin. One guest was sufficiently disturbed that he had her arrested for soliciting demonic assistance. She was excommunicated but escaped the hands of inquisitors.

Serious consequences could follow a satanic attribution, which is why citizens of that time were so practiced in dissimulation. They needed to hide anything that might attract inquisitional attention. A conjuror in Naples, Horatio Galasso, wrote a book on prestidigitation for entertainment purposes.[38] He approached Nicola Stigliola, a physician and city planner of high standing who had a printing press, who published the book but disguised it with a Venice imprint.[39] Even a politically respected man like Stigliola was afraid to openly publish a book on simple conjuring tricks.

King James I of England, who inflamed fears of witchcraft, commented on jugglers who learned tricks with cards and dice in his essay on demonology. These magicians, he declared, are easily corruptible by the devil, who will teach them wicked arts and sciences to perform counterfeit miracles.[40] One never knows if they are using trickery or devilry, so they deserve the same punishment as sorcerers and witches.[41] These were not good times for learning the double lift or a false shuffle. Nevertheless, James later became more skeptical and was proud of his ability to identify what were called popish impostures.[42]

I FOUGHT THE CHURCH AND THE CHURCH WON

Porta was first arrested by inquisitors and sent to Rome for "things concerning the faith." After some unspecified torture, Porta was forced to disband his Academy of Secrets because the group was presumed to be practicing magic—the evil kind. Again in 1580, Porta was charged with "having written about the marvels and secrets of nature." The records of Porta's trial show that his writing was deemed unacceptable. The focus of interest was not his book of secrets but his intrusion into a dispute between a French jurist and a German physician over details about the nature of witchcraft. Inquisitors voted to torture him mildly, but this time his health was deemed too precarious to survive the ordeal.[43]

The dispute was laid out as such: Johann Weir (also known as Weyer), a German physician, argued that witches were delusional about their activities, perhaps even deluded by demons.[44] Weir asserted that these accused individuals were not heretics who had made a pact with the devil but old ladies whose imaginations had been deranged. Thus, they deserved treatment rather than death. Jean Bodin, a French witchfinder, was furious with this soft-on-crime position.[45] Bodin accused Weir of excusing the blasphemous behavior of witches.

Porta entered this argument on the side of Weir, a Lutheran. Porta supported the idea that witches smeared salve on their bodies that contained an ingredient causing bad dreams or hallucinations.[46] He endorsed a natural explanation for the behavior of witches.

Bodin concluded that Weir's support of witchcraft was proof that Weir was in league with the devil. Now Porta was on that same team. Porta's opinion about witchcraft was not congruent with the conclusions of the church, and the church determined what people should believe about witches. Thus, Porta lost on this tilted playing field that was designed and refereed by inquisitors. Rome instructed the Inquisitional Committee in Naples to keep an eye on him, and he was warned about publishing any future writings without approval. This resulted in a lengthy delay of his next book on the secrets of human character as identified through physical traits, a belief called physiognomy.[47] Moreover, the inquisitional scrutiny eliminated the possibility that Porta's third edition of *Natural Magick* would ever be published. After his death, only meaningless fragments of his final work were found. As a further humiliation, *Natural Magick* was banned in both the Spanish Index of Prohibited Books (1583) and the Roman Indexes of 1590 and 1593.[48]

THE SINS OF PORTA

Porta did not get in trouble for his tricks with magnets but for more subtle reasons.[49] Porta was interested in natural explanations when others were satisfied with metaphysical and theological concepts. As a result, inquisitors saw him as a threat. The following incident illustrates how Porta's work challenged the church.

Porta read about an antidote against poisons, which he had an opportunity to try a few times after some farm laborers were bitten by tarantulas. He traced the victim's footprint on the dirt and placed some magic words inside the tracings.[50] The grateful workers recovered. However, Porta stopped this practice after a local cleric told him the ritual was sacrilegious.[51] Later he observed an acquaintance using a similar procedure with completely different associated words. Then he discovered that Epicurus and other ancient authorities recommended similar remedies applied to the foot, but they believed the foot needed to be bare before the tracing. All of these procedures had successful outcomes, so Porta concluded that the specific ritual was not the healing factor. Instead, he came to conclude that a natural process of recovery had occurred.

We pause here a moment to acknowledge that the bite of a tarantula results in a range of consequences. Few victims, most likely none die, and the remainder have symptoms that vary from mild to severe.[52] Those with minor reactions might claim that they experienced a miracle, but Cardano, our probability expert, would remind the lucky ones that they were merely on one end of all inevitable outcomes.

Were magic rituals by folk healers as diabolical as the church claimed? Sometimes the treatments by physicians were similar. Porta viewed secular acts as merely unauthorized ceremonies sometimes infused with frightful words.

> I began to look over the books of the physicians, printed and manuscripts, that were full of magical rituals and worthless words, and I discovered that the effects derived from natural causes; and by simply experimenting on these things, and arriving at the truth, I unmasked the frauds and the diabolical trickery.

The mystery, he realized, is that the eventual outcome was always good, no matter how painful the bite or how long it took to heal. Witches and physicians alike relied on the typically good outcomes that naturally followed. Ceremonies, rites, and spells were useless, maybe even blasphemous or heretical as the church declared, but withholding the truth about the mysteries of nature was diabolical. Porta said that he "rejoiced" in discovering a natural explanation that made a fraud of connecting witches with demons.

The church maintained its own panoply of healing rituals, such as making the sign of the cross, sprinkling holy water, anointing with oil, and exorcism.[53] The real question was about who had the authority to administer the procedure. But if, as Porta claimed, demons were not involved, then exorcism was unnecessary and useless. As such, the church was not about to allow Porta to tell people that its rituals were useless. The recent Council of Trent had empowered the church to define religious truth, and the church was dominated by priests committed to a narrow scholastic interpretation.[54] The Inquisition closed Porta's academy and restricted his writing.

Porta's observations and conclusions got him in trouble because church authorities had the final say regarding who could conduct rituals, which words were acceptable, and who could take credit. Porta's support of natural causes challenged the claims of miraculous cures, and the church was sensitive about intrusion into its monopolistic business enterprises. A moment's reflection on this cultural situation will help us understand the difficulties and dangers that Galileo faced in just a few years.

PORTA IS SO YESTERDAY

Porta was on the edge of science when he breathed and belched on magnets to show that they are not "at enmity" with garlic. Despite his stated intention to avoid the mistakes of alchemists, he still could not drop the idea that magnets had human intentions. In 1600, Porta was severely criticized by Englishman William Gilbert. Gilbert said it was time to give up that metaphorical language and the pretense that magnets had anthropomorphic characteristics and intentions.[55] Porta was guilty of the alchemist's folly, confusing a metaphor with reality:

> Because there is such a natural concord and sympathy between the iron and the lodestone, as if they had made a league, that when the lodestone comes near the iron, the iron presently stirs, and runs to meet it, to be embraced by the lodestone.

Porta ended his chapter on magnets by informing us that the unfaithful wife would be "attracted home" if her husband slept with a lodestone under his pillow. He believed that a magnet could attract a stone-cold heart. Porta failed to realize that the only metaphorical power of the magnet was to draw money from the wallets of the gullible.

Looking back, we see that Porta had fun playing with magnets. In contrast, Gilbert conducted research. Porta's magnetic sand particles mimicked the formation of an organized army because they formed a pattern, but he failed to see this as the magnet rearranging the environment. It does more than simply attract, as Gilbert concluded.

Likewise, Porta noticed the declination or dip of a balanced magnetized needle, a slight deviation away from the horizontal, but Gilbert went further to ask why this occurred.[56] Gilbert proposed that the Earth itself was a magnetic lodestone, which then provided him a strategy for further study. He obtained a large round lodestone to represent the Earth. He placed a needle at various latitudes along the sphere and observed the changing magnitude of the dip. This led him to conclude correctly that the same would occur at different latitudes on Earth. Thus, Gilbert conceived strategies for obtaining information not otherwise available, a tactic we see used by Galileo.[57]

Those who read Porta in later years were distressed by his inconsistencies. Sir Thomas Browne dismissed him as overly enamored with secrets and games.[58] He said books of secrets transmitted foolishness and superstition that impeded the arrival of science, and he viewed Porta's work as the apogee of looking back to the masters for answers. But Browne made the same mistake that I did when I first got Porta's book. Porta understood that ancient authorities did not always have the correct answers or the ability to move knowledge forward. He envisioned the future of independent investigations, but he did not have a path to get there. He had the idea that he should not take the word of an authority without first making his own direct investigation, but without the tools to investigate for himself, he had no alternative.

Porta was headed toward science, even if he got distracted on the way. He tinkered with old secrets because he had no concept of an organized investigation. Porta had trick drinking glasses, but he did not have the scientific instruments that would soon become available, and he did not have any notion that nature followed mathematical rules.

Porta reflects the beginning of a cultural change in which curiosity could no longer be suppressed. During the second quarter of the seventeenth century, modern scholars have noted a marked increase in the number of words expressing notions of doubt; at the same time words like "curious" and "inquisitive" lost the air of pious disapproval that they previously carried.[59] The concept of "novelty" evolved through a similar change, from something dangerous to something praiseworthy and desirable.[60] That which nature had hidden was no longer feared as a forbidden mystery. Whereas alchemists and scholars had previously hidden their ideas in metaphorical language or Latin text, Porta wrote in his native language. He exposed all of his secrets; everyone could now enjoy the natural wonders of life.

Porta opened the window to science in all fields, drawing in fresh air that universities were not inhaling. He sought out practical ideas that made life more tolerable. Historians Herbert Butterfield and David Wootton both make the point that Renaissance intellectuals had no concept of progress because of their adoration of antiquity.[61] Porta promised to make life better by challenging old wives' tales, the claims of alchemists, ancient philosophers, and the church. He probably had a bumper sticker that read "Question Authority."

A SURPRISE TRANSITION

Porta is important for an additional, unexpected reason. In the spring of 1604, Porta was feeling old and discouraged in his sixty-ninth year. English physician William Gilbert had upstaged his contributions on the magnet, and the Inquisition had deprived him of his academy. They placed Porta's book on the prohibited list and thwarted his intention to write a third edition. However, his spirits were lifted by the visit of Federico Cesi, the eighteen-year-old son of a Roman nobleman. Cesi travelled to Naples just to tell Porta how much he admired his work. The two immediately bonded. Cesi understood that Porta had conceptualized new ways of looking at the world, and Porta was the father figure Cesi did not have at home. Porta was eager to mentor the young man who appreciated his past accomplishments.

Cesi was already a skillful investigator with his own cabinet of curiosities and a botanical garden for studying plants and bees. A year earlier Cesi formed his own Academy of Lynxes with three of his young friends.[62] He got the idea of using the lynx as a name for his academy after seeing one on the title page of a book by Porta. The keen-eyed lynx was believed to possess extraordinary vision to comprehend the hidden meaning of things. When Cesi designed his academy's logo, he showed the lynx tearing its claws through Cerberus, the mythical dog that guarded the gates of Hades, or more properly, prevented individuals from escaping. His illustration represented truth struggling with ignorance or knowledge attempting to burst from its hellish confinement. There is one more important part of Porta's relationship with Cesi.

Porta prepared Cesi to recognize a lynx at the University of Padua.

REVISING THE WORLD

8

EXTRAORDINARY
VISION

*There are some persons who, by habit acquired in youth, or by
evil guidance, or foolish desire, or perverse inclination, are so
strongly attached and bound to an error or false view that they
refuse to relinquish it, and it is grievous to them to hear any
opinion to the contrary, and they have become unpracticed in
hearing the truth.*

—Aristotle, *Metaphysics*

*A man with a conviction is a hard man to change. Tell him
you disagree and he turns away. Show him facts or figures
and he questions your sources. Appeal to logic and he fails to
see your point.*

—Festinger, *When Prophecy Fails*, 1956

People have always been fascinated by an object swinging at the end of a
tether—a pendulum. Perhaps a parent placed one over the first crib ever
constructed. Before Galileo, the motion of an object (like that of a pendulum,
a rolling ball, or a projectile) was described by philosophers according to their
belief. Galileo had the idea that he could evaluate the validity of his belief if it
was based on direct observation. What a brilliant, elegant concept.

He started timing and measuring pendulums and discovered that the movement follows mathematical rules.[1] He had a strategy for gathering information
for a deeper comprehension of nature. If he had never looked through a tele-

scope, his name would be honored for creating a new breed of natural philosophers who abandoned reason as a method of exploring the world for that of direct observation and measurement.

Then Galileo started making telescopes.[2] Once again, he discovered things that had never been previously observed, things that had always been there but hidden from sight. He enabled people to "see" something in order to believe in it. Before Galileo, people lived in the center of the world. After Galileo, the secret was out: people lived on a small, ordinary planet in a gigantic universe.

Galileo was a pivotal person in providing humanity a reliable way of uncovering the secrets of nature. Rather than seeking metaphysical meaning, he emphasized observation. Earlier searchers of truth were usually constrained by religious or magical thinking. He pushed past those barriers to provide natural explanations. In this chapter, we focus on why both theologians and philosophers resisted his thinking, and we briefly attend to some who saw the value in where he was headed.

TO GET THE BALL ROLLING

Galileo enrolled at the University of Pisa at the age of seventeen to study medicine, but he was impatient with his teachers. In those early years, he considered Aristotle and his commentators the only teachers he needed, but he soon began to challenge some of Aristotle's ideas.[3] During his third year of college, he started skipping classes to study mathematics and philosophy on his own, and his father probably uttered some choice Tuscan words about the younger generation when his son left Pisa without a degree.

Galileo was sufficiently successful in tutoring, lecturing, and publishing that he was invited back to the University of Pisa at the age of twenty-five to teach mathematics. While there, he conducted experiments in a manner that forever changed science. He learned from his father that direct observation was the best teacher. His father was an accomplished "lutenist, a mathematician, and musical preceptor to the Florentine musical academy."[4] He conducted experiments with different lengths of musical string under various amounts of tension, and his conclusions were reported in a style later used by his son, namely reliance on direct observation:

> It appears to me that they who in proof of anything rely simply on the weight of authority, without adducing any argument in support of it, act very absurdly. I, on the contrary, wish to be allowed to raise questions freely and to answer

without any adulation [of authorities,] as becomes those who are truly in search of the truth.[5]

Galileo cared little about the adulation of authorities; however, contradicting them was problematic. Indeed, he received troublesome attention when he gathered information that supported the theory of Nicolaus Copernicus (1473–1543), a Catholic theologian in Frombork, Poland. Copernicus constructed a heliocentric (sun-centered) arrangement of the heavens that conflicted with everyday observation, the conclusions of Aristotle, and statements in the Bible.[6]

Galileo, the lynx, at the Museo Galileo in Florence, Italy. *Statue by Carlo Marcellini in 1674 is on display at the Museo Galileo in Florence, Italy. Getty Images*

Luther and Calvin condemned the idea when they heard rumors of it.[7] Never-theless, a young Lutheran astronomer from Wittenberg named Georg Joachim (known as Rheticus) planned a two-week visit with Copernicus in northern Poland to learn more. He stayed for two years. These two men of different gen-erations, faiths, and ethnic backgrounds cooperated in their scientific endeavor during the height of intolerance and religious zeal. Rheticus convinced Coperni-cus to publish his work despite its controversy, offering to bring his book to one of those fancy printing presses in Germany. Without those multiple copies, *De revolutionibus* might have been lost in the cathedral library of Frombork. The first printing took a month to travel back to him, but Copernicus got a copy of his book through some grand scheme of fate—or against all odds of probability—on the day he died in 1543.[8]

Copernicus was a man of the sixteenth century, so we should not be surprised to discover his use of anthropomorphic language and metaphorical images:

> In the middle of all sits the Sun enthroned. How could we place this luminary in any better position in this most beautiful temple from which to illuminate the whole at once? He is rightly called the Lamp, the Mind, the Ruler of the Universe. . . . So the Sun sits as upon a royal throne ruling his children the planets which circle around him.[9]

Historian Owsei Temkin points out that Copernicus was obligated by liter-ary and philosophic expectation to praise the sun in order for it to be worthy of being placed at the center of the universe.[10] Today, science is pursued without consideration of social expectation, magical properties, or religious conviction. That detachment did not exist at the time of Copernicus. Yet Copernicus fol-lowed what nature and reason presented in order to formulate his theory. Simi-larly, Galileo was determined to accept whatever nature revealed.

We should pause here a moment to remind ourselves how deeply science has been entangled with religion and magic. When Newton presented his theory of universal gravitation, it eliminated the need for a god to keep things in order. In response, one of his students, a professor of mathematics at the University of Cambridge, felt compelled to defend the religious tradition by writing a comprehensive tome about the six-day formation of the universe along with an explanation of all the known physical laws as being in total harmony with the Bible.[11] Even early chemists had to justify chemical reactions as natural responses that did not require divine intervention.[12] Lynn Thorndike filled eight volumes reviewing the history of science as it developed in the context of magical thinking.[13]

Bertrand Russell reminded us that modern commentators often ignore the silly remarks of great philosophers, but we have also disregarded the extent to which science was entangled with magic, religion, and the supernatural. Galileo tried to avoid those entanglements without throwing out God, but his critics were relentless.

Galileo generated conflict from the earliest days of his teaching. He had an Olympic skill in irritating philosophers who were loyal to Aristotle. Consider, for example, his famous demonstration from the top of the town's leaning bell tower, dropping two balls of different weight. Aristotle asserted the perfectly rational belief that heavier objects fall faster than lighter objects. Galileo showed what really happens: the balls hit the ground at the same time. The event was not presented as an experiment but as a demonstration to his students and the professors who were still promoting Aristotle's infallibility.[14] If you had been there to count noses, more were likely tweaked than persuaded.

Galileo's method of searching for truth, for the first time, placed philosophy and science in conflict, and, incidentally, religion and science as well. Valla, Cardano, and Porta all engaged in some local skirmishes, but Galileo precipitated a world war.

Aristotelians had difficulty accepting what Galileo presented because they considered their belief, based on reason and logic, to be more compelling. They tried to neutralize observations that did not fit. They had difficulty abandoning concepts that had lost their usefulness. It was easier to keep believing the sensible idea that heavier objects fall faster than to accept the reality of a demonstration by that gadfly Galileo.

As unlikely as it seems to us now, watching the simple fall of an object sparked religious conflict. Was an object's attraction to Earth the consequence of some outside cause or an inherent property of matter? If inherent, that would be equivalent to endowing an ordinary stone with qualities belonging to God.[15] But the word "gravity" (similar to words like "attraction," "cohesion," and "affinity") is merely a label that only *implies* knowledge unless associated with mathematical laws. Galileo ignored ultimate causes and directly measured the effects of gravity. After Galileo, any named force was expected to follow some mathematical law or there was a "danger of producing an illusion of understanding things and blunting the sense of the mysteriousness of all-natural phenomena."[16] Aristotelians claimed that an object falls to the Earth because that is its natural place, which was their explanation for gravity. So, when you drop an object and it falls down, the experiment is confirmed. But you have not learned much. Meaningful experiments establish and confirm laws that govern the full spectrum of what one expects to encounter.

Historian David Wootton argues that before Galileo the only experiments conducted were with magnets and rainbows.[17] When experience was utilized, no measurements were ever systematically evaluated or mathematically construed.[18] Wootton makes the startling claim that no medieval natural philosopher had a goal of making progress in science, and none was engaged in methodical study to push knowledge ahead. Philosophers analyzing motion discussed all types of conditions, causes, and forces; some even constructed the beginnings of mathematical analysis.[19] However, not a single European philosopher attached their beliefs to observations of what really happens. In the sixteenth century, astronomical explanations were not based on their consistency or correspondence to physical observations.[20] Galileo ignored *why* things move and measured *how* things move, a deceptively simple shift with enormous consequences.

SUPERHERO OF THE SUPERNOVA

Galileo was lecturing at the University of Padua when a blazing supernova suddenly appeared in the sky. This new light suggested something was terribly wrong up there because Aristotle said the heavens were unchangeable.[21] Although the Earth was in constant change, the sky beyond the moon was declared unchangeable and incorruptible. Astrologers concocted dire predictions of doom—famine, plague, and wars—which had a frightening impact on the public. All heavenly events had meaning in those days, and any scholar or theologian who said otherwise was exposing his ignorance.[22] The nova generated philosophical and religious controversy, especially at the University of Padua, which was established in 1222 as a center for Aristotelian studies.[23] Galileo's observations would challenge the very foundation of his university. The alumni club must have been furious.

In November 1604, Galileo lectured on the nova to a crowd estimated at more than a thousand people. He said the nova could not possibly be this side of the moon because he had checked its distance through parallax triangulation.[24] Galileo colorfully described its bright flare and extinction, well aware that he was challenging the writings of Aristotle and the beliefs of the church. Shortly thereafter, a philosopher issued a pamphlet dismissing the parallax calculations of Galileo, and he further claimed the nova had always been there but no one had noticed it. Aristotelians needed to dismiss Galileo's conclusions in order to preserve their beliefs.

Galileo responded with his own pamphlet, anonymously published in the form of a dialogue between two unsophisticated but bright citizens of Padua.

They began their discussion by ridiculing the fear-mongering astrologers, wondering why anyone would believe that an unknown star could influence them.

The location of the supernova, of course, was a major point of contention. Piece by piece, these two fictional characters dismantled the arguments of Galileo's critic.

> "Well, what does the fellow know? Is he a surveyor?"
> "No, he's a philosopher."
> "Then the guy should stick to what he knows."

The first character says that he can measure the length and depth of the field that he owns, so he presumes a mathematician and astronomer (like Galileo) can measure the heavens. It would be senseless for someone with no expertise in measurement to pick a fight with a professor at good old University of Padua who knows what he's talking about.[25]

Then they addressed the incorruptibility and permanency of the heavens.

> Does that philosopher really think he can see all the stars in the sky? There must be three or four, maybe lots more, that you can't see up there because some are small.

The other character laughs at Aristotle's idea that if one star were added to the heavens, they wouldn't be able to move.

> What a plague! This star has really made a big mistake. It's gone and ruined all their philosophy.[26]

Galileo's writing did not appeal to ancient authorities but to the common sense of common people. His dialogue allowed him to satirize controversial topics and put them into the mouths of fictional characters, and he gave abstract ideas a convincing emotional appeal.[27] His opponents had no evidence to contradict him or any strategies in their intellectual toolbox for responding, so they resorted to threats.[28]

These were difficult times. On one side, Protestant reformers were challenging the theology of the church, and now this astronomer was rearranging the furniture of heaven.[29] Meanwhile, printing presses were inexorably placing dangerous ideas into the hands of an expanding literate population. The church decided to bring out the kryptonite of intimidation against this superhero who had powerlifted the supernova above the moon.

The supernova's location was not the only concern. Although the location of the star was a bothersome fact, the primary problem was the implication. Facts were like deviant behavior that could be ignored. Galileo's implications, on the other hand, challenged deeply embedded tradition and authority. Looking back, we see Galileo, the church, and philosophers preparing to play rock, paper, scissors.

GALILEO, THE LYNX

In April 1609, small telescopes were available for sale in Paris, and a former pupil of Paolo Sarpi informed Galileo. Apparently, Galileo had not heard these rumors, or he did not believe them until he visited Sarpi in Venice during June or July. (Sarpi was in poor health due to attacks on his life for his stance against Rome's attempted control of Venice.) Galileo returned to Padua and assembled a three-power telescope using lenses he purchased in a shop. Meanwhile, an entrepreneur approached the senate of Venice with an offer to sell them a telescope for 1,000 ducats. The senate asked Sarpi for his opinion, and he expressed confidence that Galileo could do better.

At this time, our old friend in Naples, Giovanni Porta, heard rumors that Galileo was constructing telescopes, and he wrote to his young admirer in Rome, Federico Cesi: "I have seen the secret use of the eyeglass and it is a hoax." To Porta, the instrument was disappointing and unimportant; he had studied the idea and described its construction in his *Natural Magic*. Porta's telescope could only magnify about one and a half times, so it was merely a useless curiosity.[30] Old tales about spotting ships six hundred miles out to sea was a pipe dream, nothing more than a legend perpetrated by Roger Bacon.[31] Porta failed to anticipate Galileo's ability to make significant improvements to this toy.

After only six days of work, Galileo created an eight- or nine-power instrument that he brought to Venice as a gift. The senators convened in a tower where they successfully identified ships not otherwise visible, which would provide them a great advantage in preparing for hostile approaches.[32] They rewarded Galileo with a salary increase and a permanent teaching position at the University of Padua.

By the beginning of the next year, Galileo's telescopes magnified thirty times or better. He transformed this insignificant tube into a scientific instrument that Aristotle could not have imagined. Porta wrote again to Cesi, this time acknowledging that a simple idea in the hands of Galileo was now "filling the world with astonishment."[33] Porta must have slapped his wrinkled forehead and said, *"I should have known."*

Porta tinkered with lenses and suggested improvements in eyeglasses and the camera obscura. Galileo constructed a telescope to expand knowledge of the physical world. Galileo was not just a good craftsman and observer. He was sufficiently curious to look for evidence that would support or conflict with the theory of Copernicus, whom he secretly supported.

Although the Venetian senate richly rewarded him, Galileo was tired of teaching math 101 to undergraduates. He approached the Medici in Florence with the promise of even greater discoveries if they would name him professor of mathematics and philosophy *without* specific teaching responsibilities. He got what he wanted and immediately set out to develop better grinding and polishing techniques to make objective lenses with longer focal lengths.[34] But his technical skill was only part of his genius. While others considered the telescope useful for identifying ships at sea and spying on an enemy, Galileo's inquiring mind was prepared to focus on mysteries hidden since creation.

Galileo's discoveries in astronomy appeared with the speed of light. By March 1610, he published *Sidereus nuncius* (*The Starry Message*), which presented three startling findings. First, he claimed stars existed beyond the Milky Way that could not be seen by the naked eye.[35] Critics dismissed this because Aristotle had declared that nothing existed beyond the visible fixed stars.[36] Second, Galileo identified mountains on the moon, like those on Earth, which he measured by the lengths of their shadows. He said the surface of the moon was pitted, creviced, cracked, and riven.[37] Scoffers said he was the victim of an optical illusion because Aristotle described the moon as a perfect luminescent sphere. Finally, Galileo observed moons circling Jupiter, which showed that the Earth with its satellite moon was not unique.[38] Opponents said that his lenses had aberrations or that he was encountering atmospheric interference.

Galileo was particularly excited about those moons going around Jupiter. Aristotle said that compound motions were impossible. Therefore, the Earth's moon, for example, could not go around the Earth *and* follow the Earth as it moved around the sun, as Copernicus suggested.[39] Yet Galileo could see moons going around Jupiter, and Jupiter was circling either the Earth or the sun. Either way, Jupiter's moons stayed with their planet as the system hurled through space. Galileo had captured an outside view of a planetary system analogous to our own. The old Ptolemaic cosmology was now damaged goods, and Galileo was convinced that Copernicus had devised more than a mathematical theory. Moreover, Aristotle's treatise, *De Caelo* (*On the Heavens*), was now a failed myth about how God was running things up there.[40]

From our perspective today, we are ready, at this point, to conclude that Galileo is going to win this controversy. For Galileo, the outlook was not so rosy. All the planets were aligned against him.

Anyone who looked up could see that the sun was circling the Earth, and human senses ought to be capable of interpreting the fundamental features of reality. Otherwise, our senses are lying to us. Furthermore, a ball dropped from a great height would not land on a spot directly under it because the Earth would be moving during the fall.[41] For that matter, everything would whirl off a spinning Earth, especially at the equator. If the Earth rotated around every day, that would require seemingly impossible speed.

Galileo considered other ways of determining if Copernicus was correct: Venus and Mars would show phases like the moon. And the stars would presumably show a parallax variation when viewed from opposite sides of the sun over a six-month period.[42] Theologians cared little for these ideas because the Bible clearly stated that the sun circled the Earth—it even stopped one time so that Joshua could finish a battle. All of these arguments—personal perception, Aristotle's assertions, opinions of philosophers, and biblical authority—created a vast cultural inertia at odds with Galileo.

But he gained a couple of welcome friends.

YOU'VE GOT A FRIEND

After *Starry Message* was published, Galileo immediately sent a copy to Johannes Kepler by way of the Tuscan ambassador to Prague. Kepler was the most important astronomer in the world at that time, so Galileo was particularly eager to hear his reaction.[43]

Kepler's rapid response began by saying that he believed the information was not faked, which tells us how implausible Galileo's discoveries seemed at that time. Many, like Porta, did not believe in the possibility of a meaningful telescope. Kepler had read Porta's descriptions, and he concluded that a better telescope was at least conceivable. Porta was not always clear because he mixed fantasy with reality, but Kepler was certain that Galileo was capable of accomplishing the improbable. Nevertheless, Kepler admitted that previously he had doubts about whether a telescope, if ever made, would be successful for viewing the sky. He predicted that lens refraction would cause difficulty in seeing distant objects and that the air itself would cause distortion, just as clear water distorts vision. However, Kepler had no hesitation in acknowledging that Galileo had proved him wrong, and he was ecstatic about the new world opening up in front of him.

Kepler mentioned that he had poor eyesight, now further rendered defective by the accomplishments of Galileo.[44] He was acutely aware of how far Galileo had moved astronomy, and he was counting on him to push forward even further. Question after question entered Kepler's mind as he thought of many inquiries that could now be answered, and he considered specific ways the telescope could assist his own research. He was like an eager child waiting to play a new game. He was grateful that the political and religious differences between their two countries was of no concern to him when it came to gathering knowledge. He was more interested in getting the facts than in assigning credit.

Galileo found another friend after publishing *Starry Message*. Federico Cesi, the young admirer of Porta in Rome, was a prince now that his father had died. He realized that Galileo's eyes had penetrated what no one else had ever seen before. Galileo was truly a fellow lynx with extraordinary vision. Cesi's Academy of Lynxes had incorporated a principle into its charter that insisted on "freedom to philosophize in natural matters" without regard for the opinions of previous authorities, including the Bible. Cesi offered Galileo a membership in his academy, and Galileo was honored to accept. Cesi quickly changed the focus of his academy and added more members to expand their research into astronomy. Cesi maintained a respectful friendship with Porta, but the future belonged to Galileo.

Finally, Galileo gained some friends, temporarily, down at the Jesuit College in Rome. But before reviewing that relationship, we first step into a night of wonder at Cesi's palace.

TO SEE OR NOT TO SEE

On the evening of April 14, 1611, Cesi threw a grand reception for Galileo at his palace. The Lynxes spent the night looking at the stars, the moon, and the moons of Jupiter. The experience confirmed their dedication to astronomy. Here's why.

Critics of Galileo were not ready to believe that some trivial tube was providing an accurate view of the world. Scoffers argued that the human sense organs are the only direct, honest, authentic way to evaluate the world. An intervening instrument may be only creating an illusion.[45] But Galileo responded that the human eye is an imperfect instrument, and all our sense organs have "instrumental" errors that create false conclusions.

As the night ended, Galileo focused his telescope on the inscription above the door of the papal palace, which was way across Rome.[46] Above the door, the name of Pope Sixtus V was clearly legible. Cesi's assembled guests verified that the telescope, for the first time ever, had extended the power of human senses. Reading that inscription gave them confidence that the instrument could provide valid information about distant objects.

The telescope's optical limitations were not the main problem that Galileo faced. His critics were upset because his instrument contradicted established beliefs and treasured expectations. A friend of Galileo said there were some who, "in order to escape knowing the truth about the stars around Jupiter, do not even want to look at them."[47] For example, philosopher Cesare Cremonini refused to look because Aristotle had declared the moon was a perfect sphere.[48] Cremonini later expressed his hatred of Galileo by bringing inquisitional charges against him.[49] In his later years, Galileo retaliated by using Cremonini as his model for Simplicio, the dogmatic Aristotelian goofball character in his *Dialogue Concerning the Two Chief World Systems*.

STONE COLD

Galileo had another surprise for his exhausted companions the next morning.[50] Galileo opened a box and took out a cold stone that glowed in the dark. They were speechless. Everyone knew that Aristotle said this could not exist. The marvelous stone had been mined in Bologna and subsequently became known as the Bologna stone or Bolognian phosphorus, now known to be calcinated barium sulfide.[51]

When the stone went dim, they exposed it to the faint light of dawn, and it glowed again in the dark. This "light sponge" had absorbed the substance of light then emitted it. This supported Galileo's idea that light was corpuscular, which meant the light would eventually disperse when travelling through space. In contrast, Aristotelians believed light had no substance and was therefore consistent over distance. If Galileo was correct, then the light of distant, unseen stars would be expended before reaching Earth. Moreover, the moon and planets did not soak up light; Galileo was certain they were sending light back to the Earth by reflection. The implication was not lost on members of the academy. If light from the sun reflected off planets, that explained why Mars, for example, varied in brightness as it travelled closer and farther from the Earth, as proposed by Copernicus.

SUNSPOTS

The wonders kept coming that April morning. Galileo had something else to show the Lynxes that was not apparent through the human senses. As the sun rose, Galileo showed the academy members something only he had ever seen: spots on the surface of the sun. (We now understand sunspots as intense magnetic activity on the sun.[52]) Galileo confided to the group that he had been observing these unusual manifestations since the previous January, concluding that they moved in a way that suggested the sun was rotating.

Sometime later, Galileo received a letter from an anonymous correspondent claiming he had identified sunspots the previous May. Galileo had been scooped by not immediately publishing his findings, although he had coded his discovery to Kepler in an anagram.[53] But Galileo was even more concerned by the author's assertion that the sunspots were merely stars passing around the Earth in front of the sun. That theory saved belief in the sun's perfection. Aristotelian philosophers would suffer another blow if Galileo's theory were correct. Thomas Aquinas said he was pleased that Aristotle asserted the immutable sun reflected the perfect creation of God. Galileo was pleased to have discovered imperfections on that rotating sphere.

Galileo knew that the sunspots were on the surface—or very close to the surface—of the sun. In a letter to Cesi, Galileo expressed his hope that he could now write a treatise on sunspots that would result in the funeral of "pseudophilosophy." More and more observations were accumulating that showed Aristotle was no longer worthy of unquestioned devotion. Galileo was ready to drop the anvil on his skeptics. Unfortunately, Galileo's early optimism about changing minds fell with a thud.

While editing Galileo's book on sunspots, it suddenly dawned on Cesi that the Copernican theory eliminated the need to explain the sudden backward movements of the planets as seen against the fixed stars.[54] Galileo might have asked what took him so long to figure that out. Instead, he wrote back to express his pleasure that Cesi was getting the big picture.[55] In a statement we will have reason to consider again and again, Galileo's response to Cesi captured the essence of the science that Galileo was creating:

> I only want to tell you what you know much better than I, namely that we should not wish nature to accommodate itself to what seems better ordered and disposed to us, but that we should rather accommodate our own intellect to what nature has made, in the certainty that this is the best and only way.[56]

Philosophers were not accustomed to viewing the world "as it was." They understood the world as the ancients had described it, wrapped in meaning. (Remember that Valla stated he tried to record the history of what people think, not what they ought to think.) Galileo's contemporaries wanted the sun to be pure and without blemish; therefore, spots could not be present on its surface. But Galileo and the young prince in Rome faced trouble beyond the offended philosophers. The church was not ready to accept the world as it was.

A DARKER SKY

Galileo knew that the Jesuits were confirming his astronomical findings with their own telescope, but their support would soon be undermined by the higher powers of the church. About one month after Galileo's all-night gala at Cesi's palace, the students of astronomy and mathematics at the Jesuit Collegio Romano organized a reception in Galileo's honor. Noted astronomer Odo van Maelcotes, the featured speaker, praised Galileo and confirmed that the Jesuits had verified his findings. Not all the faculty accepted his interpretations, but the younger Jesuits, in particular, were determined to reconcile the observations with their faith. They did not realize that their intentions were already doomed.

A couple of weeks earlier, on April 19, 1611, Cardinal Robert Bellarmine had written a letter to the mathematicians at the Jesuit College asking for their opinions about Galileo's discoveries. They quickly responded with an affirmation for all five of his questions, although they did not unanimously agree about some minor issues. However, they confirmed their observations regarding the phases of Venus, but they did not state the profound and inescapable conclusion that Venus circled the sun. This knowledge of Bellarmine will soon become important to our story.

Although many Jesuits supported Galileo, he was having trouble with censors reviewing his book on sunspots. They rejected his scriptural quote that described the heavens as violent. Censors also struck out a passage about the "divine goodness" of the Copernicus theory. They said that Galileo was meddling in dangerous and subversive affairs that were none of his business. Who was he to claim that the incorruptibility of the heavens was false? The issue of concern here was not specifically about sunspots or their locations but about Galileo's interpretation of scriptures. Censors realized they could not contradict Galileo's scientific observations; however, they could keep him from trespassing on their territory.

Here's the rub: Galileo, an astronomer, was telling priests the meaning of the Bible. The conflict was similar to what the church faced from the Protestant Reformation. When Luther published his German vernacular Bible in 1522,[57] ordinary citizens read the text in their native language. Individuals could then challenge traditional interpretations by priests.[58] And now the church's authority to explain scriptures was challenged by Galileo. Moreover, he published his findings in his own Tuscan dialect for common people to read. Scholars usually hid information in Latin texts to maintain their aura of authority.[59] Galileo instead revealed the mysteries of the universe, usurping the authority of the church to describe the architecture of heaven.

When Galileo's book on sunspots was finally approved for publication in 1613, Cesi hired a talented German engraver whose thirty-eight illustrations dispelled any idea that the sunspots were stars.[60] Despite the censors, Galileo managed to slip in some zingers, like accusing followers of Aristotle of never lifting their eyes from books in order to understand the real world around them.

Aristotle's reputation was in trouble, and the Ptolemaic system was unravelling. But no one was hoisting the white flag, even when Galileo struck another blow. Copernicus predicted that Venus would show phases, like the moon, if it circled the sun. Venus would appear like a full moon when going behind the sun and then larger as it approached the Earth. And that is exactly what Galileo observed. Contrary to popular belief, Venus was not self-luminous.[61] But instead of eliciting cheers for expanding the understanding of astronomy, the sky grew stormy with the lightning strikes of Aristotelian scholars and the thunderclaps of theologians.

Galileo realized that at some point he would need to address passages in scripture that conflicted with his findings. He decided to go back to Rome to promote his views. But he was headed for trouble.

FALSE AND CONTRARY TO HOLY SCRIPTURE; FOOLISH AND ABSURD IN PHILOSOPHY

Galileo and Cesi believed that the acceptance of the Copernican theory would depend on harmonizing their findings with scriptures, and both were certain this could be done. Galileo said that the scriptures simply described what nature placed before our eyes by sensible experience (perception of the senses), and nature cared nothing if her hidden operations were revealed to the inquiring mind. Therefore, it was the responsibility of wise expositors to work to find the true sense of passages in the Bible that accord with those physical conclusions of which we have first become sure and certain by manifest sense and necessary demonstrations.[62]

But in 1614, Jesuits were ordered to adhere strictly to the teachings of Aristotle. The task of reconciliation became difficult if not impossible, and Jesuit scholars who were sympathetic to Galileo were prevented from supporting him. The next year, Dominicans hounded him, and one preached a blistering sermon warning that

> no one is permitted to interpret the divine scriptures contrary to the sense on which all Holy fathers agree, for this has been forbidden both by the Lateran council under Leo X and the Council of Trent.[63]

Thus emboldened, two Dominicans brought accusation against Galileo as "suspect in matters of the faith," citing his friendship with Paolo Sarpi, the Venetian priest who defied Rome in supporting the independence of Venice.[64] Meanwhile, Galileo's stay in Rome centered around his supporters, who were enthralled with his wit and eloquence; their adulation distracted him from the brewing trouble. In February 1616, the inquisitional arm of the church, known as the Congregation of the Holy Office, asserted two critical propositions: (1) the idea of the sun as the center of the world was "foolish and absurd in philosophy, and formally heretical" because it contradicted Holy Scripture, and (2) the belief that the Earth moves was similarly erroneous. Once Pope Paul V received this information, he told Cardinal Bellarmine to order Galileo to abandon Copernicanism.[65] If Galileo had not agreed, the Holy Office was prepared to imprison him. Bellarmine told Galileo in no uncertain terms that the Copernican system ought not contradict scriptures, "the interpretation of which is to be reserved to the professors of theology who are approved by public authority."[66] Bellarmine laid down rigid restrictions against teaching or promoting the Copernican theory, even though he was well aware that Galileo's finding had been confirmed by the Jesuits' own telescope. In contrast, the pope and the members of the Congregation of the Holy Office were devoid of knowledge about science, and they believed that it would never be possible to prove the Copernican theory. They obviously could not anticipate that their actions would be forever associated with a demand for blind obedience to a belief that contradicts reality.[67]

Bellarmine, however, chose blind belief. His life was directly linked to the Council of Trent, when rules were laid down for following the opinions of the church fathers. Thus, Bellarmine's interaction with Galileo was determined by rules that had nothing to do with science or reality. And, as an aside, he failed to create a proper record of the exact conditions imposed on Galileo, which became controversial in 1633.[68]

In a coda to the 1616 drama, the Congregation of the Index of Prohibited Books banned the work of a Carmelite priest that was favorable to Copernicus. The congregation also suspended until corrected the book of Copernicus, which had been published decades previously. Galileo was not mentioned, but everyone knew that he was seriously handicapped now that the immobility of the sun was declared "false and contrary to Holy Scripture." It was a double punch to the gut.

A FIGHT TO THE END

After his initial censorship of 1616, Galileo began pushing the edges of his constraint against teaching the Copernican theory. He corresponded with scholars throughout Europe who recognized the importance of his findings and could more easily ignore the restrictions of the church, being farther from Rome or in Protestant regions. Galileo began preparing another book, *Dialogue Concerning the Two Chief World Systems*, which he published in 1632. The context is a discussion among three friends about how the world is constructed. One protagonist makes a brilliant case for Copernicanism against his foil, Simplicio.

Inquisitors were displeased: "He declares war on everybody and regards as mental dwarfs all who are not Pythagorean or Copernican."[69] Through a series of innuendoes, Pope Urban VIII got it in his head that Simplicio was modeled after him.[70] Although once a friend of Galileo, Urban turned vindictive, even out of control.[71] At that time, Urban was in turmoil over so many political and religious conflicts that he became paranoid to the point of eating nothing unless first tasted by an attendant. Another blow to Galileo was the untimely death of Cesi, who would have edited out some of his more obvious jabs. And Galileo's fate was sealed when inquisitors realized he had withheld information about the restrictions demanded by Bellarmine. He was ordered to appear before the Congregation of the Holy Office in Rome, despite his serious medical limitations. If he refused, he would be taken from Florence in chains.

In what is considered his second trial in 1633, Galileo faced a staged drama of coercion held under the threat of torture.[72] Galileo was not willing to suffer the fiery fate of Bruno, so he renounced his belief in Copernicanism.[73] Popular legend suggests that after his abjuration, he rose to his feet and muttered, "and yet it moves." Most historians agree he was not that careless with his life.

The intervention of friends in high places saved Galileo from imprisonment. Instead, he lived the remainder of his life in comfortable circumstances in Florence, which distressed the pope, who watched Galileo's book become increasingly popular on the black market. Friends smuggled his next manuscript, *Two New Sciences*, into Protestant Holland for publication by Louis Elzevier in 1637.[74] Because his book contained no references to Copernicanism, the Congregation of the Index could do nothing about it. Nevertheless, historian Maurice Finocchiaro suggests that this book was Galileo's final revenge. He went back to his physics and mathematics, establishing the laws of motion that revealed the foundation for celestial movement.[75]

When Galileo died on January 9, 1642, the nephew of Pope Urban VIII wrote:

Today news has also come of the loss of Signor Galilei, which touches not just Florence, but the whole world and our whole century that from this divine man has received more splendor than from almost all the other ordinary philosophers. Now, envy ceasing, the sublimity of that intellect will begin to be known which will serve all posterity as guide in the search for truth.[76]

But Uncle Urban VIII, on the other hand, will always be remembered for his role in the trial of Galileo. During his reign, he was popular for some of his Urban renewal projects in Rome, but he was mostly known for raising taxes and draining the Vatican coffers for the benefit of his family. His papal army was defeated in a senseless military campaign,[77] and his troops were likely responsible for spreading the bubonic plague. When he died in 1644, outlasting Galileo, the news was greeted with spontaneous jubilation in the streets of Rome.[78]

In Florence, Grand Duke Ferdinand hid Galileo's body in a small chapel in the Franciscan church of Santa Croce with the intention of moving him to the main basilica according to the wishes of Galileo.[79] But Urban maintained his opposition, denying the move. Vincenzio Vivani, a pupil and personal secretary of Galileo, commissioned two sculptors to design a monument for a proper tomb. That plan was placed on hold until a Florentine pope was established in Rome, Clement XII. Finally, in 1737, Galileo's body was moved to the main part of the church, where his statue is flanked by marble statues of astronomy and geometry. On the night his body was transferred, the middle finger of his right hand was removed, which you now can see defiantly pointing upward if you walk over to the nearby Museo Galileo.

9

FIGHTING
FOR MEANING

"Faith" is a fine invention
For Gentlemen who see—
But Microscopes are prudent
In an Emergency.

—Emily Dickinson

Throughout his brief life, Federico Cesi persistently attempted to organize the chaos of botanical variety into a meaningful taxonomy. In 1616, he read a paper about his difficulties with this task to members of his Academy of Lynxes, Galileo among them. He noted that most people read books and acquire information without much sense of order or organization. As a result, few—actually very few—obtained much organized knowledge. Cesi was not satisfied.

The older systems of information were inadequate. Aristotle's description of the heavens had failed Galileo, and Aristotle's notion of botanical "differences" was not helpful to Cesi. He wanted to arrange the world based on what nature revealed, not on how men wanted the world to be. He was following the new strategies of Galileo—careful observation and all that—but he was not getting anywhere. While Galileo was racking up points, Cesi admitted that his shots were hitting the rim and bouncing off the backboard.

FUNGUS, FOSSILS, AND CONFUSION

Medieval alchemists, physicians, and astrologers believed that if you knew the true name of something, you would understand its characteristics and purpose.[1] In contrast, Galileo said the observation of a thing should direct its name and attributes. The essence should not be derived from its name, "since things come first and names afterwards."[2]

Botanists and physicians typically attempted to understand the mystery of plants by considering how they might be useful as medications or dangerous as poisons. Their perceived task was one of discovering a plant's purpose based on color, texture, odor, and form. The curative virtue of a plant was presumed to be indicated by external signs, although in some cases only an adept might be able to recognize it. No one doubted that living things had an "essence" that reflected God's purpose for creating them, but no one could decide if that was a single feature or a combination of characteristics.[3]

The next generation of naturalists focused on the aberrant forms of organisms. They believed these jokes or sports of nature were created for the amazement and education of men.[4] Drawings of a plant at that time might exaggerate certain features to emphasize how its purpose was reflected by the external qualities. Cesi learned from Galileo that he must draw nature as it was, not as it was presumed to be. He and his fellow Lynxes worked to highlight objective features that distinguished a specific plant from other plants, but even that task was surprisingly elusive and subjective. Accurate representation was difficult because a plant could vary according to the season and cycle of its life. Cesi and his colleagues realized with regret that pictures could not provide sufficient information for classification of plants or animals, but this did not curtail their activities observing the natural world and accurately representing it, even to the extent of redrawing earlier illustrations.

Cesi studied reproductive systems and mechanisms of nourishment as a way of classifying plants only to be baffled by lichens, moss, ferns, mushrooms, and fungi. He was also confounded by unknown species arriving from the New World at a dizzying pace, flora beyond Aristotle's dreams.[5] Moreover, previously unrecorded plants were regularly discovered locally. During an excursion near Rome, Cesi and some friends collected forty-one new plants in one day.[6] The task of illustrating and classifying them was simply overwhelming.

And then there were fossils. Were they vegetable or mineral? Cesi called them "metallophytes." He had no notion of geologic ages because everyone assumed a recent creation.[7]

MICROSCOPIC DISAPPOINTMENTS

The Lynxes began dissecting plants, surmising that a better understanding might arise from a deeper look beneath the surface—befitting of the mythical lynx—to discover their secrets. They studied reproduction and seeds, but they were not certain what they should be looking for. Thus, Cesi was immediately intrigued when he learned that Galileo was making a new instrument that might help his analysis.

In 1624, between his battles with the church, Galileo turned his attention to grinding lenses for an instrument that would be able to "view the smallest things as if from nearby."[8] Galileo gave one of these instruments to a couple of cardinals and another to Cesi. He hoped the microscope would provide the information necessary to allow for better classification. Everywhere Cesi looked, he saw only complexity. He found no clues with which to differentiate the plants into meaningful categories.

When that first microscope arrived, the Lynxes turned their attention to bees because of Cesi's lifelong fascination with them. That same year, they produced three documents on bees, which they dedicated to Pope Urban VIII. Urban was from the Barberini family, whose coat of arms was decorated with three bees. These documents were panegyrics to Urban that celebrated classical learning and new discovery; thus, each included a subtle message for him to accept science along with an embedded plea for tolerance and restraint. The first publication was a large copper engraving, which was the first printed illustration created with the assistance of a microscope. This appeared forty years before the most famous early book on microscopy, Robert Hooke's *Micrographia* (1666), which was the first major publication of the Royal Society. *Micrographia* became the first scientific best seller.

The Lynxes obtained a better microscope in 1626, and their enthusiasm for research resulted in greater understanding and better illustrations. Galileo had demonstrated there were more things in heaven than anyone realized, and now the microscope revealed more things on Earth than previously known.[9] Not only was the universe larger than anyone suspected, it was also smaller. But the microscope also caused disappointment. Philosophers, especially, expected to discover the occult powers hidden within plants, animals, and stones. Indeed, the word "occult" refers to something that is concealed or not visibly apparent— something hidden.

Amid his academy's discoveries and mutually supportive work, Cesi died in 1630 at forty-five years of age. Cesi had worked tirelessly to create an encyclopedic system of classification, but he was unsuccessful. An ideal taxonomy places every object in its own box, but he was frequently stymied because many objects fit into more than one box. Cesi was unable to conceptualize a satisfactory system because there were too many overlapping, borderline, and crossover cases that defied arrangement. At every turn, plants resisted straightforward classification. (The nomenclature of Linnaeus was still a hundred years away.)

During his short life, Cesi endured many disappointments and harsh events. His first wife died at the age of sixteen, and all six children by his second wife died in childbirth or shortly thereafter. He was continually on the edge of surveillance by his father and then the Inquisition. Despite some important accomplishments, he died leaving many of his goals unfulfilled. His death left Galileo bereft of protection in Rome, and the Academy of Lynxes soon fell apart without his strong leadership and financial backing.

MICROSCOPES

Although not explicitly stated, the microscope introduced many of the same religious dilemmas as the telescope. St. Thomas Aquinas had declared that no animals could exist below the threshold of our senses.[10] Any belief to the contrary would appear to contradict Genesis, wherein Adam is said to have given names to all animals as they passed before him. If something could not be naturally sensed, then God probably did not wish ordinary men to understand it. Exceptional knowledge was once thought available only through divine aid, but lenses opened once-hidden worlds to ordinary individuals. The very existence of the telescope and microscope expanded human senses. Moreover, their existence dramatically reinforced the idea that observation advances knowledge, an important divergence from past authorities. Secrets in old books, like those of Cardano and Porta, now seemed like the products of outdated technology. Porta's dream of discovering the magical secrets of nature was becoming a reality.

The publication of the academy's microscope findings was left to Francesco Stelluti. Just as Galileo had opened the heavens, Stelluti demonstrated that the lowliest things on Earth were suitable for investigation. Although Galileo's telescope immediately revolutionized astronomy, the microscope's effect was only quietly appreciated. Stelluti's book explored common insects like bees and the weevil, which infested grain harvests of Tuscany. The mysteries of these usually unnoticed insects were displayed in illustrations for everyone to understand. It took another 150 years for the microscope to be appreciated for its ability to advance medicine.[11]

Like Galileo, Stelluti wrote in the vernacular. Knowledge and science were now ready to tip from the exclusive domain of Latin scholars to general readers. The mysteries of nature would soon be accessible to everyone. The ancients were no longer the authority for explaining the heavens—or the Earth. The person to consult about the workings of nature would be someone making careful observations, perhaps through a lens, and taking measurements with apparatus not previously available. Any ship's captain might find a new island, and any careful observer could uncover a new mystery of nature.

FORGING AHEAD

It was still not clear where this new search for knowledge would lead and whether it would be sustained. But nothing gets the hometown crowd riled up more than the disdain of a rival bully. Roman suppression of Galileo's findings was a sore point in Tuscany. About a year after Galileo was condemned, some teenagers discovered manuscripts that sharpened the conflict. Nineteen-year-old Curzio Inghirami and his younger sister discovered some encapsulated manuscripts in an area where farmers regularly unearthed artifacts from the Etruscans who once lived there.[12] Over time, the siblings and others in the community discovered about two hundred related manuscripts and fragments written by a priest nearly two thousand years earlier. This priest had been prompted to preserve his knowledge because he believed survival of his Tuscan culture was at risk from an impending Roman attack. His fears were well founded, and the Roman destruction of the Etruscans was a known historical event that allowed accurate dating of the fragments found.

These recovered papers were examined by a local academy, which attracted the attention of Ferdinand II, grand duke of Tuscany. He summoned Curzio to present his findings at the University of Pisa, where his documents were received with excitement, although some declared them factitious and forged. Curzio's *Fragments of Etruscan Antiquities* was published under a false imprint, date, and location, a subterfuge around the hands of censors. The book was constructed to appear as if published in Germany.[13]

The Etruscan fragments were exciting because they contained ancient history of local families and governments, illuminating the Etruscan view of their conflict with Rome. Who said that history is written by the winners of war? There were also predictions of the advent of Christ, "after whom the years shall be numbered." Moreover, the Etruscan astronomers, of whom Galileo was a descendent, appeared to support some of his ideas. Additionally, the fragments confirmed an older belief that Etruscan priests had been in contact with Hebrew priests since the time of Noah, an idea promoted about two hundred years previously by a Dominican friar named Annius of Viterbo.

Curzio became a popular speaker because of his youth, quick wit, courtly manner, and extensive historical knowledge. He was trained in debate and oratory within the same traditions as Galileo, but, as Galileo learned too late and Curzio was soon to discover, the rules of courtly conduct were insufficient for success in Rome. As Renaissance historian Ingrid Rowland noted, the presence of an international population of scholars and religious censors created an atmosphere that "to an exquisite degree, could be quick, smart, and nasty."[14] Curzio's book attracted attention in Rome right up to Pope Urban VIII, because the theological and astronomical revelations rekindled the conflict between Rome and Galileo.

Perhaps the papal court had learned something from its loss to Valla. Now it was their turn to cry forgery. The Etruscan documents were written on paper not available during that era, and they contained anachronisms that failed to align with known historical facts. Such minor defects failed to impress Tuscan scholars; they enjoyed the fight and the version of truth being advocated.

Despite his youth, Curzio was ultimately determined to be the author of these fragments, although he may have received assistance from pals who were in on the joke. And joke it was. Hoax and ridicule are commonly employed by those of lower status against those with power.[15] Curzio's forged documents were a parody of Rome fighting Tuscany. He and his friends must have roared with laughter when constructing bogus oracles from the mouths of Roman priests, mocking their proclamations: "The vulture hath raised its voice from the face of the Locust. The Locust shall devour Lions. The stones shall sweat in horror."

Alas, Curzio was not the first Tuscan forger to hoax the Romans. That Dominican friar, Annius of Viterbo, had also forged his documents about Noah and the Etruscan priests.[16] And the tomfoolery is even more extensive. Remnants from the Etruscan civilization have been popular from the earliest times among collectors, spawning a manufacturing industry of fake artifacts. Forgers have been busy to the present day, fooling such prestigious institutions as the Metropolitan Museum of Art, the British Museum, and the Getty Museum.[17] Rome had its own manufacturing center for forgeries, and nineteenth-century craftsmen from Orvieto invented a nonexistent village where statuary was said to be unearthed.[18] As one forger told Thomas Hoving, "mankind seems to have embedded in its brain the notion that every day something miraculous will be found in an unlikely spot."[19]

Even after Curzio's forgeries, Tuscans still could not forget what Rome did to their native son, Galileo. The embers of that conflagration were still smoldering when Grand Duke Ferdinand II organized the first laboratory devoted entirely to experimental science, the Academy of Experiment. The year was 1657, fifteen years after the death of Galileo. They wanted to show what can happen when following the path of Galileo's investigative methods. The time was right to move past Aristotelian dogma and study the natural world through organized observation.

10

HIDDEN AGENDA

*Nature did not make human brains first, and then construct
things according to their capacity of understanding, but she
first made things in her own fashion and then so constructed
the human understanding that it, though at the price of great
exertion, might ferret out a few of her secrets.*

—Galileo

Galileo was not very successful in changing the minds of his contemporaries.
His voice was silenced by the Inquisition, and his few supporters at the
University of Pisa were expelled.[1] His books sold here and there on the black
market, but historian David Wootton says that science cannot be viable until a
community of individuals critically examines experiments, replicates the results,
and conceptualizes the next steps. It takes a whole village to nurture science.

Galileo may have lost his battle but he won the war. No one now doubts the
Copernican system, and we review in another chapter how that transition of
acceptance ocurred. Equally important, Galileo's method of examining nature
became the standard of science. It is difficult to overstate the importance of this
evolution. American philosopher of science Thomas Kuhn says that a "paradigm shift" is accomplished when younger scientists ignore the old restraints
and forge ahead with a new mind-set.[2] Those left behind feel betrayed, and
those moving into new territory discover that change is arduous. In this chapter two Medici brothers assemble the first community that nurtured science
through its early developmental steps.[3]

ANOTHER SUBVERSIVE ACADEMY

Grand Duke Ferdinand II helped Galileo during his darkest days even when restricted by the church. Galileo acknowledged his debt by dedicating *Dialogues Concerning Two New Sciences* (1632) to Ferdinand. Their relationship ran deeper than their inherited Tuscan distrust of Rome. Ferdinand was interested in scientific instruments, and he recognized their importance for measurement as promoted by Galileo.

Ferdinand teamed up with his brother, Prince Leopold, a man of considerable energy and extensive talent. The brothers decided to honor Galileo and sustain his memory by creating an academy to continue his method of investigation. Their first recruit was Carlo Rinaldini in 1657. Rinaldini had worked as a military engineer, a professor of philosophy, and a private tutor for Leopold's son. The brothers were impressed with his courage in lecturing on Galileo's findings at a time when it was not safe to do so, and Rinaldini promoted the ideas of Pierre Gassendi, a controversial French priest who criticized Aristotle and his followers. Rinaldini's first task was to conduct a survey of topics worth investigating.

All ten recruits for the Accademia del Cimento (the Academy of Experiment) had studied with either Galileo or one of his students. Alessandro Marchetti had boldly published seventy philosophical theses directed against Aristotle.[4] All academy members were convinced that commitment to the teachings of Aristotle thwarted the discovery of new ideas, but that belief put them at risk. They knew what they were facing. This was not a club for rich men who wanted to tinker with beakers or scare ignorant peasants using tricks with magnets or speaking tubes. Following Galileo's path led directly into unforgiving territory. The name of their academy was a double entendre that reminded them: in Italian, *cimento* means experiment; its second meaning is risk, danger, or arduous ordeal. Any endorsement of Galileo or challenge to Aristotle flagged the academy as a threat. They all knew what happened to Porta's academy and to men like Bruno and Campanella, who lost their lives for contradicting church belief.[5]

Nevertheless, these two Medici brothers offered some powerful incentives. In addition to their political and economic influence, their grandfather, Cosimo I, had created his own artisan's workshop because of his interest in alchemy, glassmaking, and artificially produced gems.[6] Therefore, the academy had the best glass blowers in Italy to produce instruments that Aristotle never imagined. The Experimenters had telescopes, microscopes, air pumps, timing devices, and thermometers that measured with numbers.[7] Their calibrated instruments enabled the scientific revolution because they could now describe things with numbers. The barometer and thermometer, for example, made it possible to "see" air pressure and temperature.[8]

HIDDEN AGENDA

All of the Experimenters believed that they could provide further evidence that Galileo was correct to heed his instruments rather than Aristotle. However, they needed to conceal that agenda. In doing so, they employed precautions and safeguards that inadvertently became a part of science. For example, they were scrupulous about avoiding mistakes. Suppose an experiment showed that Aristotle was wrong. Any error in their procedures was an opportunity to construe the blunder as an intentional misrepresentation. Thus, the Experimenters had good reason to plan each step of the design of their experiments and then execute each phase carefully to ensure reliable results. And to do it again, just to be certain. They did this to protect themselves, but of course that process is required to ensure that findings are congruent with reality.

Similarly, the academy was scrupulous about describing their findings without the intrusion of their own expectations. The correspondence of one Experimenter shows his concern for honesty in conducting an experiment. He noted that others with less reliable instruments would certainly be unable to replicate the study that they themselves had trouble completing. He said they should "confess frankly that it has not been possible to make [our] instruments work."[9] Modern scientists understand that a study is not confirmed until independently replicated elsewhere, and academy members clearly accepted that principle.

The Academy of Experiment's emblem was gold in a vessel tested by aqua regia (nitrohydrochloric acid) with the motto "try and try again," alternatively translated as "by trial and error." Aqua regia was the ultimate solution for identifying gold. The Experimenters were looking for truth and willing to subject each nugget of discovery to the ultimate gold-standard test. Fear of error was the cradle that first rocked modern science into life.

The extensive correspondence and associated documents of the academy members still survive, disclosing their interactive style and operating procedures. Members had an agreement that no one would speak negatively about a colleague in his presence, which was in stark contrast to university traditions. They were working for the Medici family, and the context of wealth, splendor, and political power undoubtedly calmed these divergent individuals, some of whom had quite different philosophical viewpoints, even to the extent of prior hatred.[10]

Their wide-ranging studies included investigations of static electricity, magnets, light, sound, heat, freezing of water, air pressure, movement of fluids, vacuums, mirrors, and lenses. Never before had comparable questions been asked or answers sought using similar procedures.

Discussions about a vacuum traditionally inflamed philosophical controversy about whether God existed where there was nothing or if a vacuum could have a purpose. The Experimenters ignored these questions and instead explored what happened when air is removed from a vessel: How did different objects fall in a vacuum compared with ordinary air? Can sound travel through a vacuum? Could an animal survive in a vacuum? What happens to snow in a vacuum? When asked previously, questions like these usually elicited responses citing Aristotle's opinion or constructing a logical argument.

Before the academy's work, the concept of experiment meant nothing more than the observation of an event, perhaps with some measurement. For example, Porta might be considered an early scientist who investigated magnets; however, his activity was something less than science. A scientific endeavor involves a meaningful organization of facts, theories, and an investigation strategy.

The Experimenters devised conditions designed to evaluate competing hypotheses. For example, they wondered whether salt dissolved in water fits into the "spaces" between the water particles or if the water increases in volume equal to the salt. To evaluate these possibilities, they designed an ingenious vessel with a small tube emerging from the side that could measure volume with exquisite precision. First, they recorded the volume of water with the still-undissolved salt. The vessel was sealed and shaken to dissolve the salt. The volume remained the same even after the salt dissolved; the salt did not crowd into the spaces between the water molecules.[11] One theory was confirmed and the alternative rejected based on observable evidence that could be replicated by anyone.

The Experimenters also created conditions that could never be observed in the natural world, like putting snow in a vacuum. It was "seen to liquefy just as slowly as it usually does in air."[12]

They conducted experiments demonstrating where Aristotle was wrong but without drawing attention to the obvious conclusion. For example, Aristotle's idea of "positive levity" suggested that light things rise because that is where they are assigned to be. Similarly, a ball did not bounce back to the height at which it was dropped because of the repugnance of a heavy body to go upward. The Experimenters demonstrated that air pressure holds heavier things down. They concluded that air has a corpuscular structure that expands to occupy a larger space but can be constricted into a smaller space, all in ways described by mathematical rules.

Alchemists often tried to control nature, like changing lead into gold. The Experimenters did not seek such lofty goals. Instead they accepted the tedious tasks of filling in the small, missing pieces of nature's mysterious puzzle. Sometimes their best observations were questionable. In one study, Lorenzo Magalotti stated that he thought he could feel an "imperceptible little breeze" from the focused radiation of ice. However, he had no confidence in this sensation because "I know that I am biased." One wonders if this basic statement of science had ever been previously uttered.

What could be done to protect scientific inquiry from personal bias? The academy provided a seminal example of controlling self-deception during its study of Saturn's rings.

RINGS AROUND YOUR SATURN

Galileo was the first to see Saturn through a telescope, but he was confused by what he observed. On either side were two small discs or protuberances that looked like handles.[13] Another time he said it looked "three-bodied" or as if there were two stars on the side. He concealed his finding in a Latin anagram to establish the priority of his discovery: "I observed the highest planet [to be] three-bodied."[14] A couple of years later, Galileo was further mystified when he discovered these "handles" had completely disappeared!

More than a decade after Galileo's death, Christiaan Huygens discovered an explanation.[15] While studying Saturn, Huygens concluded that the planet was "surrounded by a thin, flat ring that nowhere touches [the planet] and is inclined to the ecliptic."[16] He believed Galileo had been confused because the ring appeared to vanish when the planet went through phases of tilting. (This is similar to looking down at a plate on a table, which then appears to vanish if viewed from the side at table height.) Huygens dedicated the publication of his observations to Prince Leopold.

When Leopold received his copy, he decided the academy should immediately investigate Huygens's explanation. After some brief calculations, the members concluded that it would take eight or nine years to study all the phases of Saturn. Thus, they decided to test their theory by making a model of Saturn using the proportions described by Huygens. They painted the model and its ring white and placed it in a long gallery in the Pitti Palace. They illuminated the model planet with hidden torches and viewed it from different distances through both good and low-quality telescopes. They took the model outside and repeated the tests with a low-quality telescope and the naked eye.

Galileo's sketch of what he called the "triple nature" of Saturn. *Getty Images*

When the ring was tilted, the model appeared to have two handles; as the angle of the ring decreased, the rings disappeared from view. When they looked at the model through a telescope, the ring became obvious. Huygens had nailed it.[17] The academy had devised a clever strategy to investigate Huygens's hypothesis in a short period of time without waiting through the planet's phases.

Galileo's confusion was perfectly understandable. This was an excellent example of how the academy could generate a hypothesis and conduct an experiment to support or cast doubt on an idea. They did not rely on a single test but observed their target under various conditions to improve their confidence—convergent validity.

It gets better. Some of the Experimenters had second thoughts. What if they were deceiving themselves, seeing what they wanted to believe? Had their conclusion been influenced by the hypothesis they believed was correct? Were they so eager to support the conclusion of Galileo that their senses betrayed them with a misperception? The Experimenters devised an elegant solution. They enlisted people off the street, some of them illiterate, to describe and draw what they observed. These citizens had no dog in the fight, yet almost all participants agreed with the Experimenters. The hypothesis was confirmed, but more importantly the Experimenters were confident that their results were not based on self-deception.

A tendency to confirm one's own beliefs is so natural that deliberate intention is required to consider alternative possibilities.[18] Perhaps no one person has the necessary skills and sufficient objectivity to conduct value-free research. No matter how brilliant, one might be tempted to take shortcuts or even cheat. Thus, the watchful eyes of the members of the academy and the Medici brothers made the academy a success; similarly, most research today is not conducted by isolated individuals but within universities and commercial laboratories that work cooperatively and then submit their results to outside peer review.[19]

Nonscientific thinking protects tradition and offers an inherent capacity for explaining away failure. In contrast, the history of science is littered with discarded theories that are not consistent with experience.[20] However, this is not to say that the academy prevented its agenda against Aristotle from influencing its final report. On some minor issues, it failed to mention results that appeared to favor Aristotle.

FINISHING THE PAPERWORK

The Experimenters compiled their results into a book dedicated to their sponsors, Ferdinand II, grand duke of Tuscany, and Prince Leopold. In the introduction they declared that nature had been created without falsehood but had often been misperceived. Therefore, they laid out high standards for their own observations, and they made a plea for tolerance:

> we would never wish to pick a quarrel with anyone, entering into subtle disputes or vain contradictions; and if sometimes in passing from one experiment to another, or for any other reason whatever, some slight hint of speculation is given, this is always to be taken as the opinion or private sentiment of the academicians, never that of the Academy, whose only task is to make experiments and to tell about them.[21]

The Experimenters did indeed want to share their experiments, but they needed to report them carefully. They never mentioned that many of them contradicted Aristotle. To ensure its tone of neutrality, their manuscript went through at least five major revisions based on continual suggestions from individual members. The author of the manuscript was never identified; the introduction was signed only "the secretary." Thus, censors were less likely to condemn the work. As a result, the book successfully passed through four ecclesiastical censors; this was a triumph given how intentionally the church guarded against contrary opinions.[22] Apparently all censors were oblivious to the implications against Aristotle.

Because the academy was so slow in getting its book to press, Englishman Robert Boyle earned the distinction of publishing the first book based entirely on experimental research. The academy's experiments did not have a major impact on the scientific world. Nevertheless, it earned the distinction of being the first organization exclusively devoted to scientific investigation.[23] The Royal Society in England acknowledged the influence of their Italian friends, adopting many of the same guiding principles. For example, an individual experimenter was not to be criticized, only the design and procedure was fair game for discussion. The goal was finding truth, not promoting support for one's personal beliefs. Moreover, a focus on the explicit results should not elicit discussion of more distal and conjectural issues of causation, which typically created conflict. The hope was to gather consensus about what was discovered, that which existed regardless of political or religious belief. And, incidentally, the Royal Society believed that by studying the book of nature, it could eliminate mistakes in the Bible that had been corrupted by Catholics.[24]

WHY THE ACADEMY ENDED

The academy closed after a decade of work. Pope Clement lured Leopold away from the academy by making him a cardinal.[25] Personality conflicts started to erupt among the Experimenters, which ultimately destroyed the academy. One Experimenter lied rather than admit an error in judgment, and three of the more talented members departed. Giovanni Borelli was left a beggar, and Antonio Uliva was tortured so badly during an inquisition that he committed suicide.[26] I have been unable to determine whether Leopold realized that one of his academy members was being abused (how could he not know?), but he pleaded for compassion when Honoré Fabri was incarcerated for his interest in science.

SOME FINAL EXPERIMENTS

Francesco Redi (1626–1697) was a member of Cesi's Academy of Lynxes and subsequently recruited for the Academy of Experiment. Records indicate that he studied the possibility of spontaneous generation of toads from rain. The story is significant here because Redi challenged Aristotle's conclusion that many animals were spontaneously generated and that animals of one species could create new species.[27] Yet no one had ever bothered to determine whether this was true.[28]

After the Academy of Experiment published its book and closed operation, Redi continued his own investigations. His experience undoubtedly helped formulate his thinking about how to address new questions, and as Julian Corrington observed, "No better sets of controlled experiments could be devised today than the simple and ingenious ones described by Redi."[29]

Redi began by observing the decay of dead animals. He noted the arrival of flies, the deposit of eggs, the hatching of maggots, and the emergence of flies. He repeated the experiment with raw and cooked meat, concluding that all maggots came from the eggs of flies, not from the decay of meat, and that flies attracted to a specific meat later produced flies of the same species.

Each experiment produced new questions that piqued his curiosity, and he stated that "belief unconfirmed by experiment is vain." In continued experiments, he sealed meat in flasks without air, covered meat with a fine netting that allowed access to air but not flies, and buried some meat underground. He observed these three conditions during different seasons and concluded to his satisfaction that flesh not exposed to flies did not develop maggots. He also observed rotting fruits and vegetables that similarly allowed him to conclude that different foods attracted different insects, which then produced insects of the same species.

His conclusions, which seem so innocuous to us, were attacked with vigorous opposition. Those who accepted his results still refused to reject completely the idea of spontaneous generation. Their belief was transferred to smaller species. (Even Louis Pasteur became involved in the spontaneous generation debate in the 1860s.) But Redi demonstrated that the strategy of science leads to trustworthy knowledge. His carefully planned research revealed compelling evidence that science can resolve conflicting beliefs.

This strategy of investigation became sufficiently compelling that fellow academy member Marcello Malpighi adapted the procedure for his investigation into the possibility of spontaneous generation of plants. He protected soil from the intrusion of stray seeds just as Redi had screened meat, and Malpighi found no spontaneous generation.[30] He did not attempt to answer his question with a single experiment but accumulated converging evidence. The Experimenters learned how to ask questions and search for answers in ways that the world had never before conceptualized. Now people could change their mind based on objective observations instead of the subjective opinions of authorities.

THE EVOLUTION OF EMPIRICAL SCIENCE

Three major religions blended Aristotle's philosophy and his concept of the universe into their dogma. When he could observe something directly, Aristotle was usually right, but he was frequently wrong when things were unobservable. He was, of course, constrained without instruments of observation, like microscopes and telescopes,[31] so he relied on rational argument or reasoning, like a philosopher.[32] He knew less about mathematics than his contemporaries and did not incorporate it into his scientific thinking.[33] He did not conceptualize a path for experiments or for verifying information through multiple strategies. Once Aristotle defined something, his followers revered his statements as inerrant truth, which petrified science and blocked the road to further development.[34]

At the time of Galileo, clergy, lawyers, and philosophers all regularly relied on tradition to answer questions. They did not resolve questions of fact but made their case through an appeal to authority and rights. Theological arguments, legal wrangling, and philosophical discussion defend a point of view; they are strategies of confirmation, not of testing belief against objective reality.

Galileo's new strategy of investigation was promoted by the Academy of Experiment and the Royal Society.[35] They changed the thinking and language of scientific investigation. Their new science was also a challenge to authority and therefore a subversive enterprise. During a milieu of control and censorship, these new natural philosophers, sometimes called mechanical philosophers, were forced to go underground to operate in secrecy. However, they were not the only ones with opinions contrary to the official position of the church or civil authorities. Others hid their private journeys and unaccepted beliefs: occultists, astrologers, indigenous healers, alchemists, magicians, academy members, church reformers, and all those with divergent religious convictions. Marcello Palingenio, a poet whose work was on the list of banned books, said: "not to know how to dissimulate is not to know how to live."[36] Sometimes literally!

An invisible network of natural philosophers grew and flourished across Europe. Before the advent of scientific journals, these individuals developed informal correspondence that became known as "philosophical commerce," "the invisible college," or "the republic of letters."[37] Scientists in Italy and other repressive countries were able to remain informed while keeping their unacceptable beliefs concealed. Not all their findings were unpopular or censored, but the observational method itself rankled many scholastics. Some scholars tried to adopt the new science into the Aristotelian tradition rather than overthrow it completely,[38] but many were unwilling to accept modifications.[39] Inquisitors refused publication of anti-Aristotelian books for years after the death of Galileo.[40] As a result, communication of sensitive ideas was carried by trade vessels, diplomats, merchants, and travelling scholars. Sometimes their letters took a year to arrive, and some were lost at the bottom of the sea.

Matters started to change in England when the Royal Society formed in 1660, a founding that corresponded with easing political oppression during the Restoration. Robert Boyle urged a new era of investigation based in large part on the strategies and procedures of the Academy of Experiment. Writings before this time show no sense of steady intellectual advance. From this point on, science created an expectation for the accumulation of new knowledge and a deeper understanding of the world around us.[41] The days of the solitary alchemist hiding away with his lone apprentice by the sooty furnace were over. For the first time, discussion of the natural world avoided metaphorical language, and mathematical notation became standardized.[42] The Royal Society decided that experiments would be conducted in the presence of informed observers. A whole community now critically examined experiments, replicated results, and conceptualized the next steps.

How far have we come? The search and discovery of the Higgs boson by CERN involved three thousand physicists from 175 institutions in 38 countries, all working cooperatively together.[43]

REVISING
THE UNIVERSE

11

SOME
ENCHANTED EVENING

*If men will believe every Thing without examining, they will
run the Hazard of being always deceived by confounding
Truth with Falsehood; and if they believe nothing, they deprive
themselves of the Knowledge of Truth.*

—Ahlers, *The Pretended Rabbit Bearer*, 1726

The idea of the Earth going around the sun was once unimaginable. When Galileo first looked through his telescope, only a few people accepted the possibility that the Earth *might* be spinning like a top while circling the sun. Galileo's findings changed that. Although he tried to communicate his discoveries to common people, he was silenced. Supporters of Galileo were rooted out and summarily dismissed from the University of Pisa. In contrast, most people today understand that the Earth circles the sun every year.[1] Yet nothing we experience gives any clue about the true structure of the solar system. How did this change in belief become commonly accepted?

Canadian psychologist Jim Alcock likes to confound his students by reminding them that they have no *direct* evidence to convince him that the Earth is round. The concept seems so clear and obvious, even though it is not. We are unaware that we have a belief system based on someone's authority without any direct evidence of our own. Nothing we have experienced tells us that we are standing on a round, spinning ball blasting through space. Most of us "know" the actual configuration only from hearsay evidence.

Voltaire, among others, informs us that Fontenelle was the first person to popularize knowledge about the true arrangement of the solar system.[2]

POPULARIZING SCIENTIFIC FINDINGS

Bernard de Fontenelle, you will recall, was our seventeenth-century French-man who explained that the devil was not responsible for the workings of Greek oracles. His intended goal was to end the nightmare belief in ubiquitous demons. Early in his life, Fontenelle struggled to accept his own insignificance in the expanded and rearranged world of Galileo.[3] As he was coming to grips with this, he wondered if he could help other ordinary individuals avoid the existential depression that he endured. How could he transfer the knowledge of this new universe into the public domain without causing loneliness or dread? Or getting himself in trouble with the church?

Fontenelle realized that he was living in exciting times. Ships were bringing new plants and strange animals from distant lands. New tribes were discovered with such strange visage and culture that Pope Paul III was called to declare that the indigenous people of the Americas were rational beings with souls. People were excited to learn about new civilizations on Earth, so maybe they were ready to consider the possibility of rational souls on other planets.[4] Fontenelle decided to write a book that explained the new organization of the solar system. But why stop there? Maybe he could even suggest a new universe inhabited by other civilizations. It was a bold plan. Only a few decades earlier, Giordano Bruno had lectured widely about the possibility that the cosmos was filled with innumerable inhabited worlds. If our world is good, he offered, then ten hundred thousand globes may exist where God is glorified. Bruno was not bothered by the immensity that disturbed Fontenelle. Bruno said he felt "loosened from the chains of a most narrow dungeon."[5] He was now free to move about in the riches of infinite space. Unfortunately, his movement ultimately was constricted in an actual dungeon cell, and then Cardinal Bellarmine, impatient with his refusal to repent and conform his belief to church doctrine, had him burned.[6] Fontenelle decided to adopt Bruno's bigger-is-better outlook and Galileo's dialogue style.

The strange title of Fontenelle's book, *Conversations on the Plurality of Worlds*, suggested the existence of more than one inhabited world.[7] Despite its placement on the list of banned books, *Conversations* was reprinted thirty times in France and approximately the same number of times for English readers. He did not address, like others, whether moon dwellers had souls. The text is about thinking, keeping an open mind, and allowing astronomers to tell us how the world is arranged, not demanding that it conform to our expectations or desires. Fontenelle was a deeply religious man who believed, like Galileo, that theology could explain heaven but not the arrangement of what might be in the heavens.[8]

Fontenelle's dialogue entwined a natural philosopher (a scientist) and a woman of status, a marquise. At this time in France, women were at best considered unimportant. In contrast, Fontenelle had no difficulty admitting: "Most women have quite refined minds; they think and express themselves more precisely than men when they think well."[9]

Fontenelle's philosopher gives the marquise the respect she deserves as an intelligent, curious, witty, and bold woman. She holds her own in conversation with him. She challenges his assertions, expresses doubt, and sets limits on topics. Authors using this literary style usually have their characters periodically comment about their surroundings to avoid a tedious interaction. Fontenelle placed his couple outdoors looking up at the sky. Their diversions, therefore, were the stars above, which was their topic. And those interludes sparkle with romantic tension.

DISCUSSING SECRETS IN THE GARDEN

After an intolerably hot day, our philosopher and the marquise walk out to the garden of her estate. The moon has risen and is shining through the dark greenery that now appears black. No cloud hides the smallest star. Looking out, the man trembles at the beauty; turning his gaze to the woman, he is overwhelmed. "Don't you find the night more beautiful than the day?" he asks. "Yes," she answers, "day's beauty is blond and dazzling, but the night's beauty is brunette." The gentleman notes her modesty, because she herself is blond, and he declares her more beautiful than the night. Then they discuss why lovers address their poems and songs to the night. "I love the stars," she says, "which the sun overpowers." He responds by saying, "I have a peculiar notion that every star could well be a world."

That line might be high on the list of "ten things not to say on a first date." However, in this situation, Fontenelle's goal was not human romance but romance in the sense of a journey of mystery and adventure. That journey was five evenings of discussion about the possibility of new worlds (the *plurality* of worlds) and an effort to understand the implications. Galileo never mentioned the possibility of life on other planets. Fontenelle took that risk.

In dialogue, the philosopher explains that the Earth and other planets are rotating around the sun, even though that is not how things appear. He explains that the new philosophy is based on two things: curiosity and poor eyesight. If we had better eyesight, we could see things that are hidden, and curiosity makes us care about knowing. True philosophers spend a lifetime not trusting what they see and making theories about what might be happening outside our immediate awareness. (Our philosopher, you can see, adopted the thinking of Galileo and the Academy of Experiment.)

As an analogy, the philosopher explains that our view of nature is similar to watching an opera. We see only the stage effects while the wires, pulleys, and wheels that move everything around are hidden behind the scene. In nature, the wires are so well hidden that the sages of the past, like Aristotle and Pythagoras, were not able to guess how or why the cosmos was functioning so efficiently.[10] Copernicus drew back the curtain sufficiently to notice that the sun was at the center of all activity.

"But why don't we notice that we are moving?" asks the marquise. Using another analogy, our philosopher suggests that someone asleep on a boat travelling down the river would wake up in the same relationship to the boat. Nothing on the boat would suggest movement or a change in location. Only by looking out at the riverbank would one notice motion.[11] And on they go into deeper questions.

On the second evening, our philosopher declares that the moon circles the Earth and apparently is inhabited. The marquise responds to such a shocking declaration with the observation that she had never heard anyone say such a thing except as fantasy or delusion. "Someday," he assures her, "we will go to the moon to discover the truth."

On the third evening, the philosopher begins by admitting that the moon is probably not inhabited by intelligent creatures, thereby disappointing the marquise, who protests that she had already overcome her trouble accepting the idea. He cautions her never to give more than half of her mind to beliefs of this sort so that the other half could admit the contrary position if necessary. Despite reasons for not believing in inhabitants similar to humans, the philosopher endorses the likelihood of simpler life forms on the moon. As for other planets, it would be strange if they were not populated like the Earth, although perhaps with creatures considerably different in appearance and culture.

The marquise says she is pleased to live on the Earth where it is not too hot or too cold. Our protagonist says she should instead give thanks that she is young, beautiful, and French. They part for the evening after she makes him promise that his philosophy will provide her with new pleasures the following night.

On the final evening our curious couple consider that the fixed stars, like our sun, may have planets as well. The marquise then reflects the early fears of Fontenelle himself.

Here's a universe so large that I'm lost. I no longer know where I am, I'm nothing.
. . . All this immense space which holds our Sun and our planets will be merely

a small piece of the universe? As many spaces as there are fixed stars? This confounds me—troubles me—terrifies me.[12]

Our protagonist reassures her that the old world with stars nailed to a blue vault now seems small and narrow. Following the conclusions of Bruno, he says the universe has infinitely greater breadth and depth, and he can breathe more freely knowing that its magnificence is a profusion of riches worthy of admiration. Nothing is more beautiful to visualize, and the telescope reveals its infinite wonder. As they prepare to part, he asks her to think of him whenever she looks at the stars.

He also reminds her that she can decide to believe only what she chooses. The significance of this permission can easily slip past our awareness, and perhaps inquisitors missed the point because it was directed to a woman. The marquise declared herself ready to ask questions, learn, challenge, and think on her own. She is now an independent woman without restriction. She will believe what she wants and choose for herself without constraints from others.

When she went out running the next morning, her T-shirt said, "Just do it."

THE EXPANDING UNIVERSE

My summary of the philosopher and the marquise cannot capture the immensity of these ideas, which were completely beyond what people of that day had ever considered. As soon as his book was published, the public consumed it as if a door had been opened in a smoke-filled room; everyone needed this breath of fresh air.[13] From our perspective today, speculating about intelligent life on other planets seems rather ordinary. At that time, however, the topic was more than an issue of curiosity. The idea was forbidden, hidden away in a locked closet.

If Fontenelle wanted only to discuss the new science of Galileo, he could have written about air pressure or how a ball rolls down an inclined plane. Instead, he went straight to the heart of the new reality. The Earth-centered world was gone; Copernicus had reorganized the planets. From now on, telescopes tell us how the heavens are arranged, not the Bible. We can no longer accept the conclusions of our battered friend, Aristotle. There are now better ways of finding things out.

To make these points, Fontenelle presented two individuals who openly discussed their belief without consideration of ancient opinion or current authorities. The new constraints were the boundaries of science, not the restrictions of the church. Maybe intelligent beings were looking down on us, wondering if we were looking up at them. Such speculations were not merely novel, they were anathema to religious authority. His book on oracles had thrown out demons, and now he was throwing out the church's arrangement of the heavens and its authority to insist on conformity of belief.

Galileo backed away from his early consideration of plurality, and even Isaac Newton shunned the Copernican implications in his *Principia* (1687). Every topic was infested with religious implications that could be construed as heretical; debate was stifled by fear. Roman ecclesiastical authorities feared the "crisis of the European mind."[14] Inquisitors focused on the growing numbers of atomists, materialists, libertines, and atheists. As a consequence, Naples endured nine years of trials that censured the "freedom of philosophizing." As late as 1714, inquisitors were so angry with those who discussed celestial movement that one monk sadly noted "they hate them more than the plague." Another worried that the prisons of the tribunal were "full of men to be questioned; their number is always growing."[15]

I FOUGHT THE CHURCH AND I WON (SECOND VERSE)

The church and state were difficult to distinguish. Before the revolution in France, about 160 censors were on the thought-police payroll of Louis XVI.[16] Nearly a thousand writers, publishers, and book dealers were sent to the Bastille between 1600 and 1756.[17] Debates about the simplest of topics raised religious questions concerning the existence of God and his role in creation. Few topics in science were safe. To avoid untoward consequences, French writers often published their books in Holland, where Protestants ran the presses.[18] Yet Fontenelle published in Paris, followed by multiple reprints and translations into other languages.

The implications of plurality were sufficiently alarming that Fontenelle's book was placed on the Index of Prohibited Books. How did Fontenelle escape personal persecution?

Even at his young age, Fontenelle had friends in high places, including the Marquis d'Argenson, the chief of police who saved him more than once,[19] and he avoided conflict throughout his long life.[20]

What really saved Fontenelle, however, was what he did *not* say and the style of his writing. Fontenelle did not say anything about God or the church. He urged an open mind, making conclusions only when evidence was available. If humans were no longer the center of the universe, then he could celebrate that situation. But he never celebrated by praising God—or rejecting him. He did not debate; he was impelled by wonder and intellectual thirst: curiosity and poor eyesight.

Fontenelle's style with the marquise is so pleasing, easy, and elegant—affectionate—that he remained safe in a culture still amidst witch hunts and inquisitions. Using a woman in the discussion was perhaps his way of avoiding the conflicts typical of men, as in the interactions of Galileo's protagonists who engaged in debate and conflict. Fontenelle did not introduce someone like Galileo's Simplicio as a fool to defeat but rather a charming marquise who expressed innocent curiosity about the structure of the cosmos. Fontenelle did not address the concerns of free thinkers, rational theologians, and materialists. And he never mentioned the Bible.

The church dared not censure Fontenelle for writing about two people discussing the night sky. Theologians may have hated his Copernican conclusions, but who would risk charging him with heresy? The church could have, but it avoided confronting him. How could anyone say it is wrong for lovers to discuss the findings of a telescope? The popularity of his book was an indication that the church had lost this battle. People found something deeply satisfying in Fontenelle's writing, and perhaps the style of interaction set the stage for listening and understanding the new organization of space.

Sarpi complained that he was at risk when expressing his opinions, and therefore was "compelled to wear a mask; perhaps there is nobody who can survive in Italy without one."[21] Fontenelle provided a model, if only on paper, of people interacting without external intimidation.

Fontenelle's book was so popular that the idea of plurality became a topic for other writers, although not always in a way that would have pleased him.[22] You may remember that the marquise said she had never heard anyone talk about people on other planets except as fantasy or delusion. She had in mind some individuals we encounter in the next chapter.

12

MEN ON THE MOON

With eager stomach they swallowed all he said. . . . No statement of detected falsehood could counteract or weaken the infatuation.

—Anonymous, *The Fortunate Youth;*
or Chippenham Croesus, 1818

Writing about life on other planets was Fontenelle's way of getting people to stretch their minds and confront the new organization of the cosmos. The philosopher in his book convinced the marquise that the moon was inhabited. The next day, he admitted that the moon was probably not populated by intelligent creatures. He suggested that she never give more than half of her mind to such speculative ideas so that the other half could easily admit the error if the idea proved wrong.

The marquise acknowledged that she had encountered the possibility of men on the moon but only as fantasy or delusion. She was undoubtedly thinking of French author Cyrano de Bergerac. We consider him as well as some other "lunatics" because these people created a cultural path that eventually led to the acceptance of mesmerism, clairvoyance, and spiritualism. Maybe there are a few grains of truth hidden within all the wacky ideas.

FANTASY MAN ON THE MOON

Cyrano de Bergerac (1619–1655) was a man of many talents who wrote a couple of books so fanciful that some were concerned that he might be delusional. His *The Voyage to the Moon* and *The Voyage to the Sun* were initially circulated in manuscript form among friends.[1] Censored revisions were published only after his death, about forty years before Fontenelle. On his voyage to the sun, Bergerac's character lands on a "sunspot," one of several small planets circling close to the sun. He meets talking birds who put him on trial because of how they have been treated on Earth.[2] Bergerac was addressing moral issues, but he also tipped his hand to reveal his opinion about the form of the universe, even if only by describing the route he took to the sun. To make his points yet stay safe, he created situations and interactions so fanciful that only his friends would catch his meaning, while inquisitors would conclude he was delusional or writing bad fiction. Scholars within nearly every university department have subsequently taken the opportunity to enlighten the rest of us about what Bergerac "really" meant. Features of his work have been described as utopian, anti-utopian, satirical, foolish, paradoxical, hermetic, atheistic, and pantheistic, as well as employing features of science fiction, free thought, and a parody of contemporary travel literature. We learn from these sages that Bergerac's "syncretic enterprise" concerns "epistemological bodies" by using "corporeal visions" in which the "anthropocentric universe disappears."[3] Were these academics facing publish-or-perish requirements to obtain tenure, or were they eating laurel leaves and inhaling ethylene?

Fontenelle's marquise was undoubtedly thinking of Bergerac when she mentioned fantasy, but certainly she never met anyone as delusional as Swedenborg.

DELUSIONAL MAN ON THE MOON

Emanuel Swedenborg (1688–1772) was raised in a deeply religious family but turned his life to science. He became known as the Swedish Aristotle for his writings on everything from mathematics to cosmology. In his mid-fifties, he began having mystical experiences that resulted in a prolific outpouring of theological treatises written in Latin. In one of his books he said that the universe is so immense and the stars so innumerable that some have their own planets, just like our solar system. Moreover, some of these planets are populated. These statements were not much different from what Fontenelle had said or, for that matter, from what Carl Sagan has suggested. However, Swedenborg further claimed that God had granted him the opportunity to visit—or, rather, his spirit to visit—those inhabitants.

Whereas Bergerac created fantasies about creatures on other planets and Fontenelle discussed the possibility of other civilizations up there, Swedenborg claimed he visited and confirmed the plurality of worlds.[4] He conversed with spirits, angels, and humans that lived on our solar system's planets, our moon, planets of other stars, and on moons of those planets. Some of his visits lasted only a day but some were a week or even a month. He learned about the customs and religions of these habitations, which were congruent with his understanding of the God who had introduced himself to humans through the incarnation of Jesus. All the inhabitants he met were human in basic ways, although men on the moon were dwarfs and those on Jupiter used their hands for assistance in walking. The inhabitants of Mercury, however, were like a whitish flame or spirit who examined facts in his memory. He said they were shielded from the sun's intense heat by a protective atmosphere.

Swedenborg never mentioned Uranus or Neptune, because those planets had not yet been discovered by astronomers who were forced to rely on more prosaic investigative strategies like telescopes.

All of Swedenborg's planetary visits provided evidence of a universe consistent with the Bible, with the exception of hell. Swedenborg could not believe that God cast people into the abyss, although evil spirits were there because evil is its own punishment. Therefore, he reinterpreted hell as metaphor.[5]

When Swedenborg visited London in 1743, he became so disorganized in his thinking that he locked himself in his room for two days in fear of being suffocated by spirits or directed to steal or commit suicide by aliens. This brought him to the care of Henry Maudsley, a famous British psychiatrist. Without regard for confidentiality, Maudsley described Swedenborg's psychological history in his book *Pathology of Mind*.[6] He said that Swedenborg believed he was a messiah sent to be crucified for the Jews.[7] He was suffering from visual and auditory hallucinations with delusions of taste, odor, and bodily sensations. Maudsley said Swedenborg had episodes of mania along with fits that suggested temporal lobe epilepsy, a condition sometimes associated with intense religious experiences. Maudsley's revelations provoked passionate criticism from Swedenborg's followers, who did not want him considered psychotic. Therefore, Maudsley withdrew mention of Swedenborg in the second edition of his book, and Swedenborg's disciples recast his hallucinations and delusions into dreams and visions.[8] They had good reasons for reframing his pathology. They found something helpful in Swedenborg's writing that they did not want painted with the mental illness brush.

In 1992, I was walking down a London street when I saw a large house with a sign indicating the headquarters of the Swedenborg Society. I asked if they had a copy of his book on planets, and the pleasant attendant disappeared into the basement before emerging with a crisp, new reprint. I noticed that the cover was bound upside down from the text, and she apologetically offered to find another. I insisted on having that copy, convinced it was emitting metaphysical vibrations into the surrounding ether. I was right: she gave me a 50 percent discount.

I wanted to ask her how the society now evaluated Swedenborg's belief in people living on other planets, but I could not think of a way to inquire without sounding provocative rather than curious. I think I understand now. An articulate Swedenborgian would likely reply that the planetary visits were merely his context to tell a greater truth. Like a myth, the details might not be historically accurate, but the message is important.[9] As one of his followers said, "It is not as a seer of Ghosts, but as a seer of Truths, that Swedenborg interests us."[10]

Swedenborg's message infused charity into Christianity, in contrast to Luther's insistence that "faith alone" was the door to salvation. Swedenborg believed that faith was the force that allows one to live a life of charity. His plurality made the world a bigger place, an idea that had comforted Fontenelle. Swedenborg's followers considered his views worth keeping whether he had sometimes been insane or not.

Not everyone admired his wild speculations about other planets. In 1785, a mining engineer in Cornwall, Rudolph Raspe, wrote a chapbook about the exaggerated and self-aggrandizing lies of a character named Baron Munchausen.[11] Noted for his ability to tell outrageous whoppers with a straight face, the baron provided bizarre descriptions of inhabitants on the moon and the Dog Star, also known as Sirius in the constellation Canis Major. This was understood to be an obvious ridicule of Swedenborg.[12]

During Swedenborg's lifetime, only about fifty people appreciated his doctrine.[13] Nevertheless, his writings eventually influenced a surprising spectrum of poets, philosophers, and theologians. One would think, however, that his embarrassing belief about contact with people on other planets would be quickly dismissed and ignored. But that did not happen. We turn now to a German theologian who wove Swedenborg's views of heaven with features of mesmerism to concoct a belief that one's soul could contact life on the "other side."

JOHANN JUMPS OVER THE MOON

Without focusing on celestial bodies, the occult meandering of Johann Heinrich Jung-Stilling (1740–1817) also contributed to thoughts about the possibility of contact with a spiritual world. Jung-Stilling was a highly regarded German scholar and theologian who also practiced as a physician. He cobbled together old German myths, Swedenborg's planetary travel stories, and the emerging ideas of Mesmer. The result was his own idiosyncratic theological goulash. The complex admixture was surprisingly influential given how bizarre all of this appears today.

Everything had significance for Jung-Stilling, but it was hidden or disguised. He discovered meaning in life, but it emerged only through esoteric knowledge.[14] For example, he endorsed the idea that the soul can "see" without using the sense organs of the body because when a person dies, the soul goes into the invisible world with a recollection of its earthly life. According to Jung-Stilling, experiments with somnambulists (those deeply mesmerized) convinced him that human souls can become en rapport with souls previously departed. Even when alive, one's soul may leave the body and connect with another through the ether. This ether penetrates and pervades everything, which allows transmission between the visible and invisible world. These propositions, Jung-Stilling said, "are sure and certain inferences which I have drawn from experiments in animal magnetism."[15] By "experiments," Jung-Stilling meant his reading of published writings (no matter how obscure) and his knowledge of stories told by individuals of unquestioned character. And herein lies a thread that we continue to follow: we begin to encounter honorable individuals who bring us observational evidence of things unseen. After rejecting the opinion of ancient authorities, "honorable people" became a reliable source of knowledge, a problem addressed later.

Jung-Stilling said that people living "holy lives" were often able to develop a faculty of "presentment" that allowed them to give messages to those in need of warning. He conducted experiments in which individuals in a somnambulistic state were contacted by angels and spirits who provided information otherwise unobtainable.

He acknowledged that some of these presumed contacts were not genuine spirits. Some individuals suffered from "delusions of magnetism" and were not really connected with the other side. (He reported an unintentionally funny story about one patient who was cured of seeing angels after being given a laxative.) Even worse, some somnambulists made contact with evil spirits who falsely presented themselves as good guys. Satan and his hosts easily produced false signs and wonders to deceive the faithful; therefore Jung-Stilling repeatedly insisted that all communications be tested for authenticity. He was determined not to be misled or deceived.

Because there were true and false prophesies, Jung-Stilling emphasized the importance of making certain that claims were "minutely and rigidly examined by the Word of God and sound reason."[16] He chided the materialists with their sophistry who dismissed these spirit contacts when any scientist could conduct similar experiments to obtain the truth. The true sage, he said, understood the difference between divine magic and the black arts.

While repeatedly suggesting that claims be challenged, Jung-Stilling wildly raged against those who doubted his conclusions. For all his trust in "research," the bottom line was that Jung-Stilling himself was the judge of "true" revelations. He was the only one with sufficient knowledge to separate those with true vision from those who needed a laxative.

No story that reached him sounded absurd or ill-founded if consistent with his expectations and belief. Despite his guideline that reliable seers were pious people, he recounted the story of a drunken gravedigger who had the ability to tell people when they would soon be in need of his services. The local parson insisted he stop these predictions because they might frighten some of the sick and elderly to an early death. Jung-Stilling was apparently unconcerned that his drunken gravedigging prophet was more profligate than pious; the man still had a spiritual gift.

The belief system of Jung-Stilling emerged from Germany's sordid entanglement with black magic, witchcraft, and the intrusion of devils. Nevertheless, he tried to represent himself as part of the Enlightenment. Yet his biography shows how deeply he was involved in occult beliefs. His disturbing childhood was characterized by social isolation, and his father was given to vivid hallucinations of angels and departed friends. He said he was often followed by a guardian angel and was sometimes seized by beings who tempted him.

All this moon jumping was rather speculative and theoretical. In the next chapter we review individuals who claimed direct experiences with life in heaven, on the moon, and on distant planets.

⓭

SPACE RACE

The traveling pilgrim, by virtue of his repeating his tales, finished by first half-believing them, then entirely, and his voice thence took that accent of truth which alone can produce conviction in the audience.

—Jusserand, *English Wayfaring*, 1889

The previous chapter reviewed individuals who created a philosophy based on belief in a universe inhabited by spirits and denizens of other planets—with other esoteric ideas thrown in. This chapter begins with individuals in the early part of the nineteenth century who claim more direct encounters with life in heaven, on the moon, and on distant planets. Here our credulity is put to the test. Just how much aberrant behavior and wild claims are we willing to believe? Fontenelle would be wondering how his discussion of plurality had been hijacked to such an extreme position.

The first encounter in this chapter is with doctor Justinus Kerner, who describes the life of a patient with abilities (or symptoms) and claims (or delusion) beyond what anyone had previously encountered. Yet Kerner and his patient started a "space race" in which other mesmerists and their subjects battled to see who could journey the farthest into the uncharted ether of psychic ability or psychopathology, whichever it was. Moreover, all of these psychic and paranormal activities within mesmerism later created a welcome environment for the emergence of spiritualism.

SYMPTOMS JUMP OVER THE MOON

In 1826, a young woman appeared in the German village of Weinsberg. She was nearly dead, wasted to a skeleton, and unable to stand without assistance. She was in a "magnetic sleep" and asking for the local physician, Justinus Kerner. Kerner (1786–1872) was a well-educated man assigned to practice in this remote area after experiencing multiple personal tragedies and family deaths.[1] Thus, he was sensitive to questions about the hereafter. Kerner became known for writing poetry and his civic role in restoring a local castle, complete with a gigantic aeolian harp whose plaintive harmonics reverberated through the valley. He also became widely respected for his therapeutic use of mesmerism, metallotherapy, and exorcism. And, thus, the appearance of the unfortunate young lady at his doorstep.

Kerner's unsolicited guest was Frederika Hauffe, the daughter of a vineyard keeper in Prevost, who became known as the Seeress of Prevost. Kerner kept a diary of his work with her over a period of three years.[2] Her life was so sensational that visitors thronged to see her, and I wish I could have been among them.

Kerner unintentionally revealed what happens when a seriously disturbed patient meets a physician with unlimited credulity. He found someone on whom he could project all his occult beliefs. Thus, his personal theology distorted his interpretation of her symptoms, all of which he recorded for us.

The seeress could fall into a deep trance or sleep at any time through Kerner's mesmeric passes or her own mental actions, which were sometimes intentional, sometimes not. He said that while mesmerized, she was so in tune with nature that she identified the curative properties of plants and minerals, often prescribing her own treatment. She predicted her own symptoms and even her own death. She said she did not have the strength to live on her own, so she took energy from the eyes and fingertips of those in her presence. Some said they felt the transfer of vitality from their own body into hers. She also gained strength by drinking mesmerized water and knew exactly how many passes had been made over any glass offered to her.[3]

The seeress could change her weight and defy laws of gravity. She had prophetic dreams, out-of-body experiences, and travelled to other worlds, where she talked with angels and spirits. She was uneducated but could speak in poetic verse or in languages previously spoken by sages. The knowledge she shared, Kerner said, was the intuitive wisdom once known to prophets but now mostly lost. She associated numbers with words, which allowed her to forecast the future. Kerner compared her wisdom with that of Pythagoras and Plato. She understood the moral laws that circumscribe reality after death. She could see with the pit of her stomach, but she could also perceive the past and future, a talent known as second sight. She communed with ghosts and warned people of coming dangers. These ghosts created strange knockings without any physical action on her part.

Kerner's report became an international sensation, and the seeress pushed the boundaries for somnambulists and clairvoyants who would soon become the prized patients of mesmerists everywhere. As a reminder, somnambulism in this context refers to a deep trance state, and clairvoyance refers to the apparent ability to see past or future events and things outside the vision of others. Even today, Kerner's book is regularly cited, but I wonder how many have actually read it. I suspect its cultural impact is so deep that authors presume knowledge of its content by reputation, like that of Herodotus.

Kerner and his seeress opened the door for others to visit worlds beyond our own. For example, about three years after her death, another seeress experienced similar somnambulistic flights to visit the moon and planets.[4] Other young women in magnetic somnambulism similarly reported visions that "support the claims of Swedenborg as the divinely authorized revelator."[5] To keep up with the Germans, a French mesmerist introduced eight of his patients who visited heaven.

ECSTATIC SOMNAMBULISTS JUMP OVER THE MOON

Alphonse Cahagnet (1809–1885) identified himself as an ordinary working man with a gift for inducing the mesmeric state. He kept a journal as he experimented with eight "ecstatic somnambulists" who looked so directly into heaven that they provided specific descriptions of what awaits us there.[6] These eight interacted with about seventy or eighty spirits, including angels, demons, philosophers, priests, and ordinary people who had recently died. His sixth ecstatic was described as so powerful that she could answer any theological, metaphysical, or psychological question. When provided the names of deceased individuals, she could see and converse with them, providing exact details about their lives. Her contacts had visited the moon, stars, and other places in the universe. Cahagnet was scrupulously fair in his investigation: "In order that the reader may judge my impartiality I give an account of all my questions and the answers obtained, without, in the least, perverting their meaning."

Cahagnet's book was translated into other languages, all generating wide admiration. Frank Podmore declared that "these trance utterances are at once amongst the most remarkable and the best-attested documents on which the case for Spiritualism depends."[7] In vivid detail we experience the bliss of angelic beings clothed in white, walking on beautiful lawns among surrounding beds of sweet-smelling flowers.

The visions of Cahagnet's ecstatics were so congruent that he claimed their unity proved the reality of heaven and confirmed the theology of Swedenborg. His rhetorical question was: who would still dare treat Swedenborg as a madman after so beautiful a description?[8] Those who laughed at Swedenborg, Cahagnet asserted, were now forced to admit he was right.

Actually, it was not quite that easy. If one looks closely at the statements of Cahagnet's ecstatics, they had disagreements that he did not notice or care to acknowledge. Reverend George Sandby, vicar of Flixton and an ardent believer, reluctantly identified many of these discrepancies.[9] Sandby noted that Cahagnet's devotion to Swedenborg made it unclear how much his clairvoyants were influenced by leading questions and expectations. The subjects were likely clairvoyants, he admitted, yet how much of their visions were only fantasy? One ecstatic witnessed the battle in which her uncle had been killed by a cannon ball. Was that a vision of an actual event or a picture concocted in her brain?[10]

Sandby noticed that one ecstatic's answers improved as sessions progressed, which might have resulted from practice rather than from better clairvoyant vision. He believed that some subjects got carried away and tried "to prove a little too much."[11]

Even more problematic, Sandby noted that their statements were devoid of meaning. Their notion of paradise was impoverished, like those of boys and girls who gathered their ideas partly from fairy tales and partly from Swedenborg. Sandby was dismayed by their melodramatic descriptions of flowers. He concluded by saying that if these clairvoyants had fallen into the hands of a mesmerist of a different belief, then ideas of someone other than Swedenborg would have emerged. They were dreamers who had deceived themselves, yet their wildest flights of imagination were accepted as holy inspiration. That possibility got Sandby wondering. Maybe Swedenborg himself was a self-mesmerized clairvoyant who worked himself into such a frenzy that he believed his own fantasies. Maybe it was all a cerebral illusion.

Despite Sandby's insights about confusing belief with truth, we later see that Sandby was unable to make critical distinctions about his own experiences.

Cahagnet was also a worthy target in *The Zoist*, the first English journal of mesmerism, which typically promoted clairvoyance. Fred Hockley said that Cahagnet's book did not support Swedenborg at all. In fact, Cahagnet was not even a Swedenborgian as he had proclaimed because his ideas were those of a pure theist. Hockley argued these finer points of esoteric theology then enlightened readers about the true nature of the soul: "the soul is the luminous material atmosphere which surrounds the body, described by many somnambulists as appearing like a lambent flame."[12]

Hockley then expressed his irritation with clairvoyants for testing his patience. He complained that the famous American seer, Andrew Jackson Davis, claimed he had gleaned extensive philosophical information from his visits with inhabitants of the planets and the moon.[13] But Davis stopped right where his revelations would be most useful and convincing—at the limits of our present knowledge. Therefore, Hockley expressed his doubts about whether Cahagnet would produce the book he had promised about alchemical secrets. If you do write this book, he demanded, make certain you reveal something not previously known. Then you will have more converts than Mesmer and all his disciples.

I have searched for Cahagnet's promised alchemical treatise on the dusty shelves of forgotten bookstores. Maybe Cahagnet never wrote that book because he got distracted by his attempts to obtain ecstatic experiences for himself. He tried occluding his carotid arteries, hoping to enter a trance by depriving his brain of blood, though he achieved better success drinking coffee laced with hashish.[14]

YOUNG JUNG JUMPS OVER THE MOON

Carl Jung (1875-1961) was a psychiatrist whose views on occult philosophy were shaped by Jung-Stilling, discussed previously and to whom he was related.[15] Jung was raised in a family with strong roots in unorthodox religious belief, and he believed he possessed second sight, inherited from his maternal grandfather who conducted séances.

Jung obtained many of his ideas about psychic reality from Kerner's book on the Seeress of Prevost, whom we met earlier in this chapter. Before he was twenty years old, Jung was directing séances with women, most of whom had strong emotional feelings for him. One was a maternal cousin nicknamed "Helly" who was not quite fourteen years of age. After some mysterious table tipping, Helly went into a trance and began transmitting dramatic messages from spirits. After three séances, Helly's father refused to allow her to attend any more. Jung was disappointed if not furious. He returned to his medical studies but diverted considerable time to books on spiritualism and occult philosophy. He was also impressed by French psychiatrists and mesmerists like Charcot, Pierre Janet, Alfred Binet, and Theodore Flournoy.

After two years without holding séances, Jung reintroduced them when Helly was able to attend. She told her father that she was going to the Jung home to work in their garden. She was now much better prepared to act as a confident spirit medium. Helly introduced to the group the spirits of many deceased individuals including the Seeress of Prevost, much to the delight of Jung. The seeress related marvelous tales, told through Helly, about travels between the stars and visits to Mars. She described intimate details about the martians and their highly developed civilization. Jung-Stilling had warned his readers about the sin of using trances to foretell the future or to amuse oneself by contacting the spirits, but Jung had long since cast off any concerns about religious taboos.

The group finally broke up when everyone realized that Helly was in love with Jung. She was bright and bold in faking her mediumship as a way of making an impression on him and keeping his attention. She had learned how to produce spirits from books that Jung had sent her to read, including Kerner's book on the seeress. Jung believed her whole song-and-dance routine until the group was finally forced to face the reality of what was happening.

When Jung began his medical dissertation, he described his experiences of these past séances as a way of explaining the pathology in spiritualism. In 1903, he published *On the Psychology and Pathology of So-called Occult Phenomena*.[16] Jung reinterpreted the séances and dressed them up with the popular ideas of French psychiatrists of the time. He cast Helly as suffering from hysteria, automatisms, and distortions of reality. He described her splitting into multiple personalities and finding spirits within her unconscious.

The publication of Jung's work scandalized his mother's side of the family. Helly and her sister were ostracized, and she never married. After a period of depression, she succumbed to tuberculosis and died at the age of thirty.

Behind the facade of Jung's current image, biographer Richard Noll documented a consistent history of fact-twisting that tarnishes the reputation of the man who popularized mythology and meaning in symbols. For Jung, the unconscious would always be a source of higher knowledge beyond the confines of time and three-dimensional space. He believed that one could establish a personal relationship with the voices and images of one's unconscious, one's inner "land of the dead."[17]

Like Kerner and Jung-Stilling, Jung promoted a wide range of ideas that have no foundation in reality and are of interest in academic psychology only for their historical consideration. Nevertheless, many of his concepts still have a lingering cultural presence: the archetype, the collective unconscious, synchronicity, introversion-extraversion, and dream analysis. Jung still retains a band of dedicated supporters, some of whom are highly defensive about the psychological test they commonly promote that is based on his theories, the Myers-Briggs Type Indicator. A National Academy of Science study concluded the test has too many methodological inadequacies for use in career counseling, its primary intention.[18]

SPIRITUALISTS JUMP OVER THE MOON

In most cases, mediums claimed to be in contact with spirits who informed them about spiritual life on the other side. Native Americans were popular spirit guides[19] as were angels.[20] More direct experience was claimed by a young, attractive Swiss medium. Helene Smith (a pseudonym) boasted a large throng of supporters who were enthralled by her ability to talk in the martian language and paint pictures of her planetary experiences. Here was the chance to prove the reality of planetary visitation.

Psychologist Theodore Flournoy (1854–1920) stepped up to investigate the truth of her claim. Flournoy clearly conceptualized his assessment question: were Smith's martian contacts an authentic revelation of affairs on the planet or a fantasy of her imagination? His question implies that he would give both possibilities a fair evaluation; he was not merely looking for information to confirm his private beliefs.

Flournoy made careful observations during séances with Smith, but he also checked into her past to see what experiences she brought to the situation. He discovered a history of childhood reveries, daydreams, visions, and hallucinations. She had the ability to create mental images of fantastic proportions that sprang from her with startling proficiency. These fantasies laid a pattern for her adult personality and were consistent with her lively conduct in séances.

It should be no surprise, then, that Smith's visions quickly became problematic for Flournoy. She could not answer the most basic questions about Mars, like the structure and purpose of its canal system or the sociopolitical arrangements that made such projects feasible. Her descriptions of life were nothing beyond what already existed on Earth or could be easily anticipated. She said, for example, that martians eat on square plates with a furrowed trough for gravy. Flournoy called such responses a childish fancy of a "profoundly infantile character."[21] She projected onto martians nothing more than a simple variation of us.

The martian language and writings of Smith consumed a large chapter in Flournoy's book. His tedious work was necessary, he said, only because her followers believed in the linguistic authenticity of her presentations. They deserved to know the facts that led to his conclusions. Her supporters were astonished by her ability to speak and write in an unknown language. But Flournoy was not that impressed; he viewed her performance as simplistic fantasy. In his analysis, her linguistic production was transparently silly because she merely twisted sounds of the French language. The writing was something like that created by a child with an exuberant imagination.[22]

Smith painted pictures of her recollections of Mars, mostly while in her normal state although sometimes in a semitrance, with spirits apparently controlling her hand. Again, Flournoy rejected these as representations of actual martian creatures because they appeared to be earthly forms dressed in eastern garb. The buildings were like pagodas with hanging gardens. She depicted Persian carpets, people in tunics, and Indian tapestries. Flournoy said this was from her earlier fascination with oriental culture, and he dismissed it outright. He left no doubt about his conclusions, and he repeated his criticism in many ways with little constraint or qualification. However, the divergence of opinions between Flournoy and Smith's supporters is worthy of consideration.

As we have seen, one's belief dramatically influences perception. Seekers at an oracle felt encouraged about better things to come when told "There is in the land of Trachis a garden of Herakles, full blooming, well watered, where fruit is gathered every day." But the impact was quite different when Oenomaus discovered that many others received the same boilerplate message that day. Alphonse of Aragon undoubtedly trembled simply knowing about the existence of The Donation of Constantine, but Lorenzo Valla perceived it as such an obvious forgery that he blurted out "O sancte Jesu." Porta laughed while peasants cowered, fearing the presence of the devil when sand moved across his table. In the same way, people were impressed with Helene Smith's pictures of Mars, but Flournoy knew she had been drawing pictures of people in tunics since she was a little girl.

Spiritualists reassured themselves that future explorers would prove their case, and they sneered at skeptics for doubting. Spiritualists said scientists got nothing new from the spirit world because of their doubt. Hiram Corson said that some scholars who had passed into the spirit world were studying the laws of nature for the betterment of our world.[23] Apparently, they were no more advanced than scientists on the Earth, and their concern for our physical welfare was pathetically passive. Maud Weatherhead said she was told by spirits about the scientific advancements of people on Venus and Mars, but she failed to provide any details.[24] Francis Smith[25] channeled the spirit of Sir Humphrey Davy,[26] England's great scientist and inventor. Davy told Smith that he would not reveal the secrets of science because men on Earth must discover such information for themselves.[27]

If new scientific knowledge was not forthcoming, spiritualists were certainly not short on inspirational lessons and moral platitudes. Even when laying out the absolute moral foundation of the universe, spiritualists provided little more than what mothers teach their children about the importance of being good. Flournoy's question separated him from Smith's supporters. They asked for evidence of her travel to Mars, and she gave them the martian language, paintings of her experiences, and exciting stories. Flournoy asked if these responses could have arisen "from the fantasy of her imagination." Once he looked for evidence to support an alternative possibility, he discovered a clear developmental path from her youth to her séances.

And, finally, progress in our understanding of extraterrestrial space has shown that the spiritualists were not merely wrong; they were dramatically, outrageously, and ridiculously wrong.

HOAXES JUMP OVER THE MOON

In August 1835, readers of the *New York Sun* were informed about an article that had appeared in the *Edinburgh Journal of Science*. The *Sun* was a penny newspaper that attracted working-class readers with lurid headlines reporting colorful neighborhood crimes. Readers were not likely to have much interest in a report by an assistant of John Herschel, a British astronomer working in South Africa. Nevertheless, there on the second page was his article on the history of telescopes, some technical features of their construction, and the currently understood structure of the universe. The most newsworthy part mentioned that Herschel had developed a new instrument measuring twenty-four feet across and weighing nearly seven tons. And it magnified forty-two thousand times.[28]

The next morning, the paper carefully corrected its previous report, explaining that it mistakenly stated the cost of the telescope as seventy thousand American dollars, but it was actually British pounds. This small touch, as Pooh-Bah said, "gave artistic verisimilitude to an otherwise bald and unconvincing narrative." The story soared into a description of Herschel's discoveries on the moon. His telescope projected images onto a large screen. In exquisite detail, the article described pastoral meadows, verdant forests, and wonders never seen. Moving their view further along, they observed lunar bison and agile goats.

The circulation of the paper soared. The third report created such demand that the presses could scarcely keep up. The story included volcanoes, beautiful birds, exotic animals, and a humanoid beaver that was probably utilizing fire.

The fourth installment reported the appearance of four-foot creatures like men who flew in the air with nearly transparent wings that could be folded behind them. These man-bats lived in peaceful bliss, eating fruit, bathing, and enjoying their leisure. The details were so enthralling that everyone had to have a copy. The *Sun* had prepared twenty thousand reproductions of the article and later sold sixty thousand in a month.[29] Other newspapers quickly reprinted the story and raised their prices, selling newspapers as fast as they came off the press.

The fabricated story was created by Richard Adams Locke, an American educated at Cambridge who had moved from London to New York only a couple of years previously. He quickly became the *Sun*'s head writer after producing a sensational piece about "Matthias the Prophet," an insane man who murdered someone near New York City. The editor urged Locke to try something new, and his mind wandered back to an article he had read in the *Edinburgh New Philosophical Journal* by Thomas Dick, an amateur astronomer who speculated about the possibility of corresponding with the inhabitants of the moon. Dick had many unusual ideas, such as God punishing evil with natural disasters, so Locke decided to write a satire or hoax based on Dick's unrealistic expectations of his own telescope. (William Griggs says that Locke intended his story to be a satire, but he quickly played the hoax game once the article was accepted as

Inhabitants of the moon as purportedly discovered by John Herschel's telescope. Richard Adams Locke, *The History of the Moon, or an Account of the Wonderful Discoveries of Sir John Herschel* (London: printed by B. D. Cousins, 1835). *From the author's collection*

real.[30]) Locke centered his fiction on Sir John Herschel, son of famous astronomer William Herschel, who was at that time mapping the southern hemisphere
skies in South Africa.[31]

A group of scientists from Yale who believed the story asked Locke for a
copy of the *Journal* and were sent on a wild goose chase from printer to editor
in an endless loop. During a press conference, one man claimed he had a copy
of the *Journal* and that the *Sun* had reliably reported its content.[32] Another man
asserted that he was present when Herschel's seven-ton telescope was loaded
on the ship bound for South Africa. When Herschel was informed of the story,
he had a hearty laugh at the joke. Eventually the story fell apart, but the *Sun*
remained popular and Locke continued his successful career. Eventually the
joke got old for Herschel, who was continually pestered about whether he had
really identified men on the moon.[33] People threw away their cheap pamphlets
in disgust, making them now rare and expensive.[34]

ASTRONOMERS JUMP OVER MARS

The work of Italian astronomer Giovanni Schiaparelli inadvertently set an expectation for intelligent life on Mars in 1877 when he publicized his discovery of
multiple *canali*. The Italian word *canali* refers to channels, ditches, or grooves;
therefore, the word was reasonably translated into English as "canals." We now
know that Schiaparelli was born near the city of Canale, Italy, which was near
several major transportation canals.[35] His mention of ditches or channels may
not have been inadvertent.

This fervor of canals inspired Camille Flammarion, a French astronomer and
committed spiritualist with the skills of a popular writer. His scientific writing
about Mars, which started in 1892, was difficult to distinguish from his science
fiction.[36] His *Urania* infused scientific discoveries with a cultural agenda that has
been described as an early example of the misappropriation of scientific research
on Mars.[37] Flammarion decided that we would someday communicate with
that planet through psychic waves.[38] An American cartoonist called him Flim-
flammarion. Nevertheless, his mystical presentation of martian society gained
enormous popularity and sent science fiction writers flying off to the red planet.

More cautious individuals trusted the opinions of American astronomer Percival Lowell (1855–1916).[39] Lowell was born into a distinguished Boston family, which provided access to a Harvard education and the privileges of wealth
and travel. He had no special training in astronomy but established his own
private observatory outside of Flagstaff, Arizona, where observations would

not be disrupted by clouds, city lights, and air pollution. That location also put Lowell out of reach of Harvard astronomers who frequently told him his belief about Mars habitation was pollution that distorted his vision.

Although the public and press loved him, astronomers began to identify many problems with Lowell's conclusions. For example, Blackwelder enumerated several of his "astonishing and disastrous" misunderstandings of geology in his comparison of the Earth and Mars.[40] Then, when Mars appeared close to Earth in 1909, even the astronomers who had supported him realized that his canals had now dissolved into a complex of spots and irregular details.[41] Nevertheless, Lowell maintained his belief until his untimely death in 1916.

The telescope improved Lowell's vision, but failed to help his poor eyesight.

WHAT GOES UP MUST COME DOWN

Over the years I have acquired books and pamphlets by individuals who describe their contacts with extraterrestrial beings, mostly from the UFO craze in the 1950s and 1960s, although one is from 1922.[42] Some of these books are scams and hoaxes,[43] of course, but most have a common theme. They promote a moral imperative that the rest of us must acknowledge and immediately heed or suffer some disastrous consequences. Like those books about heaven, these space travelers are selling salvation. Those who reject the good news are doomed to live in the hell we have created here on Earth.

Authors select aliens as the harbingers of their messages for reasons that are easy to understand. An ordinary person demanding that governments abolish boundaries, live in brotherhood, and clean the environment would sound impertinent. Thus, the mandate is delivered by an extraterrestrial civilization offering a cosmic plan to solve Earth's problems.

At first glance, these stories have a fascinating attractiveness. On the other hand, most are pathetically simplistic. The authors are convinced that scientists are wrong while blithely ignoring the limitations of physics. Spaceships cannot travel "several times the speed of light,"[44] even with the advanced skill of the craft's operator.[45]

Are all these writers merely kooks?[46] Belief in extraterrestrials is alive and well in the minds of those who attend UFO festivals across the nation. At the same time, scientists at the SETI Institute are searching for intelligent life on other planets. How would we know if aliens have tried or are trying now to contact us? What evidence would we need to be convinced?

Some were convinced by grainy photos or the testimony of people who saw something in the sky they could not explain. Perhaps we should listen to Fontenelle's philosopher who suggested we give only half of our mind to a belief so that we can change it when new information arrives. Those grainy photos did not provide new information.

SECRETS OF CHEATS

POSTER BOY
FOR CLAIRVOYANTS

When once the minds of a people are prepared with a solution
for every event, there will never be wanting events adapted to
the situation.

—Salverte, *Philosophy of Magic*, 1846

Alexis Didier was a young Frenchman who provided entertainment for the rich and famous in 1844 Victorian England. But his amusements created controversy that raged within the medical and scientific journals of the day. The editor of a British medical journal said of him, "He came, *saw*, and conquered." Didier *saw* what no one else could see because of his clairvoyance, his ability to perceive the past, future, and circumstances outside the range of ordinary vision.

Didier amazed his admiring public with demonstrations unlike anything ever witnessed. After being deeply mesmerized, Didier claimed to become clairvoyant to the extent that he could read words sealed in envelopes and peer into the lives of those in his presence. He described past events and distant details that he could not have known. Historian of hypnosis Alan Gauld said that if a case for clairvoyance cannot be made for Didier, then no case can be made for anyone.[1]

I always considered clairvoyance a quaint historical belief, now the domain of mental entertainers and television psychics who write foolish books for publicity.[2] But Didier was different. I first encountered him in Alison Winter's excellent book on the cultural impact of early British mesmerists.[3] She did not claim his clairvoyance was real; her goal was not to evaluate his veracity but to document his enormous influence on some of the best minds in England.

My quick searches on the internet only reinforced Didier's reputation. For example, an essay by Bertrand Méheust[4] said the French Academy of Medicine had been afraid to test him. What quickly drew me into the story, however, was his claim that Didier's ability had been endorsed by French magician Jean-Eugene Robert-Houdin, the father of modern magic, whose name was later

Alexis Didier (1826–1886), poster boy for clairvoyants. *Purland scrapbook, National Library of Medicine*

appropriated by Harry Houdini. Robert-Houdin was in a position to evaluate the clairvoyant ability of Didier because he had perfected a second-sight routine with his son, which they performed with acclaim in his Paris theater. In my opinion, an endorsement of Didier by Robert-Houdin was more powerful than the opinion of the French Academy. The *Illustrated London News* was not exaggerating when it described Robert-Houdin as "the sole monarch of the world of wonders; all other conjures and wizards, from whatever point of compass they arrived, sink into insignificant imitators before him."[5]

The first English translation of Robert-Houdin's report on Didier was written by Alfred Russel Wallace, the cofounder of the concept of evolution with Darwin. Wallace offered his translation as a rebuttal to Frank Podmore's conclusion that Didier "might have been, and probably was, a clever impostor." Wallace reminded readers that Didier's clairvoyant vision was too incredible for trickery. The tests conducted by Robert-Houdin offered final proof. Robert-Houdin carefully bandaged the eyes of Didier, yet Didier named the ten cards that Robert-Houdin dealt for a game of écarté, a popular card game of the day. Moreover, he named cards in Robert-Houdin's hand with "accuracy and infallibility." Robert-Houdin was "convinced of the utter impossibility of chance or conjuring having been responsible for such marvelous results."[6]

I was left with one certain conclusion: Didier was the poster boy of clairvoyants. I started wondering about Alan Gauld's statement that the case for clairvoyance is made with Didier. How solid was the evidence for Didier's clairvoyance?

I spent a couple of days wondering if certain individuals might have evolved special sensitivities, just as some people have unique physical, mental, or musical abilities. Alan Gauld suggested that maybe telepathy does not break the rules of science.[7] Perhaps under certain conditions some individuals can be in contact with past or future events in ways that have not yet been explored or understood.

A SMALL CRACK IN A LARGE WALL

As I thought about that essay by Méheust, some of his comments suggested that something was amiss. Méheust mentioned an examination of Didier by a noted literary figure, Reverend Chauncey Townshend, who practiced mesmerism with his friend Charles Dickens.[8] Townshend held an open book, *Jocelyn*, which Didier read with his eyes closed. Then Didier asked if Townshend wanted him to read pages further into the text, and Townshend suggested he read eight pages ahead. Didier traced invisible words with his finger while "reading," "*a déchiré d'un trait toute ma sympathie.*"[9] Townshend counted down eight pages and discovered that Didier read correctly with the exception of a single word, exactly where his finger had traced. Didier had said "*déchiré*" instead of "*dévoré.*" Townshend declared that "human incredibility began to stir in me, and I really thought perhaps Alexis knew *Jocelyn* by heart."

Incredibility began to stir in me, as well, but of a different kind. If Didier was truly reading, I wondered, why did he misread a word? Moreover, if he could read through eight pages, why did he read only a simple phrase instead of a complete sentence or a full paragraph? An alternative explanation was that he had caught a glimpse of a few words that he gave back incorrectly. The most likely explanation was that he was *pretending* to read, which caused Townshend—and Méheust—to be caught up in the excitement of an apparent miracle. Townshend accepted mesmeric lucidity as the explanation instead of considering a natural solution, which he could have tested by asking Didier to read more. Could Didier read seven pages down? Nine pages? In reality, Didier did not read eight pages ahead; he misstated a simple phrase. Townshend failed to think of ways to rule out cheating, in contrast to Flournoy, who accepted the responsibility for deciding whether his subject had visited Mars or had constructed a fantasy from her imagination.

Magicians have devised dozens of "book tests," in which apparently random words are mysteriously divined. In my own collection I have several gimmicked books—even a gimmicked dictionary—as well as several books and pamphlets describing ways to accomplish this feat even with randomly chosen books. But none of them goes back to 1844. This poster boy was certainly worthy of serious investigation. My initial bewilderment turned to eagerness as I wondered if there might be any remaining traces of DNA to crack this cold case.

My confidence in uncovering the tricks of Didier—if they were tricks—did not last long. I consulted *The Zoist*, the first English-language periodical devoted to mesmeric arts, which provided extensive coverage of Didier's 1844 appearance in England.[10] Reading those contemporary descriptions made me realize why people were so overwhelmed by him. I could see why it was difficult—or maybe impossible—for the brightest men of the day to offer alternative explanations for his incredible feats of apparent clairvoyance.

WHO WAS DIDIER?

Alexis Didier was born into a poor working family in Paris in 1826. His parents sent him to a mesmerist for treatment of epileptic seizures, and there he demonstrated an ability to enter a deep somnambulistic trance. During his teenage years, Alexis volunteered as a subject for a demonstration of mesmerism conducted by Jean Marcillet, a former military officer who was part of the underground revival of mesmerism alternatively known as animal magnetism. The convergence of a powerful mesmerist and an excellent somnambulist created perfect conditions for exploring the mesmeric state.

Marcillet was the owner of a successful transportation firm, and he hired Didier as a clerk. The two sometimes travelled together to give demonstrations of mesmerism. Then in 1844, when Didier was eighteen or nineteen years old, the two abruptly left for England where Marcillet's adult son was living. Despite not speaking English, they quickly drew the attention of elite British society. Of particular importance was Dr. John Elliotson, a supporter of all things mesmeric. Elliotson observed Didier perform at his home in June, and the July issue of Elliotson's journal, *The Zoist*, contained two letters of praise for Didier. Two issues later, January 1845, Elliotson printed multiple reports endorsing Didier, and the French lad was on his way to fame.[11] From these extensive reports and many other sources, I was able to reconstruct an outline of Didier's typical performance.

MESMERIC MIRACLES

Marcillet began his demonstrations by bringing Didier into a deep somnambulistic state through a series of the standard mesmeric passes. This resulted in a stiffening of Didier's body, as commonly experienced with this procedure. Then Didier sat on the edge of a chair with his legs extended while a gentleman stood on his unsupported thighs without noticeable change. After this, Didier was carefully blindfolded by audience members. Pads of cotton placed on his eyes were held by bandages around his head, and he proceeded to play écarté. Didier separated out the low-value cards, as required for play, and he seemed to know the cards in his hands even without turning them over. At times he knew the cards in the hands of his opponents.

After the game, Didier read the titles of books handed to him. Then with the blindfold removed, he read passages covered by paper, sometimes reading lines a few pages further into the book, often identifying the very place on the page where these words would be found. He also read words written within a folded piece of paper and words in envelopes and boxes. Didier often ended his demonstrations by mentally travelling with someone into their past or to their home where he described the location of doors, windows, the fireplace, and household furnishings. He described pictures on the wall and ornaments on the mantel. This visual activity became known as "remote travelling" or "travelling clairvoyance" and is currently called "remote viewing."

In one situation, Didier instinctively knew that a man was seeking his service to locate 20,000 francs embezzled from his business.[12] Didier identified the thief who he said was hiding in Brussels at the Hotel des Princes, gambling away the money. Elliotson said that Didier could identify criminals by name as easily as he could identify a word written on a card and sealed in a pillbox.[13]

As a typical example of his "remote travelling," I provide here the report of Henry Sims, presumably a complete stranger.

> Didier took my hand. I asked him if he could tell me where I lived. After a good deal of hesitation he said, "North-east of London;" and gave the distance very correctly in leagues. He then said, "There is a railroad which leads to your part of the country. There are two branches to this railroad, and your house is situated on the left branch; and on the right side of that branch." He then called for a piece of paper, and began to draw a map of the part of the country he was describing. He delineated the railway with great correctness, marking the branch which turns off eastward at Statford, and continuing to a point where he said there was a station. He gave a very minute account of the position of this station, answering in all points to that of Roydon; the river running nearly parallel to it, and the bridge immediately in front: and he also described with much truth the general character and appearance of the surrounding country, and said that the railroad extended only three or four leagues from this point, which is the fact. He then marked on his chart another station, a few miles farther on, and gave exactly the relative distance and position of my house with these two stations. He then said, "Now let us go to your house," and proceeded to give a sketch of the road with its various turnings. As he approached the house he was more minute, and described with singular correctness the sudden descent; the brook about half as wide as the room, the steep ascent on the other side, and the gateway on the right hand of the road. He gave the distance of the house from the gateway very exactly, mentioned a piece of water on the right with ducks upon it, and described the position of the stables, etc. The perfect accuracy of the whole of this minute description was truly astonishing.[14]

After an hour, Mr. Sims stated that Didier "began to make mistakes, and I would not suffer him to proceed, being perfectly satisfied with what he had already done." Reports like this were peppered throughout the pages of *The Zoist*, and they were overwhelming.

Yet I wondered why Didier started making mistakes with Mr. Sims. Was he tired or did he exhaust the information someone provided to him? Our homeowner, Mr. Sims, was convinced of Didier's power, so mistakes were no longer of interest to him. But mistakes and failures are important in scientific investigations when exploring the parameters of an unknown phenomenon. The more I thought about Didier's failures, the more I became convinced that they would lead to an understanding of his methods. And at the beginning of Mr. Sims's report was a clue suggesting how he might have known the local geography, a point I return to later. But first I needed to develop a strategy.

I started by identifying seven different categories of effects Didier commonly performed: (1) stiff-leg endurance, (2) ability to see while blindfolded, (3) ability to play écarté blindfolded, (4) ability to read covered words in a book, (5) ability to read hidden words, (6) various mind-reading feats, and (7) remote travel.

My goal was to identify a possible method for each effect. Any magic trick can be conceptualized in terms of its method and effect. The effect is what the spectator sees or experiences. The method is what the magician does to accomplish the effect; it's the secret that must be kept hidden.

A magician may think of an effect that he wants to accomplish, like floating a lady in the air or getting a signed card into his wallet. Then he tries various methods that might fool the spectator. Sometimes a magician knows a method or constructs a gimmick, and he imagines various ways to create an effect. For example, how many different tricks can you perform with magnetic sand? Or a magnet inside a floating wooden swan?

For each of Didier's effect, I wrote down the circumstances of any failures, situations he avoided, and conflicts about restrictions. Refusing to accept a limitation would suggest a method. The task was to discover whether Didier had available methods to accomplish his effects.

I was surprised by how much my first readings of Didier had blocked my critical thinking. On rereading the reports on him, I noticed all sorts of clues that I did not catch the first time. Nearly all of Didier's fans acknowledged his mistakes and limitations, but they were trivialized as if to magnify the importance of his accomplishments. Perhaps the authors were attempting to depict him as merely human (we all make mistakes), but his fallibility never piqued their curiosity about his range of limitations or the possibility of trickery. My adventure quickly turned exhilarating as I attempted to untangle the mysterious and occult claims from an understandable, entertaining performance.

Before discussing my findings, I need a moment to explain why I lay out so much background for each of the seven categories. Scientific articles typically begin with what is commonly called "a review of the relevant research." Authors describe existing research—what is already known—to provide context for their current study. Similarly, I searched the relevant literature to find similar effects in which trickery was the understood method. That allows us to see whether Didier was ever successful when blocked from using a specific method. You will notice, incidentally, that most of my examples are from performances after the time of Didier. That means his audience would not have ever seen a similar accomplishment. With no previous experience, they would have no way to judge whether he was displaying a mystery of clairvoyance or exploiting the secrets of a cheat.

I begin with the first demonstration that Didier usually exhibited after he was mesmerized, the stiff-leg effect. It is easy to throw that piece of hokum out the window.

1: DIDIER'S STIFF-LEG ENDURANCE

Marcillet reinforced the idea that Didier was deeply mesmerized by having someone stand on his stiff, outstretched legs. Standing on a stiff body suspended between two chairs became popular with early twentieth-century hypnotists who called this "catalepsy" or "the human plank." Many of these performers described the procedure in their pamphlets and training manuals, some admitting the stunt had nothing to do with hypnosis. Any fit person can perform this feat, although some recommended using shills who had practiced the routine.[15]

The human plank becomes more interesting when a boulder is placed on the subject's chest and broken with the blow of a sledgehammer. The pounding has little effect on the person beneath because the large mass of the rock absorbs the force of the stroke.[16] Burlingame created a steel corset for a female subject who held a block of cement on her stomach.[17]

This sideshow chicanery goes back to the third century, when a Syrian recruited some followers by demonstrating his strength. With an anvil on his suspended body, his assistant forged iron with hammer blows. Historian John Beckmann said the anvil trick was displayed all over Europe in the seventeenth century by a German strongman, John von Eckeberg, and remained so popular there that it was no longer as startling to audiences.[18] Didier and Marcillet, his manager, may have known about this and adapted their own variation for display in England, where it was not known. Magicians and other performers have amassed similar secrets using what is often called "body magic."[19]

I found only one person who questioned the supernatural interpretation of Didier's stunt. Dr. John Forbes stated that someone *partially* stood on Didier's outstretched legs, and he concluded that any fit person could do this with

Catalepsy or the human plank: a popular demonstration by travelling hypnotists. Eugene F. Baldwin and Maurice Eisenberg, *Leaves from the Note-book of a Hypnotist, Including the Revelations of a Mahatma* (Chicago: Baldwin, Eisenberg & Flint, 1897). *From the author's collection*

practice.[20] Forbes did not specify what he meant by "partially," but then I have never seen a description of exactly where and how the person stood on Didier. The performance could also be understood based on principles similar to what is today identified as "applied kinesiology."[21] Some quack practitioners use this as a test for various diagnostic and prescribing purposes. The patient holds one arm parallel to the ground and the doctor pushes down at the wrist. Then, for comparison, some critical substance considered beneficial or toxic is placed under the tongue. Now the wrist either goes down more easily or remains rigid, depending on some undisclosed factors that determine the outcome. For example, the expectations of the person testing and of the subject alter the response, as does pressing the arm down faster or more slowly. Slightly altering *where* one presses on the arm also has an effect. However, the quack has the biggest advantage over the subject by knowing *how* to press. Pressing down and toward the person causes the arm to go down easily, but pressing down and away gives the subject the sensation that he is stronger in resisting. This very subtle modification of direction can produce miracles, best illustrated by a fifteen-year-old Georgia girl, Lulu Hurst, the Georgia Magnet.[22]

Lulu travelled across the United States in 1883, throwing strongmen around the stage without any apparent effort, giggling all the while. For example, she challenged the strongest man in the audience to push her over by exerting pressure on a pool cue that she held in front of her. She said the wooden stick insulated her from his power, and in multiple situations men were helpless against her. She was examined by a committee of scientists at the home of Alexander Graham Bell, and none could fathom her ability.[23] Lulu earned a fortune during her year of touring. Then, to the dismay of her parents and stage manager, she suddenly announced her retirement. She enrolled at the University of Georgia to study physics, where she gained understanding of leverage, center of gravity, and vectors of force, which had given her an invisible advantage against powerful opponents. Once she had studied the "relevant research," her powers were no longer mysterious. She had intuitively utilized principles of physics to develop an entertaining program of parlor and stage tricks.

Didier was exploiting physics, not mesmeric powers.

2: DIDIER'S ABILITY TO SEE WHILE BLINDFOLDED

After his stiff-leg demonstration, Didier was usually blindfolded so that he could demonstrate his ability to play a card game, read the titles of books, or identify things handed to him. Those who blindfolded him were certain he could not see, but our review here shows that magicians and pretenders have devised multiple methods for defeating the most careful attempts to obstruct vision.[24] The bold performer can claim "eyeless sight" even when constrained in a seemingly impossible condition.

Before the appearance of Didier, at least two individuals in England were tested for claims that they could see with their eyes bandaged. In 1804, a two-year-old girl in Staffordshire, Catherine Mewis, lost her vision after a bout of scarlet fever.[25] Then by some miracle—interpreted as a sign of end-times—her vision returned every Sunday. Her whole demeanor changed on that day, and local doctors confirmed that she could identify the letters on book titles despite the bandages on her eyes. Similarly, in 1816, Margaret M'Avoy of Liverpool read from books with her eyes carefully bandaged.[26] When she came to longer words, she placed her left finger at the beginning of the word and her right finger at the end; she properly identified the word when her fingers came together. She claimed that she had difficulty reading when her hands were cold and that she was better between the hours of ten and twelve on alternative days. She was successful through several rigorous tests, and she identified color with the back of her hand. However, she could not see in the dark, and she refused to allow any shield between her eyes and the object of identification. Despite wide endorsement of her powers, one critic concluded that many "went to be pleased, and to be imposed upon."[27]

The most impressive blindfold act ever exhibited was certainly that of Kuda Bux (1905–1981). Born in Kashmir, Bux immigrated to England in 1935, where he volunteered for the first walk on fire ever investigated by scientists, which I described earlier. However, Bux was primarily known as "the man with the x-ray eyes." He requested doctors and nurses from his audience to blindfold him by placing coins and then bread dough over his eyes. Then multiple strips of folded cloth were tied behind his head so that only the tip of his nose was visible. In this apparently impossible condition, he copied or traced writing on a blackboard or read whatever was handed to him. In Liverpool, he walked on the narrow ledge of a roof some two hundred feet above ground, causing a sensation among those who witnessed this feat, especially those who had blindfolded him.[28]

Bux eventually settled in Hollywood, where he became a regular star at the Magic Castle. After repeated observations of his act, Magic Castle founder William Larsen declared that his act was simply impossible to explain.[29] The secrets of Bux were revealed as a tribute to him by Burton Sperber in a limited-edition book of mentalists' secrets. Sperber said that Bux was so amazing in his routines that "like Houdini, there will probably never be another performer like him."[30] Martin Gardner said that the high forehead of Bux with his sunken eyes endowed him with the perfect facial structure for his blindfold trickery.[31] He was able to foil any attempt to block his vision by utilizing multiple outs if someone tried to vary his regular routine.

It does not take an accomplished magician to amaze those unacquainted with methods of defeating the blindfold. Reverend Roscoe Ronald Coyne (1943–1994) performed his blindfold act at Pentecostal meetings across the states. Coyne was blinded in one eye after an accident when he was seven years old, and he pretended that God gave him the ability to see through his glass eye or through his empty eye socket—his claim varied. Coyne declared his sight was restored at a healing campaign when he was eight, and his mother took him preaching throughout the Southwest when he was ten.[32] Believers were convinced he could still read with his good eye bandaged, but his method was crudely amateurish. On a YouTube video, you can see that Coyne holds each item at exactly the same spot in relation to the slant of his nose.[33] You can get a sense of how he thwarted his blindfold by cupping your hand over your eye. The smallest crack of light down your nose enables vision, commonly known by magicians as "the peek down the nose." However, Coyne apparently had difficulty seeing that his four-hundred-plus pounds of body weight would contribute to his death at the age of fifty.

A pastor once asked Coyne if he could see out of the eye socket all the time or just when the Holy Spirit allowed him.[34] Coyne responded: "If I get in the presence of someone who absolutely refuses to believe, the Holy Spirit can become so grieved that I begin to lose my vision through the empty socket." In other words, if someone sets conditions that prevent him from cheating, then the Holy Spirit refuses to bail him out.

IT'S ABOUT THE MONEY, STUPID

Bertram Méheust, the man who enthusiastically endorsed Didier's powers,[35] claimed that the academy refused to investigate Didier because it knew from his demonstrations to the English that he could pass its tests. This misrepresents the history I discovered. Indeed, the French Royal Academy of Medicine had a record of embarrassing mistakes, like its rejection of vaccination, its dismissive assessment of the steamboat, and its belief that meteors could not reach the Earth.[36] And yes, it was reluctant to investigate mesmeric vision, but its decision was based on experience with claimants.

In the early 1820s, mesmerists demanded the Royal Academy of Medicine revise the 1784 report that rejected animal magnetism. Responding to pressure, the academy appointed a commission in 1826. For five years the committee bungled along with little direction or thoughtful assessment.[37] Its 1831 report endorsed some aspects of mesmerism (like analgesia and clairvoyance), but historian Margaret Goldsmith said its conclusions were as silly as its experiments.[38] Alan Gauld described the report as "hopelessly inadequate."[39]

Another controversy created the establishment of a commission with the initial intention of examining the clairvoyant patient of a Dr. Berna.[40] After four sessions, the commission realized no claimant of eyeless vision would accept the limits necessary to test the matter. After the commission's 1837 report, the academy concluded that magnetizers were prone to exaggeration and somnambulists were prone to cheat.

An academy member made a two-year offer of 3,000 francs to the first person who could demonstrate an ability to read without use of the eyes, and an Englishman offered a similar £100 prize.[41] Of the thousands of clairvoyants in France at that time, only two applied for testing. They failed completely once the possibility of deception was eliminated.[42] Therefore, the academy adopted a resolution stating that it would no longer respond to requests to examine claims of eyeless sight after the offer of prize money expired.

Méheust said the academy's refusal to test Didier served "a strategic function, that of limiting, by tacit convention, an obscure domain of experience, thus stopping the flight of thought into the unknown."[43]

The academy did not deny Didier because it was concerned that he would prove it wrong. Marcillet and Didier could have sought the prize money, but they chose to exploit their tricks in lecture halls and private séances where, most likely, Marcillet exaggerated Didier's powers while Didier cheated.

Méheust said that Didier's ability confounded intellectuals, leaving "an immense polemical and heuristic impact." Therefore, the academy's response was to damn him, flattening out modern science into a bleak and narrow surface. They issued the "official truth" by simply dismissing Didier's lucidity as a myth. In contrast, Méheust said he preferred the "common or popular level of truth" conducted by informal researchers. These informal researchers accepted the accomplishments of Didier as true, in contrast to those in power, who by their decree created a vast forgetting of reality, a massive cultural repression, and a denial of our human potential. Méheust said that science did not need to take this rationalistic and materialistic path but by doing so rendered what was once possible impossible. By dismissing Didier, the academy threw out all possibility of mystical experiences and maintained its scientific materialism. As a result, nature retaliated against science by throwing paranormal events at us, like prescient dreams and flying saucers, in order to mock us and shock us out of our narrow thinking.

Méheust can accept whatever alternative facts or flavor of truth he personally prefers. For my part, I simply wanted to know if Didier cheated when he was blindfolded. Then I discovered the reports of Dr. Horatio Prater, who directly examined Didier for the possibility of cheating.

TESTIMONY OF AN EXPERT WITNESS

Horatio Prater spent his life challenging his colleagues to base their care on better scientific standards. He was dismayed that physicians had given up their reliance on Galen, the ancient Greek physician, only to cast their lot with Paracelsus, a sixteenth-century physician whom he called a quack.[44] Prater accused physicians of pretending their profession was a science where no science existed and of asserting an attitude of superior knowledge over the gullible public.[45] He encouraged physicians to study chemistry and physiology as a guide to treatment, and he offered suggestions about how to conduct meaningful investigations in medical practice.

Prater investigated ether as a general anesthetic, but he was concerned about the side effects, which were sometimes fatal. Therefore, he studied an alternative anesthesia—mesmerism—with every hope that it would also produce additional benefits. However, he left England for an extended tour of Greece. On his return, he noticed that mesmerism was more popular than ever, but its acceptance was accompanied with careless investigations. He believed scientific truths should be based on observable facts that transcended political and religious differences. Therefore, he dusted off his old lectures on mesmerism for publication as a way of reminding people that research, not belief, should determine facts.[46]

Prater's philosophical leanings were not materialistic. He believed that people had a spiritual power within that might be sufficiently sensitive to perceive some natural cues that the self could not detect. However, his belief in a spiritual dimension—a vital force—did not restrain him from conceptualizing experiments that could sort things out one way or the other. For example, he observed that mesmerists usually got the results they expected. This caused him to wonder if subjects were aroused from trance by the "life force" that mesmerists typically breathed on their subjects or if they were responding to the cue of the associated breeze. Using himself as a subject, he directed his mesmerist to try several methods of diverting his breath so that he could not feel anything. Under these restricted conditions, Prater failed to wake up.

Similarly, Prater tested whether mesmeric passes were necessary and responsible for the trance state. He blindfolded subjects, taking care that no words or signals were leaked. When the passes were delayed, his subjects entered the mesmeric state when they *believed* passes were being made. He went on to describe the essential prerequisites for testing all sorts of mesmeric phenomena, such as sympathy of taste, insensitivity to pain, secret influence of the will, identification of diseases, and description of rooms. Prater was convinced that adequate controls had not been used in previous research despite everyone's insistence to the contrary. Based on the results of his own testing, he found no situation that required an appeal to an occult force.

After his first encounter with Didier, Prater had a strong belief that Didier did not have clairvoyant powers, but he did not feel justified in saying that with certainty. Therefore, he decided to pay Didier in order to test him, money that Didier got whether he was successful or not. Prater devised a number of specific tests that he introduced with as much graciousness as possible over several séances.[47] In one, Prater took the bandages that Didier claimed he could see through and laid them over the page of a book Didier had been reading while blindfolded. Didier's success immediately stopped. Another time, Didier was asked not to touch the blindfold once placed. Nevertheless, he persisted in trying to shift the bandages by muscular movement of his face so that he could obtain a peephole.[48] When Prater tried to introduce controls, Didier behaved as though he were mistreated or mistrusted. In annoyance, "the bandages were thrown off altogether." This was like the Reverend Coyne, who said the Holy Spirit would not help him if nonbelievers blindfolded him in a way that prevented cheating. However, Prater did not rely on one strategy to make his conclusion about Didier's vision.

In another test, knowing that Didier had not been adequately blindfolded, Prater held the book directly in front of his eyes, not down low where Didier could peek. When Didier moved his head upward, Prater moved the book upward. (This certainly looks as if Didier is using the strategy of a cheater.) After five minutes of trying, Didier admitted failure. Prater also held a card directly in front of Didier, which he was not able to identify. Didier requested the card be placed behind his head, hoping to catch a glimpse as it was moved, but Prater was careful not to expose it.[49] Didier eventually gave up.

Marcillet, Didier's mesmerist, no longer allowed anyone to hold a book for Didier while blindfolded.[50] This suggests that Marcillet was not a dupe of Didier's trickery but a confederate.[51] We now know that Marcillet, Didier, and Méheust kept a secret from the eyes of scholars like Alison Winter, whose insightful book introduced me to Didier. She was not an expert able to evaluate the finer points of blindfold trickery. She understood her role as a witness of fact who could testify only about the cultural effects of Didier's visit. But Prater, as an expert witness, tells us that Didier was cheating.

What bothered me most about Méheust's huffing and puffing of polysyllabic philosophical verbosity was that he denied Didier's failures and cheating. Méheust falsely stated: "Didier was never caught cheating—he was not even suspected on the basis of tangible facts." There were, of course, some direct catches, and suspicions are rampant if one understands cheating. Speaking of huffing and puffing, consider the following description of Didier identifying a card while blindfolded: "Sometimes, when he found a difficulty in ascertaining the card, he would beg me to breathe on it; and, when I had done so, he would tell it directly."[52]

Perhaps Méheust can explain why blowing on a card helped Didier identify it. My explanation is that the card was moved down from eye level, where he could more easily peek at it. Thus, I suspect Didier of cheating on the basis of tangible facts.

Authors of articles and letters to *The Zoist* rejected Prater's testing. Sometimes seething with sarcasm and anger, Didier's supporters wondered when these materialists and atheists would finally acknowledge the existence of something unexplainable, something transcendent. This history shows overwhelming acceptance of Didier's visual accomplishments. Horatio Prater was a rare individual who had sufficient curiosity to wonder about alternative explanations and the ability to devise adequate testing to answer his question.

In reading further, I discovered clairvoyants in France were commonly using blindfold trickery when Didier came to England, which might explain why he and Marcillet crossed the channel. Some of those French clairvoyants were also playing card games while blindfolded.[53] Didier could have learned his trickery from one of those pretenders, but more likely he knew the writings of Edme-Gilles Guyot, a French mathematician who exposed blindfold techniques and card manipulations in 1740.[54] If Didier was acquainted with Guyot's writing, he would certainly know the methods for cheating at écarté. We now turn to card trickery.

15

MIRACLES OR ENTERTAINING TRICKS?

[Conjuring serves] as a most agreeable antidote to superstition, and to that popular belief in miracles, exorcism, conjuration, sorcery, and witchcraft, from which our ancestors suffered so severely.

—Beckmann,
A History of Inventions and Discoveries, 1817

The dream of every card player and crooked gambler is to know the cards in an opponent's hand, and you can bet that cheaters thought about possible methods when the first deck was introduced. Reference to card cheating goes back to 1408, when several cunning men were caught in their dishonesty.[1] Discussions of cheating regularly appeared from that time on, but the earliest descriptions of card-cheating methods appear to be from our gambling-addicted physician, Gerolamo Cardano. In 1550, he described how a card can be surreptitiously marked with the fingernail and later retrieved. Since that time, the publications on card magic have continued to expand.

Few would guess how voluminous this body of knowledge has become. Robert Giobbi[2] identified sixty different themes of card deception. It would be fair to say that David Berglass[3] needed 391 pages to teach the skills necessary for one basic card trick, and Bob Farmer[4] filled 400 pages describing all the different methods available to perform a single trick, the "Bammo ten-card deal." Steve Forte recently published more than 1,100 pages revealing gamblers' sleight-of-hand methods.[5] Keep this in mind as we now consider Didier's card playing.

3: DIDIER'S ABILITY TO PLAY ÉCARTÉ BLINDFOLDED[6]

Didier often played his card game (écarté) while blindfolded but sometimes not. Is it possible that he cheated while playing his game? We have discussed blindfold trickery and mentioned the literature on card magic. Can an ordinary person fool us with an ordinary deck of cards?

Suzie Cottrell was an attractive twenty-year-old blond from Meade, Kansas, who claimed she could name all the cards in a facedown deck with a success rate of forty-eight out of fifty-two.[7] Her favorite demonstration, however, was predicting the name of the card a person selected from a deck she spread face down on a table. After she baffled Johnny Carson on the *Tonight Show*, her father arranged for her to be tested in the famous James Randi challenge. (Her father said they would refuse Randi's $10,000 reward if she won.[8])

Suzie signed a document certifying she had not used trickery in the past

A 1901 catalog for spiritualists offers a method for reading cards when blindfolded. Ralph E. Sylvestre, *Gambols with the Ghosts* (Chicago: Sylvestre, 1901). *From the author's collection*

and would not resort to deception even under pressure of the current situation. With a protocol established by Randi and Martin Gardner,[9] Suzie was successful in her preliminary rehearsal. Once they observed her method, Randi foiled her subsequent attempt by the simple act of cutting the cards just as she began the official testing. In magician's parlance, she used the top-peek, false three-way cuts, top retaining shuffles, the Schulein force, and multiple-choice forces. So much for that innocent Kansas girl.

Suzie was the last person one would suspect of cheating. Did she learn her tricks from an unknown skilled magician or discover them by herself? The point here is that Suzie—and Didier—could cheat despite the apparent innocence of their youth and presumed lack of conjuring skills.

A couple of years after Cardano described how to mark cards with the fingernail, William Gilbert revealed a full spectrum of card-marking strategies and tricks of secret identification. In addition to visual marks on their backs, cards were prepared for recognition by touch alone. Cards were altered by slight modifications of their size or shape or with impressed dimples that only the cheat could detect. A new deck of cards could be marked during play by producing crimps or bends.[10]

Sir William Crookes was fooled by a young spiritualist, Anna Eva Fay, who developed her own method for reading the backs of cards.[11] Similarly, a gambler stumped no less than the early clairvoyant performer Stuart Cumberland. He watched the swindler name the face color of cards taken from an unopened deck.[12] After paying considerable money for the secret, Cumberland learned a clever way of switching in a marked deck. An 1877 magic book for card gamblers explains two different methods for identifying the color of cards while legitimately blindfolded,[13] and another nineteenth-century effect explains how to select face cards while blindfolded.[14] Clettis Musson revealed many card identification tricks that he used during séances for entertainment purposes, including several with amazingly clever methods.[15] For a simple example, Musson could identify a card handed to him behind his back. The secret: he previously put a known card in his back pocket, which was switched with the one given to him. Today, electronic devices can identify cards at a distance.[16]

Is there any evidence to support the hypothesis of trickery when Didier played cards? The records show that Didier was indistinguishable from a cheat. In one situation, Didier stopped the game when efforts were made to prevent his cheating. If he was truly clairvoyant, he should be successful under any circumstance. Instead, Marcillet turned the blame on the person seeking more control, which got him labeled an adversary. Moreover, Dr. John Forbes pointed out what others usually failed to acknowledge, namely that Didier repeatedly failed "and made glaring mistakes."[17] After Forbes observed Didier (and later his brother) in multiple games, he concluded that they played wonderfully compared with a man who could not see but "rather badly for a man who did."[18]

Usually it takes an expert to spot clever cheating, and even casinos are vulnerable despite proactive precautions.[19] However, when we look thoroughly through the many reports on Didier, it is clear that he was sometimes caught, and he failed when restricted. There is no reason to believe that Didier was playing an honest game of écarté, and other clairvoyant pretenders in France at that time were also caught cheating at cards.[20]

Was Didier also cheating when reading covered words?

4: DIDIER'S ABILITY TO READ COVERED WORDS IN A BOOK

Didier was known for reading words in a book that were covered by a piece of paper, and sometimes he read words several pages ahead of where the book was open. Bertrand Méheust, our contemporary French philosophy professor, reported how Reverend Chauncy Townshend was so impressed when Didier read eight pages ahead.

Was Didier able to peek at the text before the words were covered? If he surreptitiously obtained glimpses of words on later pages, he would be able to accomplish apparent miracles. Here I provide one possible method.

To understand this, note that the word "page" has two meanings: (1) one side of a piece of paper and (2) a single "leaf" in a book (consisting of both sides).[21] Books are always numbered such that the first or front right-hand side of a leaf is numbered "1" and the reverse side is "2." If a book is open at page 50 and 51, for example, the task would be to peek at pages 53, 55, and 57. Odd page numbers are more visible. Glimpsing pages 52, 54, and 56 would require opening the book wide and turning the pages completely over, which would make the cheating obvious. But those even-numbered pages can be ignored. When a page number is called out, the clairvoyant says the words he had glimpsed. Then he counts pages or leaves, starting on whatever page gets him to the selected location.

We know that Didier's audience experienced his reading covered words in a book as a dramatic example of clairvoyance. Most saw nothing to question because he successfully demonstrated the clairvoyance that they expected; they were not concerned that he "read" only a few words or a phrase. Anyone with doubts about his performance would have hesitated, knowing that a request for more control would be interpreted as a personal challenge, an effrontery, especially if the séance was in a private home. In contrast, science requires controls and intrusions. These restrictions are not designed to frustrate the subject but to assess the parameters of the phenomena being studied.

Sometimes after failing to correctly read covered words, the clairvoyant requests another task, and the examiner concludes that the covered-word test is over. The clairvoyant waits for an opportunity to peek in the book when not being monitored, and later insists he feels stronger and asks to try again with the same book.

Townshend assured readers that Didier never touched the books he was given for testing, but a magician knows that an audience will forget he had a deck of cards in his hands if he says, "you hold these cards because I don't even want to touch them." Most spectators will later report that the magician never touched the cards even though they were obviously in his hands when he made the statement and passed the deck.

Did anyone see Didier peeking or cheating when he read covered words? Was anyone sufficiently curious to test the parameters of Didier's ability? We find a clue in the middle of Méheust's awkwardly worded declaration that Didier was never caught cheating on the basis of tangible facts. Méheust said that all of Forbes's arguments were based on the assumption that such phenomena were impossible. Therefore, Méheust felt justified in throwing out whatever Forbes said. I wanted to know what Forbes observed, not what he believed.

Even if Forbes assumed clairvoyance was impossible, that would not necessarily rule out an unbiased evaluation. The issue is not what Forbes believed but whether he looked for evidence to contradict his belief. Thus, I wondered if Forbes *observed* anything that Méheust failed to report. Or did Forbes find evidence of apparent clairvoyance that he could not explain but continue to assert there was no such ability?

ANOTHER EXPERT WITNESS

Dr. John Forbes (1787–1861) was one of three individuals I discovered whose keen observations helped me unravel the mysteries of Didier. Forbes started his career as a navy surgeon during the Napoleonic Wars.[22] At the age of twenty he was dressing nasty battle wounds, which forced him to consider which services of a doctor were helpful and which were not. For the rest of his life he rejected bleeding and purging, prevalent practices at that time, and he championed rational methods of care.[23] After his naval duties, he established himself in Penzance, Cornwall, where he was introduced to a controversial new medical instrument, the stethoscope. After initially rejecting the gadget's usefulness, he changed his mind and promoted it. (What he believed did not prevent him from making an objective evaluation.) Forbes moved to London to edit the *British and Foreign Medical Review*, which gave him an opportunity to promote better medical education and practice. Forbes had criticized his colleagues for their harsh rejection of mesmerism. Some skeptics even accused patients of mesmeric surgery of pretending. In his *British and Foreign Medical Review*,[24] Forbes said that his colleagues had not engaged fairly with honest mesmeric practitioners. He experimented with mesmerism as an anesthetic for surgery, but in 1847 he witnessed the first surgical procedure in England using ether. After the operation, he hurried to his office to write his report on its success, predicting ether would replace mesmerism as the anesthesia of choice.

In London, Forbes became friends with Dr. William Carpenter, the third person—after Prater, already mentioned, and Forbes—who was helpful in my investigations. Forbes wrote a letter to Carpenter expressing his excitement that at last he had discovered an individual demonstrating genuine clairvoyance, and the two of them began attending private and public demonstrations by Didier.[25] Contrary to what Méheust said, Forbes did not begin with a mindset rejecting clairvoyance. Forbes cautiously stated over and over with unnecessary repetition that his critical comments did not disprove clairvoyance, but only that he had discovered alternative explanations for Didier's demonstrations.

I was now at a loss to generate an excuse for Méheust's misrepresentation of Forbes, this being only one of many unjustified claims. Méheust was obviously promoting his own belief with serious distortion of historical facts.

Forbes expressed excitement about the possibility that Didier might be clairvoyant, and Carpenter considered himself "much more nearly a believer" after experiencing their first séance.[26] Over time, however, Carpenter was impressed by the insightful observations made by Forbes. Forbes detected Didier feigning disinterest while casually flipping through the book that would later be used to demonstrate his clairvoyance. Forbes concluded that Didier already knew some words, the page number, and where those words were located before starting his "test." Didier created the impression that he was reading a specific area by pointing, and his audience gave him credit for being correct even when his pointing was considerably wrong. Additionally, Forbes observed that Didier did not read the text word for word but made up phrases and sentences from whatever words he had captured, just as Townshend had also noted a misread phrase. Those errors are significant clues.

To keep Didier from catching any words, Forbes inserted a full piece of paper into a book and then carefully opened it without exposing any words.[27] He asked Didier to find a word under the paper and write that word exactly on top of it. Didier cleverly tried to solve this impossible task by writing a common word and then looking for it on the page. He attempted to claim success no matter where that word was located. If he could not find the word, he looked on other pages. Didier had a creative ability to find extemporaneous solutions, but he was never successful when glances were prohibited.

Maybe Didier could read covered words, but that was never established. Far more likely, he was repeating words previously glimpsed. He was cheating.

5: DIDIER'S ABILITY TO READ HIDDEN WORDS

Didier was noted for reading words that people had written on a piece of paper and then folded over so that they could not be seen. Sometimes a single word was written and placed inside an envelope or box. In a previous chapter, we considered multiple ways that Greek priests opened sealed messages, so we know to be cautious before claiming clairvoyant powers for Didier simply because he could read a word in an envelope.

As noted above, Didier pretended to read through eight pages in a book. If he were clairvoyant, not a trickster, then reading a word inside an envelope would be even easier because there is only one intervening sheet, not eight. Horatio Prater was sufficiently perceptive to notice that Didier took much longer to read a word in an envelope. The reason, most certainly, is that Didier had already glimpsed the words he needed to read from the book, but in most cases he did not know the word inside an envelope. The clock of expectation started ticking as soon as that envelope was handed to him, but he needed time to capture the word. He could best obtain that time by entertaining his audience while creating tension over the possibility of failure.

We have discussed magicians' attention to effect and method, but equally important is the presentation. A modern magic consultant, Bruce Bernstein,[28] said that psychic entertainers know that they must craft performances that get the audience on their side, rooting for them to succeed at what appears to be a difficult task. The public has been trained by these performers since the time of Didier to enjoy this routine of struggle if presented with artistry and showmanship. Bernstein presumed that a real psychic, if one existed, would probably be dull: "You are thinking of the number 47, and your child will drop out of college and go to art school." Or "I can read through this envelope, and your note says 'Paris.'" Not very entertaining.[29]

Thus, performers must give the impression of working hard, even if they already have the needed information. They do this by straining to "get in touch," revealing general impressions that are refined into more specific details that lead to the final answer. This process has become understood and expected as the natural process of clairvoyant vision. Even Reginald Scott recognized in 1584 that the magician must make an easy trick "shew of difficultie." Houdini often escaped his bonds in seconds but stretched his struggle to the maximum tension an audience could tolerate.

Prater and Forbes realized that reading a word on subsequent pages in a book and reading a word in an envelope were the same effect (i.e., reading through paper) but each required a different method. Thus, they watched carefully to see how Didier obtained the information he needed while the remainder of his audience was distracted by his charming interactions.

If someone wrote on a folded piece of paper, often called a billet or ballot, Didier sometimes tried to catch a glimpse by opening it as he moved it between his head and stomach.[30] (The movement blurs the opening and distracts the observer; the fingers also hide the action of opening.) The action itself did not raise suspicion because of belief in "transposition of the senses." People at that time considered that hearing[31] or vision[32] might transfer their sensory mechanisms to the stomach or the back of the head. Again, the larger motion of moving the billet covered the smaller manipulation of opening the fold. The clairvoyant might ask someone to place a word into an envelope that has a secret slit. He can obtain a glimpse and announce his answer. Then, as if eager to determine whether he is correct, he tears open the envelope and destroys evidence of the method.[33]

If a word was sealed inside an unopenable or impenetrable box, Didier needed yet another method. He started by thinking out loud about "impressions." Forbes realized that Didier was searching for clues from his spectators, and he relabeled Didier's rambling self-talk as "fishing" and "pumping."[34] By suggesting possibilities out loud, Didier attempted to glean information from his subject and the gathered audience: "From the owners of the sealed packets, [he asked] various questions, as to number of letters, etc. . . . Alexis made several guesses unsuccessfully, and then gave up."[35]

For those unfamiliar with the concept, "fishing" appears to be part of the struggle. For example, after audience member Piers Healy put a word in a box, Didier took his hand and asked if the word had an "a." Yes. An "s"? Yes, again. Did the word consist of five letters? Yes. Didier then correctly guessed the word "Paris." We are not told if he proposed any incorrect letters, but "a" and "s" commonly appear in words and are likely guesses. "Paris" was a common word in both French and English, again making this a likely guess. And why, Forbes asked, would a clairvoyant need to know the number of letters in a word if he could see inside a box?

Forbes noted that Didier's audience was often so eager to see him succeed that they could not prevent themselves from helping him. Forbes successfully stopped a lady from giving Didier cues about the word she had sealed in a box, so Didier simply looked to the rest of the audience for their unconscious assistance. Forbes was forced to conclude Didier had a full house of unsolicited confederates. All Didier needed was an audience that wanted him to succeed.

I discovered a humorous example of this "unsolicited collusion" solution.[36] A prominent London society journalist, Mrs. W. P. Byrne, reported that a friend of hers invited someone to attend a séance with Didier. He declined her offer but said "I shall write down a word—orchestra, for example—seal it up, and give it to you. If Didier succeeds, have him inscribe the word on the back of the envelope."

After the séance, the lady triumphantly announced Didier's success. Her friend then opened the envelope. Inside, he had written "humbug."

Like any good magician, Didier used multiple methods for reading words inside enclosures. At one séance, Didier asked those attending to write a word at the top of a slip of paper. Carpenter wrote the word "Paris" and then, at the direction of Didier, folded the paper to hide the word. Carpenter held the paper to the light to confirm the writing was completely invisible, and he avoided giving any unconscious clues. To his surprise, Didier did not fish, yet he named all the words written on each slip of paper. The demonstration was powerful, and Carpenter was stunned. Maybe sometimes Didier did possess clairvoyant powers.

A few days later, Carpenter was visited by a friend who had accompanied him to the séance. This friend, Mr. Ottley, informed Carpenter that his sister took home some of the slips of paper that Didier had used. After some experimenting, this curious young lady discovered that she, too, could read words hidden under folds of paper. She noted that holding the slip up to light would not reveal the word, but it could be read if held at an oblique angle to the light. Carpenter realized that his expectation of testing Didier with a single experiment was hopelessly inadequate.

Even more methods of reading words were available to Didier. He may have monitored his audience to identify people whispering or mouthing words. He may have watched the ends of pencils as people were writing, a skill known as pencil reading.[37] In the 1880s, French psychologists studied clairvoyants who could identify the title of a chapter or the page number of a book opened with the cover toward them. They were confused at times, however, and numbers were often reversed, for example, 213 rather than 312. This suggested the use of a mirror, which led to the realization that subjects read the reflections from the experimenter's cornea![38]

Forbes brought boxes and envelopes to Didier's séances that he had sealed at home so that no one would know the words inside. Didier would not go near them despite encouragement.[39] Forbes secretly gave a sealed envelope to a woman, but Didier stopped trying to determine the word when he discovered she did not know what was inside. He refused to proceed until she had opened the letter and read the word herself.

Forbes concluded that he had not ruled out the possibility of reading hidden words by clairvoyance, but he had shown that the foundation for these marvels was built on shifting sand. Didier knew what situations to avoid: those in which there was no possibility of leaks, sneak peaks, or cheats.

6: DIDIER'S VARIOUS MIND-READING FEATS

One evening, Didier arrived at the home where he would provide the evening séance, and the hostess welcomed him with a flower. "All the guests are here," she said, "and one of them brought this flower for you. Can you tell who?"

Didier greeted everyone, holding the hand of each person for a moment or so. Then to the amazement of everyone, he correctly announced the owner of the flower. His success inspired him to try similar experiments with rings, brooches, and pins. You can appreciate that his guests believed he was reading minds.

To introduce my hypothesis about *another* mind-reading method of Didier, we must move ahead in time about thirty years to America.

We begin with a child named John Randall Brown, who enjoyed the challenge of finding hidden objects while blindfolded and holding the hand of another person to guide him to prevent him from running into objects. In this game, observers give the seeker hints like "hot," "cold," or "getting warmer." Brown developed an ability to play this game better than anyone else, gaining a well-deserved reputation in his Iowa hometown. Eventually, he learned to locate hidden objects with no hints from his playmates.

In 1873, when Brown was twenty-two years old, he went to Chicago to buy some equipment for his brother's machine shop. While in a local saloon, his travelling companion urged him to show the assembled crowd his ability to find hidden things. Among the patrons were local reporters hanging out while waiting for their next story when suddenly they watched a miracle unfold in front of them. Brown bet a prominent Chicago resident that he could find a pin hidden anywhere, and he won.[40] The next day's papers praised Brown's impressive ability, and he suddenly realized that he could make considerable money exhibiting his skills at larger venues. He presented himself as a scientific lecturer, not as a performer or entertainer, which had the advantage of allowing him to conduct demonstrations on Sundays without competition.

Brown expanded his repertoire so that he could identify someone's birthplace, ascertain a stranger's name, identify the dates on coins, and seemingly read the thoughts of his audience. Sometimes he pasted letters of the alphabet on a wall, and while holding the wrist of his subject, he could reach out with his other hand to touch the proper letters to spell a name. Brown's unexplained powers caused a furor when he was examined at Rush Medical College in

Chicago, and a professor of mental science at the University of Michigan declared that Brown "sees through the brain and eyes of another."[41] At Yale University, more than a thousand people gathered as Brown was screened for trickery. He was successful in seven of eleven experiments and later in another six of eight experiments. Brown's performance generated a variety of opinions, but most agreed that he possessed something beyond normal human sensibilities. One professor said he would stake his reputation on the authenticity of mind reading because a muscular-action explanation entirely opposed the facts he observed.

Two people were particularly captivated by Brown's abilities. The first was Dr. George Beard, a Yale-trained physiologist who did not think Brown was mind reading despite having no alternative explanation. Brown also made a profound impact on Washington Irving Bishop, a young man who had travelled as the promotional manager of the spiritualist named Anna Eva Fay (mentioned earlier), only to write an anonymous exposure of her act in the New York *Daily Graphic*.[42] Bishop realized that he could get rich if he could draw an audience like Brown,[43] so he travelled with Brown long enough to learn the secrets so that he could open his own show.

Dr. Beard, who had been studying how the mind acts on the body, recognized that Brown must have discovered something far more subtle than anything Beard had previously observed. The good doctor understood unconscious responses of the body, but he concluded Brown was in contact with something deeper. By experimenting, he discerned that *thinking* about a specific action causes minor muscle movements, and in 1874 Beard coined the term "muscle reading," a term still used today along with "contact mind reading." He credited Brown for attending to these signals with such precision that, like Mesmer, Brown had introduced a new concept to science.[44] It appears that Didier used muscle reading thirty years previously. He held the hands of his subjects without them realizing they were giving him signals.

Muscle reading can be considered a variation of those unconsciously expressed clues previously mentioned such as head nodding, facial grimaces, and other body language signals. Observers can spot these responses, and senders usually acknowledge them when pointed out, even though they might have been unaware of making them. Muscle reading, on the other hand, is expressed with such subtlety that it is undetectable by third-party observers, and the sender is unaware of creating any response. When questioned, subjects insist they have given no clues to the clairvoyant. Only a person in direct contact with the sender notices these micromovements, and then only if anticipating them.

Beard planned a public test of his mind-reading versus muscle-reading hypothesis, but Brown failed to show up despite his previous agreement. With no one to test, Beard encouraged his audience to go home and experiment for themselves. Within a few weeks, he had hundreds of muscle-reading subjects to study, although they were not as skillful or as dramatic in their stage presentation.

Because of Beard's lectures, people recognized muscle reading when they saw it, so Brown's travelling partner, Bishop, left for England in 1878, where fewer people understood the phenomenon. Bishop lectured on spiritualism, but he also presented himself as a mind reader—the real deal. To confuse his audience, he employed a ghostwriter to create a pitchbook exposing two-person code systems.[45] Giving away traditional mind-reading methods made his own accomplishments seem even more impossible. Bishop also employed Stuart Cumberland as his manager, who in turn studied Bishop's tricks in order to secretly build his own program. Just as Bishop left Brown to exploit England, Cumberland travelled to Europe to entertain royalty, presenting himself as a mind reader in his performances and his books.[46] Only on page 315 of *Thought Reader's Thoughts* does Cumberland finally disclose muscle reading as his method and admit that thought reading "is sheer fudge."[47] Extraordinary results claimed by mind readers and members of their audience, he asserted, are unconsciously or willfully exaggerated and seldom, if ever, produced under the conditions claimed.

Bishop and Cumberland both created ever more daring effects using muscle reading, but Bishop was a genius at concocting outrageous claims through false advertising. He mastered the art of misquotation, false endorsement, and misrepresentation. For example, Bishop often presented himself as a man of means who had no pecuniary interest in displaying his abilities, which was all a clever act.

When tested by investigators, Bishop knew how to avoid precautions that would foil his efforts. As a result of his spectacular publicity and masterful presentations, the premier issue of the *Proceedings of the Society for Psychic Research* was devoted to Bishop and mental feats.[48] Bishop taught his methods to Douglas Blackburn, a Brighton newsman and performer who also fooled the society, a blunder it never openly acknowledged.[49]

Magic historians credit American John Randall Brown as the first person to introduce contact mind reading, but Didier almost certainly was using it thirty years earlier. Perhaps Didier accidentally discovered this secret while desperately trying to identify the owner of that flower. He may have been searching facial expressions only to discover he was holding the answer in his hands. He was sufficiently perceptive to identify involuntary muscular movements and wise enough to exploit them. Records of his séances in France mention hand holding without attaching any meaning to it, so perhaps Didier was already practiced in muscle reading before he arrived in London.[50]

Good magicians employ multiple ways of presenting the same effect, and we cannot leave this topic without noting that another clairvoyant created a variant mind-reading strategy that astounded and entertained audiences.

BACK TO DELPHI

"Professor" Samri S. Baldwin, also known as the White Mahatma, was born in Cincinnati, Ohio, in 1848.[51] While in medical school, he followed the famous Davenport brothers to learn their spiritualistic tricks. Then he started his own show, which featured exposures and "experiments" in supernatural responses to questions. After his marriage, Baldwin and his wife evolved the first full-evening program of answering audience questions through clairvoyance.

During a typical evening show, members of the audience wrote down questions, which they held in their hands.[52] Mrs. Baldwin sat on stage, blindfolded with her back to the audience, and called out the names of people, described their clothing, their occupations, and so forth, and then answered their questions. Sometimes it was a question held only in the mind. An observer ruled out the possibility of individuals secretly cooperating with them because Mrs. Baldwin addressed forty to fifty people during a performance, and the "trick" hypothesis was precluded by the circumstances. "I am sure . . . it was a genuine case of thought-transfer." Another reviewer confirmed that "collusion was impossible."

The Baldwins travelled around the world several times, and Samri wrote two fascinating books that recount his adventures and explain the secrets of unusual native customs.[53] He also promoted his own accomplishments and paraded his superior knowledge of magic. He exposed charlatans and revealed stage illusions, spiritualistic tricks, and methods of second sight. He described himself as a "deceptionist" and entertainer, but he leaves the reader believing that his wife possessed clairvoyant powers, all while rejecting supernatural ability. Baldwin also gave private readings as a psychic, despite denying knowledge about the nature of his power. Whatever it was, he left no doubt that trickery was not involved.[54] At the end of one session, Baldwin suggested that no medium in all of England could have provided better results, a sentiment his client endorsed. From the description of this particular encounter, however, Baldwin was obviously using a standard billet-peek routine while pretending to consult the spirit of his client's deceased brother.

Even though Baldwin exposed secret codes, he used them in his performances without being caught. He embedded his code in the natural flow of his performance in such a way that no one recognized what was happening. Baldwin took his routine right out of the Delphi Chamber of Commerce Promotional Plan.

You may remember that Delphi priests could open sealed messages and startle the uninformed through speaking tubes in statues. They exploited vessels, altars, automata, and all the latest technological gadgets from Alexandria. However, Delphi's darker secrets included the employment of spies, who kept them informed of news and significant events. Additionally, the whole city of Delphi eagerly assisted their priests because they were dependent on a steady stream of visitors to remain economically viable. Guides and innkeepers informed priests about casual conversations, aspirations, and expectations of seekers waiting for their consultations.

Baldwin relied on these strategies as well. His main source of information was obtained the same way as those Delphi priests: informants.[55] Baldwin sent his advance team ahead to arrange publicity and venue details, but they also spied on likely attendees by consulting newspapers, visiting graveyards, and taking in local gossip. Ticket takers noted the seating locations of guests and identifying apparel. Attendant ushers who passed out pencils and paper for writing questions were able to peek at what was written down. Sometimes individuals even discussed their concerns with these ushers. Those details were passed to Baldwin by that standard code, and he then sent the information to his wife. The audience experienced an acting masterpiece. Fontenelle mentioned that the hidden wires, pulleys, and wheels that create the effects of the opera are out of sight. In the Baldwin show, individuals passed information invisibly, albeit in plain sight.

Audience members who discussed their questions with ushers were not aware that they had revealed any information afterward. They were oblivious that ushers were confederates of the visiting professor, and therefore had no way of following the path of transmitted information.

One of the convincing arguments for Didier's clairvoyance was the claim that he could not have used confederates because he spoke no English, and different people attended each performance. We have already mentioned that the people who wanted him to succeed were as good as confederates by signaling with smiles, nods, and verbalizations that are universally understood. He also held hands with unsuspecting confederates. But our knowledge of Baldwin provides yet another perspective.

Didier could have received information from Marcellet in all sorts of situations. Townshend said that Marcellet preferred to speak in French, although he spoke some English. In England, he may have pretended not to understand the idle chatter of entering guests and then passed messages to Didier in code. Maybe Didier himself understood more English than he admitted.

Brown, Bishop, Cumberland, and Baldwin were tested by scientists from time to time, but they preferred the publicity obtained by fooling skeptical newspaper editors. Eventually, a whole spectrum of mind readers was entertaining the masses. Scientists were no longer interested in marching into battle against them because their secret weapons were revealed everywhere in books, magic catalogs, and even in a boy's magazine.[56] Nevertheless, mind reading became thoroughly engrained in the mind of the public as authentic.

Remote visual travel, a variant of clairvoyance, is the final feature of Didier's performance, which I discuss in the next chapter.

16

REMOTE TRAVEL

In a historical enquiry of this kind, it is more important to establish the fact that a certain thing was done than to prove how or when it was done.

—Linn, *The Story of the Mormons*, 1923

At the time of Didier, clairvoyance was believed obtainable only in a deep mesmeric trance, often called the somnambulistic state.[1] Clairvoyance means clear vision, specifically referring to seeing into the past, future, or something occurring outside the possible range of sight. I designated the final effect in Didier's performance as "remote travel." He was apparently able to travel into situations that allowed him to describe the past and current lives of other people.

7: DIDIER'S REMOTE TRAVEL

When I first started reading about Didier, his remote travel into the lives and locations of other people was the most difficult feat to understand. But one statement by Harry Sims gave me pause and started my reconsideration. The report by Sims, which I quoted extensively in a previous chapter, began with a curious introduction: "Didier took my hand. I asked him if he could tell me where I lived."

You should recognize here the possibility of muscle reading; Didier commonly connected with people under the pretext of being en rapport with his subject. I'm not positive, of course, that Didier was reading the micromovements of Sims's hand. However, muscle reading could have helped him assemble information about directions and distances. Reading the Sims report more carefully, we note "a good deal of hesitation" before Didier gave his responses.

From the reports of similar séances, we surmise that Didier spent time "thinking out loud" to gather cues from his audience.[2] Moreover, it took Didier a full hour to produce that one paragraph of information. At the end of the session, Sims tells us that Didier started making serious mistakes, but Sims dismissed them because his mind was already made up that Didier was clairvoyant. Thus, the final report by Sims omitted information about Didier's stops and corrective restarts, any near misses accepted as true, the questions Didier asked, and the failures that ended the session. What we get instead is a paragraph of sanitized, smooth, readable prose. Today we would request a full, verbatim report (as Prater recommended) or a video recording.

A BRIEF HISTORY OF REMOTE VISIONS

Different styles of obtaining remote and hidden information have been practiced since the beginning of human history as recorded in the stories of priests, prophets, seers, scryers, shamans, popes, and saints.[3] During the seventeenth century, the term "second sight" became associated with certain Scottish highlanders who were considered able to perceive future events, locate lost or stolen items, and identify persons guilty of crimes.[4] We begin by considering an individual who popularized the idea of clairvoyance.

Duncan Campbell (1680?–1730) was one of the first persons generally known for second sight. Although deaf from birth, he was tutored by mathematician John Wallis, a pioneer in teaching the deaf to read and speak. Even as a child, Campbell was consulted as an oracle. Orphaned at twelve, Campbell set out at age fourteen for London, where he gained a following of maidens, gamblers, ship owners, and others seeking knowledge of their future. Unfortunately, Campbell was unable to foresee that his own high living would result in serious consequences. He was imprisoned as a spy in Rotterdam, captured by pirates at sea, and jailed by Catholics for telling fortunes. Returning to London, he married into wealth and rejuvenated his profession of prognostication. In a presentation similar to Didier's, Campbell informed a disbeliever that he would inherit a house that Campbell described:

before his eyes, that tho' he had never actually seen it, nor been near the place where it stood, he had seen it figuratively as if in exact Painting and Sculpture, that particularly it had four green trees before the door.[5]

And it all came true. At least that was the claim of Campbell's biographer, who some say was Daniel Defoe.[6] The style is similar to Defoe's *Plague Year*,[7] in which he compellingly described events that occurred before he was born. The colorful if not completely fanciful description of Campbell's life gives today's reader the impression of fiction rather than biography. However, this book transformed vague notions about second sight into specific belief for the general public of that time.

Scottish highlanders commonly accepted certain individuals in their communities as gifted with second sight. The author Insulanus[8] (Donald MacLeod) promoted this ability as fact to English readers who were willing to believe anything might be possible up in that inaccessible wilderness.[9]

Another dimension of remote travel emerged into public awareness in the middle 1700s with Emanuel Swedenborg's extensive visions and then with other seers we have previously discussed. But it was mesmerism with its emphasis on somnambulism that got the clairvoyance ball rolling. If somnambulists could visit distant planets, then certainly they should be able to see through impenetrable walls and inside the human body to determine medical disorders.

Dr. Edwin Lee, a physician with impressive credentials, anticipated that the lucidity of clairvoyants would facilitate the internal examination of patients to help physicians make diagnostic decisions.[10] William Gregory, a professor of chemistry at the University of Edinburgh, envisioned a mesmeric brain scan, whereby a clairvoyant would perceive the precise part of the brain involved with every manifestation of thought, sensation, memory, and muscular action.[11] He also foresaw the day when clairvoyants would recover missing or stolen goods and historians would discern the truth of past events.

For John C. Colquhoun, an Edinburgh lawyer, mesmeric lucidity explained the proclamations at Greek oracles, ecstasies of Christian saints, and prophesies by men of God.[12] He provided examples of this lucidity throughout history and across cultures. Apparitions, spirits, and ghosts were all beings perceived through visions of lucidity.

Believers expected clairvoyants would soon send messages back and forth over long distances. Then, surprisingly, the dream of distant communication became a reality with the invention of the telegraph by Samuel Morse. In 1842, he petitioned Congress for $30,000 to construct a telegraph line between Washington and Baltimore. A congressional representative from Tennessee submitted an amendment mandating that half of the appropriation go to experiments in mesmeric telegraphy.[13] His measure failed.

When Didier landed on English soil, people were amazed though not surprised when he described the lives and dwellings of people he had never met. Didier's performances confirmed their expectations. Like Sims, people noticed some mistakes and failures, but that was just pocket change falling from the riches that Didier offered them. When critics mentioned this, believers were furious.

EXCUSES, EXCUSES

Reverend Chauncey Townshend provided spectacular examples of Didier's clairvoyant powers and concluded that mesmerism has all the "certainty and fixedness" that science demands. Yet Townshend demonstrated little evidence that he knew how to play the science game. For example, he boldly asserted that "our opponents have not one intellect amongst their ranks capable of investigating the shadowy realm where Mind and Matter meet."[14] He classified those who obtained negative results as opponents. He said these critics who sham civility fool no one. They go to confute, not ascertain; their minds are made up. They set up tests merely to trip up the clairvoyant, which gets him into a fidget, makes him irritated, and ultimately destroys the experiment.

Townshend declared that nobody tells Faraday how to run an experiment on electricity, foolishly concluding, therefore, that no one should tell mesmerists how to run an experiment on clairvoyance. (Townshend completely rejected Faraday's experiments that explained how tables mysteriously started tipping.) Townshend said the clairvoyant must have tests that are not "perversely allotted." He admitted that sometimes Didier could not see into a box or words under his blindfold when placed on a page. Why? "The thing is so. One must accept a fact as one finds it." Townshend went on to say that just because Didier failed once does not mean he will always fail, which is true, of course. However, Townshend could not see (or admit) that Didier was helpless when conditions favorable for cheating were eliminated. Townshend construed experimental controls as "preventive influences" and "counteracting causes" that ruin success. In the final analysis, our Reverend Townshend believed that "mesmerism is a boon granted by God to confirm our faith and to cheer us on our way."[15] He was determined to keep his faith in clairvoyance at the cost of giving up his soul for truth. His faith blocked any intrusion of doubt like an opaque blindfold.

At one particular séance, Prater realized that Didier's failures in remote travel were having no effect on his audience. Didier's ability to entertain overshadowed their critical thinking. Prater could only shrug when he encountered an American who said he focused on what was correct and ignored mistakes.

One supporter said that Didier needed cooperation to comprehend what was being transmitted but that people had difficulty clearing their thought to concentrate on a simple object. "The constant vagueness of the thoughts occurring in some people, which perhaps they cannot help, misleads the somnambulist."[16]

In article after article I discovered comments about Didier's failures that were dismissed amid praise for his successes. Believers quickly provided explanations for his blunders. One person noted, "I have often seen him fail to accomplish what at other times he would do readily."[17] Elliotson's first report on Didier in *The Zoist* described him as strong and in good health, but he complained that Didier's master (Marcillet) worked him two, three, even four times a day. No wonder Didier made mistakes, Elliotson complained: Didier was exhausted. Elliotson expressed surprise that Didier did not fail even more frequently. Can a mathematician calculate, he asked, or a poet compose amid persons bothering and disrupting him? Others said that Didier's "delicate powers" were disturbed and annulled by annoyances in the same way a fine vocalist or orator was likely to fail if those around him acted impudently.[18] As one mesmerist stated, doubt of sincerity greatly disquiets the clairvoyant, and hostility to magnetism makes him lose the ability.[19] Didier's supporters labeled any barrier to trickery as harassment. I found no examples of harassment by Prater, Carpenter, or Forbes.[20]

But not everyone was passive in their observations. In one case, Elliotson said that a physician openly revealed his disbelief in Didier's ability, which caused Marcillet to predict that Didier would not succeed with such a person nearby.[21] Indeed, when this skeptic handed Didier a sealed envelope, he finally had to tear it open in failure. Marcillet implied this was not Didier's failure; rather, the blame was on the skeptic who offered a sealed envelope to someone who claimed to read words in sealed envelopes. Why should Marcillet be upset when Didier was asked to do what he claimed he could do?

In another situation, Didier was unable to identify a word placed inside a pillbox. Elliotson blamed the lady's handwriting, which he presumed was not legible to young Didier because he was more acquainted with the French style of handwriting. In another séance, Didier failed because he was asked "in a way as purposely to mislead him."[22] How does a misleading question block someone's ability to read a word in a sealed envelope?

Many who believed in Didier responded with anger toward those who found fault with his performance. Letters to the editor of *The Zoist* described

Forbes as being "perfectly reckless, and profoundly ignorant of the science."[23] If Forbes had only asked, any mesmerist could have told him "that for days together patients will fail, and that the cause of such failure may be by no means apparent." Therefore, the rejection of Didier by Forbes on the basis of some minimal failures was inexcusable. "How is it that Dr. Forbes should be the only one to detect the legerdemain of Alexis" among the hundreds who have been convinced of his powers?

Forbes was accused of misrepresentation when he said that Didier could see when blindfolded; the author knew better because he had carefully made certain the blindfold was secure. "Dr. Forbes labours under a natural defect which prevents him from becoming an acute observer." Another believer asserted that Forbes attended the séances of Didier in order to find fault, not so that he could learn. All were certain that science was on their side. Dr. John Ashburner lamented a particular séance at which Didier faced a hostile crowd. Ashburner knew they were not supportive because he shrewdly judged the phrenology of their heads.[24]

Reverend George Sandby, vicar of Flixton, was reverentially charitable in his ability to forgive.[25] When Didier failed to describe a specific room, Sandby said he was guessing instead of employing his clairvoyance. Sandby, Townshend, and Elliotson all inadvertently raised questions about how to determine whether a declaration was true or false, whether the opinion was clairvoyant perception or calculated guess.[26]

We should pause here a moment to acknowledge that the way contemporaries looked at Didier was quite different from the way we would view him today. Perhaps we can understand this best if we switch from considering the evaluation of clairvoyance to the task of evaluating the treatment of disease. At that time, there were few effective medical treatments, and doctors had not even used simple counting outcomes to determine if bloodletting was more effective than doing nothing at all. Physicians relied on their training and anecdotal evidence. Oliver Wendell Holmes was sufficiently honest to admit medicine's "tendency to self-delusion."[27] Audiences were willing to give Didier credit for success merely because he tried and sounded confident.

Perhaps this is why Dr. John Forbes continually enters our story with insightful observations. Forbes, you may remember, began his career dressing battle wounds, which resulted in his rejection of bleeding and purging. He observed that even the worst conditions can eventually heal naturally over time if the physician does nothing. Thus, from an early age, Forbes looked at life through eyes that penetrated deeper.

THE EYES OF THE LYNX

Dr. John Forbes did not forgive mistakes as generously as Reverend Sandby or Mr. Sims; he called a failure a failure. He said the remote travel performances of Didier were a compound of hits and misses. Forbes said that articles about Didier, even those in professional journals, appealed more to the excitement of fancy than to the accuracy of observation.[28] He wanted an uninterrupted reading by Didier to establish whether his remote travel was real clairvoyance or a clever routine. Time after time, however, the audience thwarted Forbes by responding to Didier's fishing. Spectators filled hesitations with unsolicited clues.

George Beard provided an example of how this happens. Beard said that the secret of the clairvoyants consisted mainly in the art of

> making their victims unconsciously reveal, by word or look, facts of personal history, and then, at the proper time, in re-imparting the information to them. In this way they gain the credit, even among persons of keen intellect, of being endowed with divine powers.[29]

Dr. Beard checked out this theory for himself. He told a series of patients that he would diagnosis their physical and mental ailments through clairvoyance. He requested that they not give him any information about their symptoms, where their distress was located, or any history. In most cases he was completely successful in his assessment because "nothing I could do prevented them from telling me, although I asked them no questions; unintentionally and unconsciously, they would guide me at every stage of the interview."

Beard convinced patients that he had clairvoyant powers, just as Didier persuaded his London audiences. Beard noted that clairvoyants performing with an audience gathered information by listening to their words, sounds of approval, growing excitement, murmurs of consent, and even applause. Didier could convince an audience of intelligent people that he was visualizing something remote when he was merely monitoring signals quite close by.

When Horatio Prater tested Didier's remote travel, he wrote a French word in large letters on a sheet of paper that he placed on top of a table in his bedroom. Didier could not see it. Just that simple. After this disappointment, Prater began working with a young lady who had tolerated a painful operation under hypnosis without the slightest evidence of sensation, and she had "half-frightened" a room full of people by her detailed description of places she visited clairvoyantly.[30] Prater reasoned that she, of all people, would be an excellent somnambulistic candidate to test. To his surprise, she was never able to describe the appearance of a specific room to his satisfaction despite her glib presentations that so impressed others. When examined carefully, her clairvoyance just disappeared.

William Carpenter described Didier's remote travel as so vague and general that the description of his house would have applied to the domicile of anyone in that audience.[31] And Didier made some glaring errors. For example, Carpenter had no pictures in his house at that time, but Didier presumed that everyone had pictures on their walls. Moreover, he failed to see Carpenter's pipe organ with its large gilded pipes that were impossible for a visitor to ignore. This was not like missing a word on a piece of paper. At first, Didier's remote travel appeared to be a marvel; then it just disappeared like a mirage. The real marvels of science, Carpenter said, stand the test of time and prove even more marvelous when explored critically.

Remember that Prater recommended rules for testing claimants of eyeless sight. He offered similarly rigid guidelines for assessing clairvoyants.[32] To begin, the clairvoyant should write down what he intends to describe, for example, the exterior appearance of a specific house, the contents of a room, or the floor plan. With that delimitation, the owner should privately write down all the things one would expect the clairvoyant to describe. Then, *everything* the somnambulist says must be written down. In that way, Prater reasoned, one can avoid the fallacy of attending only to correct responses that would occur with guessing. In his experience, this "severe scrutiny" showed that the best clairvoyants did not get more than one thing right for every four or five things wrong—about the rate of any good guesser. Clairvoyants commonly presented a confident or enthusiastic delivery that convinced listeners that they really observed what they described. A transcript helps an evaluator ignore the emotional impact and focus on the content.[33]

Prater was unaware that a study meeting his requirements was published that same year in America's *Western Journal of Medicine and Surgery*.[34] Somnambulistic clairvoyance was also creating controversy in America, and a young lady had excited all the good people of Louisville, Kentucky, with her ability. Nine scientists devised a study that would have satisfied Prater or any modern peer-review panel. Each experimenter wrote out a specific object or scene that he would try to send the young lady. Eliminating verbal and visual cues, each experimenter concentrated on his image, and all responses were carefully documented. With these limitations, the clairvoyant's remote viewing was an overwhelming failure. The experimenters were impressed by how few words the clairvoyant uttered, especially compared to performances when she was not being tested. Moreover, she usually prefaced her remarks with expressions like, "resembles," "appears as if," and "seems to be." Because her responses were written down, these qualifying words jumped off the page. In more than one hundred trials, her vocabulary was described as barren or eminently vague and obscure. The authors of this study concluded that when cues were eliminated, mesmeric clairvoyance "is very much shorn of its glory."[35]

ALL FALL DOWN

Hidden inside all the praise and support for Didier are the clues showing that he was a performer who used the methods of an entertainer. The critical comments were buried but not dead in the growing garden of his presumed accomplishments.[36] When I began studying Didier, I believed he was likely the first to perform some of these psychic effects. I discovered, however, that he was copying other French somnambulists who played cards while blindfolded and exhibited remote travel. These feats were pioneered within the community of mesmerists, not stage performers. Thus, no written record of their methods had yet appeared. However, many magnetizers and their subjects began confessing their collusion, which is a likely reason why Marcillet and Didier crossed the channel.[37]

Didier was a gifted performer, not a somnambulistic clairvoyant, and Marcillet brought the act to a welcoming audience. They had no commitment to search for truth. Marcillet was a businessman working with a lad eager to escape poverty. Marcillet might have been fooled when they first started working together, but he was undoubtedly a colluding confederate, not a self-deceived enabler. Forbes, Carpenter, and Prater were unhappy spectators who did not understand that they were watching theater, not scientific demonstrations of a new phenomenon.

What fooled scientists in 1844 became parlor games and stage entertainment within a decade. Mesmerism was not required. All the wonderful promises of clairvoyance remained unrealized. There were no psychic CAT scans, no crimes solved, and no transatlantic telegraphy. Newspapers in New York did not collect headlines from London. No one directed miners to a storehouse of the Earth's minerals. Clairvoyants went to the moon, but each returned with information consistent with the current understanding of that globe. If some had visited distant planets or peered into heaven, no one offered solutions for the problems on Earth other than moral platitudes.[38] Maybe Swedenborg had a vision of the 1759 Stockholm fire as it occurred, but the twenty thousand people who lost their homes would have appreciated a warning.

Bold pretenders exploited public gullibility by claiming diagnostic powers and selling secret remedies. A parade of mesmerists and phrenologists gave lectures to physicians on both sides of the Atlantic, often leading their horses to waters that gullible professionals were eager to drink.[39] By continually exposing the public to fantastic claims, mesmerism opened the door for spiritualism, which became fashionable in the 1850s. Spiritualists eagerly used all the tricks that clairvoyants had devised to convince their followers that they were in contact with the other side.

The wonderful reports about Didier illustrate the inability of bright people to catch cheats, although maybe the lesson here is that few intelligent people considered the possibility of cheating—or cared. More disappointing, maybe entertainment is more convincing than science.[40]

But one thing still bothered me: the endorsement of Robert-Houdin. Did Didier really fool the father of modern magic?

Emile performing "ethereal suspension" with his father Jean-Eugene Robert-Houdin. *Photo by Hulton Archive/Getty Images*

17

WHO IS
BEING TRICKED?

*To know the things that are not, and cannot be, but have
been imagined and believed, is the most curious chapter in the
annals of man.*

—Godwin, *Lives of the Necromancers*, 1834

For me, the most puzzling endorsement of Alexis Didier was that of French magician Jean-Eugene Robert-Houdin. According to the translation by Alfred Russel Wallace, Robert-Houdin declared he was unable to explain the clairvoyance of Didier. It seemed extremely unlikely to me that Robert-Houdin would be fooled by some young pretender. Robert-Houdin had developed a second-sight act with his son, and I thought maybe Didier's impressive performance inspired Robert-Houdin to create his own variation. When I checked the dates, however, Robert-Houdin's act was already on stage when they met.

Perhaps I need to explain my high regard for Robert-Houdin. Understanding his accomplishments shows that he caused more wonder than anything Didier ever attempted.

A CONJURER CAN IMITATE ALL OF THEM

Jean-Eugene Robert-Houdin is called the father of modern magicians because he elevated conjuring from street and fair performance into sophisticated theater.[1] Eric Weiss held Robert-Houdin in such esteem after reading his autobiography that he thereafter called himself Harry Houdini. Stage performers before the time of Robert-Houdin presented themselves as wizards or professors of natural wonder, but Robert-Houdin's self-designed auditorium gave patrons the feeling that they were guests invited into his drawing room for personal entertainment. Robert-Houdin dressed in fashionable evening clothes instead of a costume. He rejected the sleazy tricks of the day like pulling the heads off birds; instead he enchanted his audience by silently producing white doves from a red silk handkerchief. Robert-Houdin was a celebrity among the rich and famous of Paris in the mid-1800s.

In one of his famous illusions, Robert-Houdin suspended his son horizontally at the end of a broom handle. The effect was called "ethereal suspension," implying that a mysterious chemical enabled him to float.[2] Wafting a little ether toward the audience gave the trick a patina of authenticity. Robert-Houdin was also famous for his construction of automatons, one of which wrote answers to questions. After displaying this android at the 1844 Universal Exposition in Paris, P. T. Barnum paid a fortune to bring it to his museum in New York.

Of his many accomplishments, Robert-Houdin was most proud of the second-sight act featuring his son, Emile. Emile was only twelve years old when his father introduced him onstage.[3] A poster of Robert-Houdin's program of November 18, 1845, announced:

> To complete the séance of conjuring, the son of M. Robert-Houdin, gifted with a penetrating second sight, will perform a completely new and fascinating experiment with his father.[4]

Emile was brought onstage where he sat on a small stool while his father blindfolded him. His father then walked into the audience, asking people for objects from pocket or purse.

> "Emile, say what I have in my hand."
> "It is something the lady takes with her from the house, and it is white. . . . No, you have in your hand something of a different color. . . . You have something blue. A blue silk handkerchief, but it came from her white handbag. Yes, her handbag is white, but her silk handkerchief is blue."

A man produces a coin, and Emile states that it is a Dutch guilder from the year 1830. As the act continues, a physician steps forward with a small box he knows will test the clairvoyance of Emile. In the box is a small surgical instrument. Emile names it and declares it was made in Berlin. He pauses and then names the manufacturer. The physician is dumbfounded; he tells the audience there is no possible way the boy could have known this information. There are only three or four of these devices in all of France and only in the hands of surgeons. Truly, the boy has the gift of second sight, he proclaims. The audience goes wild as Robert-Houdin returns to the stage to remove the bandages from the eyes of his son. Together they take a bow, and the evening performance is concluded.

We can excuse Robert-Houdin for claiming in his broadside that he presented "a completely new and fascinating experiment." Performers are not noted for their accuracy in advertisements, especially those who announce the performance of miracles.

Clairvoyance and second sight as entertainment dates back to at least 1584, when Reginald Scot described a simple two-person code in his *Discoverie of Witchcraft*. "By this means (if you have anie invention) you may seeme to doo a hundredth miracles, and to discover the secrets of a mans thoughts, or words spoken a far off."[5]

Second-sight routines appeared in the presentations of Philip Breslaw,[6] a German magician who spent most of his life in England. (We met Breslaw and his magnetic duck earlier.) A few years later, Italian magician Joseph Pinetti developed a second-sight act with his wife. Thus, second-sight acts existed prior to Robert-Houdin. However, Robert-Houdin brought this performance idea to new heights with his son.

In his autobiography, Robert-Houdin stated that his inspiration for the second-sight routine emerged from observing a woman who remembered in exquisite detail the clothes of another woman who had just walked by. This got him thinking. After a night of rumination, Robert-Houdin took Emile to a toy store the next morning. Together they glanced at a window display; then they walked a few steps farther. With paper and pencil, both wrote down every object he could remember seeing. Each day they played this game at shop windows in Paris. Robert-Houdin reported that Emile could "often write down forty objects, while I could scarcely reach thirty." They always returned to verify their observations, and Emile rarely made a mistake. And he was only twelve years old.

Emile probably had what psychologists today call an eidetic or so-called photographic memory. Although this aptitude is somewhat common among children, most—but not all—rapidly lose the ability as they mature.[7]

Next Emile and his father learned a secret language or code that allowed them to communicate numbers and words with an absolute minimum of speech. Many mind-reading teams have developed such codes, but Robert-Houdin and his son Emile enjoyed an uncanny rapport. The general methods of communication in code are well known; however, even knowing the technique can only increase one's wonder at watching the skill displayed between individuals.

The general idea is along these lines, although the variations are unbelievably extensive.[8] First, maybe ten categories of objects are constructed. Category one includes money, category two pertains to clothing, category three is jewelry, and so forth. Certain words are used to clue the subject to the category. For example, when the performer says "I," he communicates one; "go" means two; "can" is three; "look" is four, and so on. Now the pair have a method of communicating categories, numbers, and letters of the alphabet. For example, if the performer says "Go look and tell me how old this man is," the psychic quickly identifies the age as twenty-four.[9]

French mathematics professor M. L. Despiau suggested having a spectator write the name of an animal on a piece of paper, which a confederate burns and then grinds in a mortar with a pestle.[10] The subtle sounds of knocking and crushing were designed to be detected and interpreted by the performer sequestered in another room, which he then revealed with dramatic flair. The communication methods are limitless.

As partners practice with one another, the process becomes natural, and a few clues communicate much. The receiver can stretch out the revelation for a dramatic conclusion. Near-misses can be as theatrical as direct hits because consistently quick, correct answers suggest ease—or cheating. The psychic communicates the idea that he is working hard to perceive the target. Anyone who says "I get the impression of . . ." with hand on the forehead is not reading your mind. Bet on it.

Others tried to duplicate Robert-Houdin's second-sight routine, but audiences were startled when Emile named objects in his father's hand when, without a spoken word, Robert-Houdin simply rang a small bell. Robert-Houdin never revealed their secret; however, they likely communicated through an electric telegraph, which was cutting-edge technology familiar to Robert-Houdin. His home was rigged with an extensive alarm system that greeted guests and guided them inside. (Robert-Houdin produced an electric light years before Thomas Edison created the commercial lightbulb.) Emile could have received signals generated offstage from a floor plate or from a signal inside the stool on which he sat. Several methods of communicating without speech have been devised.[11]

Emile also gleaned information before the show by scanning the audience from behind the stage curtains. He memorized the seating of people, what they were wearing, and other subtle details.[12] Thus, Robert-Houdin could "randomly" select the second person in the third row and his blindfolded son could provide remarkable descriptions of everyone nearby. Remember, he could write down forty details after a brief glance in a store window! His second-sight miracles were a display of astounding memory, an encyclopedic knowledge base, and preperformance peeking through stage curtains.

Skeptics and unbelievers flocked to Robert-Houdin's shows with unusual items to stump the boy, and royalty requested private audiences. Father and son did not rely only on clever codes and snappy patter. They continued to study, committing to memory the characters of languages including Chinese, Russian, Turkish, Greek, and Hebrew. They learned the names and uses of instruments of trades, artifacts of museums, and the names of minerals and precious stones. They learned to identify foreign coins and memorized the names, events, and dates from history. And they communicated all these matters in code. With all of this knowledge at his disposal, Emile, using small cues, could piece together fragments of information that enabled him to discourse with astounding sophistication.

Perhaps Robert-Houdin inflated his claim of inventing this method, but he could hardly have overstated his audiences' reaction. Even if spectators knew some methods of secret communication, his act was still nothing short of bone-chillingly beautiful. A famous author of that time, Theophie Gautier, recorded his impression after witnessing an early performance of Emile's second sight.

> We saw Virginie and Alexis [Didier], the phoenix of this genre, who read a sentence hidden inside three gray paper envelopes. These extraordinary and inexplicable results greatly preoccupied us, without convincing us, however. . . .
>
> You will speak of confederates, but an entire theatre cannot be filled with confederates of Robert-Houdin; I for one, was no confederate, but nonetheless the object we lent was immediately identified. The most lucent sleepers [mesmerized clairvoyants] are nothing next to this. How does this feat work? That is what is impossible to conceive. The nebulous explanation of mesmerism can serve no purpose here because the child is perfectly awake; under his black blindfold there can be no mirror effects, or acoustic ones because the miracle takes place in any room at random; never any hesitation, never any error! It leaves one speechless.
>
> Of what value are these mesmeric experiments that have so excited the imagination of savants and poets, now that a conjurer can imitate all of them and even surpass them? Who is being tricked here?

Indeed, who fails to understand the implications? Gautier asked the question that should have stopped scientists from wasting their time searching for occult powers. Can Didier be taken seriously as a clairvoyant if twelve-year-old Emile surpasses his accomplishments? Robert-Houdin and his son openly declared themselves entertainers using secrets of their craft. Gautier raised a point that no thoughtful person, one would think, could ignore. No one could meaningfully examine the powers of Didier unless acquainted with the methods that Emile was utilizing.

ROBERT-HOUDIN'S ENDORSEMENT OF DIDIER

If the performances of twelve-year-old Emile were more startling than those of Didier, why were scientists not flocking to study Emile? And why would Robert-Houdin tell the Marquis de Mirville that he was stumped by Didier? The answer is located in a limited edition, two-volume biography of Robert-Houdin sold almost exclusively to magic historians.[13] French author Christian Fechner said that the Marquis de Mirville was an influential friend of Robert-Houdin who maintained a strong belief in mesmerism. When Mirville asked Robert-Houdin what he thought about somnambulists, Robert-Houdin said that he had seen some and regarded them as entertainers.[14] Nevertheless, Mirville insisted that he meet Didier and write back to him about his experience. Robert-Houdin knew all the tricks of Didier, but he did not want to embroil himself in a conflict with an old friend.

Robert-Houdin was caught in the middle because Didier and Marcillet were earning their living by entertaining the wealthy, as Robert-Houdin himself was. Being a gentleman, Robert-Houdin did not want to tip the methods of Didier, so he wrote a couple of brief letters to Mirville, which Fechner described as a response of masterful irony. The marquis, Elliotson, Wallace, and Méheust were simply too uninformed about magic and mental entertainment to catch Robert-Houdin's sidestep.[15]

Robert-Houdin told Mirville that he could not discern Didier's method of card identification. Robert-Houdin said that all of his own cards were "faked, marked, often of unequal sizes, or at least artificially arranged." But when testing Didier, Robert-Houdin brought his own unopened deck, implying that trickery was thus impossible. This was my first clue that Robert-Houdin was leading Mirville down a path even before I read Fechner. Magicians know that laymen believe card tricks are performed with gimmicked decks, so most magicians avoid such subterfuge. Indeed, Robert-Houdin had described gimmicked decks in his books on magic secrets[16] and crooked gambling,[17] but he may have never used one himself. Most certainly, not all of his cards were faked.[18] Moreover, the best card magicians in the world are sometimes fooled by a new trick, but none has ever declared an effect impossible to accomplish through trickery. I don't care what you see with a deck of cards. You were fooled by trickery. We covered that already, so believe me.

And there are other reasons to know that Robert-Houdin understood the blindfolded play of écarté by Didier. In his memoir,[19] Robert-Houdin identified a magician named Comus as the inventor of one particular trick called "blind man's game of piquet." It was a trick that Robert-Houdin performed, one not that different from Didier's game of écarté. I would not be surprised if the two had compared methods, with Robert-Houdin promising Didier that he would not spill the beans in his letter to Mirville.

Robert-Houdin said Didier was not a juggler (using the old term for magician) but a thousand times more mysterious. Mirville then said he could take Robert-Houdin to see ten other clairvoyants who produce similar results, but Robert-Houdin declined the offer. Really? Robert-Houdin had seen a miracle a thousand times more mysterious than what a magician can do and he declines a visit? This is clearly nonsense. Robert-Houdin begged off because he had better things to do than waste time with second-rate pretenders.

Robert-Houdin told Mirville he was familiar with "second sight" but Didier was demonstrating "first sight." Here again, Robert-Houdin was spinning a tale. Robert-Houdin's second-sight performances were more startling than anything Didier could ever accomplish, and he knew Didier's blindfold was part of the act.

Finally, I realized the report about Robert-Houdin's endorsement of Didier was written by Mirville. This was not the original message by Robert-Houdin. Robert-Houdin sent only two brief letters to Mirville. Everything else is Mirville's interpretation of what Robert-Houdin believed, not what Robert-Houdin wanted the public to know. In fact, Robert-Houdin's first letter specifically stated he was anxious that his signature should not be considered a declaration for or against mesmerism. Yet believers in Didier have ignored Robert-Houdin's anxiety about his private communications with Mirville. Mirville's book did exactly what Robert-Houdin feared: his personal letters to Mirville were represented as an endorsement.[20]

Let the record be clear. Robert-Houdin never publicly endorsed Didier. Robert-Houdin's own act was light-years better than anything Marcillet and Didier had ever presented. Robert-Houdin never went on stage and told his audience that his son could not perform that night because skeptics in the audience were projecting negative influences. Robert-Houdin never claimed he was in touch with a new force of the universe, unknown to science. Not once did he ever quote the occultist's favorite line from Shakespeare: "there are more things in heaven and Earth, Horatio, than are dreamt of in your philosophy." Sometimes "more things" are lies or trickery, which the philosopher never dreamed might be hidden from his view.

I was convinced that all of Didier's effects were created through natural means even if I could not explain everything attributed to him. However, the reports of his effects are not necessarily accurate descriptions; the problem of eyewitness testimony is addressed in my final chapter. I could understand why Didier's act was so convincing at the time, and even now his methods are quietly held within small enclaves of those involved in entertainment or exploitation. The documents I have cited are uncommon and not generally available. As a result, the possibility of clairvoyance has continued to simmer in popular culture.

CLAIRVOYANCE GETS AN UPGRADE

Didier left a deep impression on the minds of fashionable intellectuals of Victorian England. His abilities strengthened interest in mesmerism and raised hopes for the utilization of clairvoyance. Historians of magic have overlooked his pioneering contributions to their fraternity. And Didier was one of the first to stimulate inquiring minds about how one might sort out innate psychic powers from mental tomfoolery.

John Forbes, Horatio Prater, and William Carpenter did not make themselves popular with their penetrating questions. They were not trying to disturb the clairvoyant but attempting to determine if clairvoyance could stand the test of critical analysis. They were among the first to think about how a newly proposed mental ability might be sorted from wishful thinking, personal bias, perceptional contaminants, and deception. They were following the standards of science as illustrated by an incident involving Herman Helmholtz. This German philosopher and scientist was visiting England when his host explained that one of the Creery girls could hold up a card on one side of a door and her sister on the other side could name it. Helmholtz appeared to show little interest. Then he was asked if he would believe if he saw it himself. "Certainly not," he answered.

> In my investigations, if anything peculiar appears, I do not accept it on the evidence of my eyes. Before any new thing can be even provisionally accepted I must bring it to the test of many instruments, and if it survives all my tests, then I send it over here to [Professor John] Tyndall, and to investigators in other countries. No, I would not believe any abnormal phenomena on the mere testimony of my eyes.[21]

Ultimately, Helmholtz's skepticism was justified because the Creery sisters were soon thereafter detected using clever codes, and they confessed their trickery, much to the embarrassment of the Society for Psychical Research.[22]

Some of the society's members were eager to demonstrate thought transference as a way of breaking the grip of "bleak physiological determinism and materialism."[23] They hoped to find something unaccountable, however fleeting, beyond codes and tricks, but they were unknowingly investigating an ordinary family using the common methods of stage performers.[24]

For more than a hundred and fifty years, psychic researchers have studied then rejected the findings of previous investigators only to place their hopes on another phenomenon. Each is eventually abandoned. Someone might see this like Edison evaluating filaments for his lightbulb, expecting that eventually something might work. But maybe this is more like trying to locate the pot of gold at the end of the rainbow. Is clairvoyance a hidden power or simply a chimera?

The possibility of clairvoyance has such deep personal appeal that it has been difficult to let go. In the 1930s, J. B. Rhine brought the study of clairvoyance into controlled laboratory conditions. Previous psychical research focused on naturally occurring events like ghosts, spiritualistic manifestations, dreams, and, of course, clairvoyance. Rhine labeled his work under the general heading of parapsychology, a term he popularized. Clairvoyance became telepathy, precognition, and remote viewing.

BRING ME YOUR BEST SHOT

What secrets have we discovered using controlled laboratory techniques and sophisticated statistical methods? For purposes of brevity, my comments focus on the critiques of Ray Hyman, now professor emeritus at the University of Oregon. Since his days at Harvard, Hyman has been in a unique position to evaluate parapsychological claims because of his extensive knowledge of statistics, experimental psychology, magic, hypnosis, and mentalism. Over the years he conducted numerous investigations for governmental agencies to evaluate the possibility of fraud, deception, or unconscious errors regarding claims of the paranormal.[25]

In 1954, Hyman concluded (though not without some criticism) that British parapsychologist S. G. Soal had made a compelling case for extrasensory perception.[26] However, years later, someone attempting to support Soal's work discovered—in data that Hyman did not have—overwhelming evidence that Soal had misrepresented his findings. Reflecting on this, Hyman realized that the apparent high quality of a report can hide an array of defects.

Hyman later dedicated more than six months to the study of forty-two "ganzfeld psi" experiments given to him by Charles Honorton as representing the best evidence for parapsychology. Ganzfeld experiments place subjects in a low-stimulus environment so they can be more sensitive to the transmission of thoughts by a sender. Judges then determine if the recipient obtained an image of what the sender tried to project. If this sounds like the 1849 experiments in Louisville, then you have the correct picture in mind.

After careful analysis, Hyman identified only three studies that were free of statistical defects, and every study had either a statistical defect or procedural error. Hyman wondered how many more issues might be hidden behind the public presentation of the data.[27]

In 1990, Honorton and others[28] published the results of eleven new experiments that they claimed met the standards that Hyman had specified. Again, Hyman disagreed after discovering "sensory leakage" and other problems, and his report only created more controversy.[29]

Meanwhile, Hyman and professor Jessica Utts were given the task of evaluating the program on "anomalous mental phenomena."[30] Hyman admitted that he did not have a ready explanation for some of the reported effects, but this was still a long way from convincing evidence for anomalous cognition. The most challenging problem in remote viewing is deciding what constitutes a valid match between the verbal description of the subject and the sender's image.

Hyman told a conference of parapsychologists that the scientific community will not view them in a new light merely because a few studies show positive results. Anomalous cognitive ability is not confirmed by a correlation between a subject's description and the sender's projected thoughts. Everyone knows that correlation does not indicate causation; a match between a subject's response and a target does not confirm clairvoyance any more than the sighting of an unidentified flying object confirms aliens from space.

Students in any particular field of science can be assigned the task of replicating one of the primary studies of their discipline. In contrast, "parapsychology . . . does not have even one exemplar that can be assigned to the student with the expectation that they will observe the original results!"[31] Even if one believes that tests of clairvoyance have been successful, there is no hypothesis that arranges the findings into a body of knowledge or follows meaningful rules to provide an underlying explanatory theory.[32] And there is no growing body of knowledge that emerges from one positive study that leads to conceptualizing new research that confirms additional facts.

HONESTY IS SUCH A LONELY WORD

Elliotson and his supporters published report after report that endorsed clairvoyance as real. These articles are sincere and compelling, written by honorable men. I have asked myself whether it is responsible to just throw them out.

While I was thinking about this, my eyes diverted to the surrounding articles that Elliotson published in *The Zoist*. On page after page, studies once considered meaningful are now in the scientific junkyard. Elliotson published articles on phrenology that confirmed the shape of a head correlated with personal characteristics. Nearby are proclamations for mesmeric cures of epilepsy, tuberculosis, breast cancer, pleurisy, sciatica, deafness, ophthalmia, asthma, pulmonary disorders, persistent dermatological swellings, rheumatism, sprains, muscular atrophy, kidney disease, lumbago, and a spectrum of psychiatric disorders. These authors did not recommend caution, consider limitations, or call for further study. Out of curiosity, I reviewed other journals of this period, particularly those promoting mesmerism. They contained the same kinds of assumptions, assertions, and downright nonsense.[33]

Promoters of these cures had no sense of evidence-based medicine, and Elliotson was printing whatever some hack mesmerist wrote about his experiences with family and friends.[34] In his defense, he had few tools to winnow out the nonsense from any nourishing grain of truth.[35] Most scientists and practitioners of that day did not understand the most basic principles of the double-blind study, and believers were willing to ignore or deny the most blatant violations of basic research.[36] In most cases, however, common sense should have constrained Elliotson to be more cautious about what he printed.

No reasonable person today believes any of these wild claims of cure. The deepest somnambulistic state cannot prevent epileptic seizures or make someone sensitive to mesmerized water. Similarly, no one should feel inclined to believe the eyewitness endorsements of clairvoyants, no matter how honorable the experimenter and subject. Elliotson was furious with those who would not follow his parade, but he was leading a band marching in place.

Similar to Elliotson's reports, Dr. Edwin Lee provided hundreds of examples of clairvoyance, prevision, luminous emanations, transposition of the senses, somnambulistic lucidity, and intuitive medical diagnosis by mesmerized subjects.[37] But some of his examples were from performances he witnessed at the Egyptian Hall, a London performance center managed by magicians. Lee did not even understand that he was watching the tricks of mental magic. He thought this was real clairvoyance. Clearly, Lee cannot be trusted to distinguish truth from trickery.

The reports of believers like Elliotson and Lee fail to provide information about the process by which the clairvoyant obtained results. In contrast, Carpenter, Forbes, and Prater informed us, for example, about whether an envelope was ever in the hands of the clairvoyant, whether he fished for information, and how long he needed to divine the hidden words. As noted, these are the clues that provide hints about possible weaknesses in the conclusions. Believers leave out this information because the interactions seem so natural that they appear irrelevant. They leave out important details not to deceive, but because they are unaware that anything has been hidden. Therefore, they perceived no need to provide what appears to be extraneous information. They do not realize that the methods of tricksters are hidden within the most innocent context. Forbes was correct when he said that even the medical journals appealed more to the excitement of fancy than accuracy of observation.

Forbes said he knew eight mesmerists whose subjects were secretly faking clairvoyance. The overwhelming evidence of cheating came as a complete surprise to three of these mesmerists, who had no prior suspicion of their subjects' subterfuge. All three acknowledged the situation when informed. A Mr. Houblier expressed grief and shame that he had been duped for four years. A Mr. Howes refunded all the money he had taken for an evening exhibition and returned all money received for previous performances. A Mr. Freeman considered it a point of duty to relate the facts of deception after his clairvoyant was caught cheating. The remaining five mesmerists were informed of their clairvoyants' deceptions, but they failed to acknowledge the cheating.

Currently, we have a vast body of research on clairvoyance with no compelling evidence that the phenomenon really exists. Instead, we know many secrets about how telepathy can be faked on the stage[38] and in laboratories.[39] Elliotson eventually acknowledged this deplorable failure. In 1856, he ended publication of *The Zoist*, closing the last pages with an essay. He declared that mesmerism is more than imagination, yet he acknowledged a persistent difficulty:

> imagination often plays a powerful part in mesmeric phenomena, and has thus proved a stumbling-block to investigators from the time of the first commission of enquiry in Paris in the last century to the present hour.[40]

This was certainly a bold admission given his own history of publishing articles devoid of skepticism. Even more, he further admitted that testing clairvoyants was seriously problematic because of deception.

we know that gross imposition is hourly practiced in regard to it by both professional clairvoyants and private individuals considered to be trustworthy but influenced by vanity and wickedness. The assertions of a clairvoyant may be heard, but should be believed in scarcely one instance out of a hundred—nor indeed ever believed unless they are free from the possibility of lucky guess or trickery and are verified by ascertainment of the facts.

Think about it. One of the most ardent supporters of clairvoyance concluded that only one out of a hundred clairvoyants could be trusted. Notice that he never said Didier was "verified by ascertainment of the facts." Elliotson must have finally realized that Didier was just another pretender practicing gross imposition. Thus, as Alan Gauld said, if a case for clairvoyance cannot be made for Didier, then no case can be made for anyone.

There is probably a good reason why Didier has remained the poster boy for clairvoyance.

THE ALCHEMIST HIDES A SECRET

Charles Poyen was a student of medicine and mesmerism in France, but his search for relief from a stomach ailment resulted in his travel to the United States in 1834. After learning English, he started lecturing and demonstrating mesmerism throughout the Northeast. Despite some ridicule from the press and the medical establishment, he successfully recruited followers including Phineas Quimby, who later treated Mary Baker Eddie. Another student of Poyen was Mr. William Andros, whose wife became a gifted clairvoyant. Poyen's extraordinary claims attracted the attention of C. F. Durant, a man of boundless intellectual curiosity and physical adventure, being one of the first Americans to make balloon ascents. Durant was not impressed when he met Mrs. Andros, who was then considered the best clairvoyant in the country. She named the value of currency placed on her stomach, but her success stopped when Durant inserted a handkerchief below her nose that blocked her peeking.[41] Mr. Andros explained that sometimes she did not answer correctly when she was not feeling well and put a soothing hand on her forehead.

Durant concluded that clairvoyance was nonsense, and he was deeply disturbed that believers refused to consider alternative explanations. Therefore, he devised a devious plan to see if he could teach people to think and ask meaningful questions.

Durant's strategy was as convoluted as the plot of a mystery novel, with twists and turns so clever that even modern historians have missed his point. In brief, Durant presented himself as the promoter of a new theory of clairvoyance. He said that subjects were obtaining information from the minds of their magnetizer, not from visions of distant places. He was parodying the theory promoted by Poyen, but Durant pushed his explanation to the absurd by claiming a magnetic cord passed from the brain of the magnetizer to that of the subject. He gained the confidence of Mr. and Mrs. Andros by always confirming her responses or by explaining away failures in a face-saving way. He told them they would all become rich by touring the nation, but first he wanted to conduct a few more tests to prove another aspect of his theory.

Consider this experiment. Durant placed a tumbler in one hat and an inkwell in another. He showed this to Andros and then told him to go into the next room, mesmerize his wife, and ask her to clairvoyantly visualize what was in each hat. As soon as Andros turned his back, Durant replaced the tumbler with his comb. Mrs. Andros saw the tumbler and inkwell, demonstrating that she was not clairvoyant. Durant proclaimed that his theory was confirmed: Mrs. Andros got her information from the brain of Mr. Andros, her magnetizer. With every new test, Durant was praised by his audiences even though the outcomes contradicted his theory. If anyone questioned, he provided straight-faced explanations worthy of a comedy award. For example, he once said that he could not easily explain the cause and effect of things because it was like the familiar proverb, "when the old cat is gone, the kittens are at play." He spoke nonsense, but people thought he was brilliant. Eventually, Durant confessed to Andros that he did not believe in their clairvoyance. Durant explained that Andros's wife was trying so hard to please Andros that she made up answers, obtained information from nearby conversations, and pretended to be clairvoyant. He told Andros to go home and revive his neglected manufacturing business.

Durant wrote a book about his experiences and then walked away from further involvement with mesmerism. He investigated seaweed, then wrote about silk production, and won awards for other intellectual accomplishments. He likely paid no attention to the scathing criticism that Poyen heaped on him.[42] Poyen said that Durant's book was "the most inconceivable exhibition of self-conceit, impudence, and vulgarity" ever presented to the public. Poyen was absolutely certain that critics of mesmerism, like Durant, were not merely wrong but were committed to gross misrepresentation of the truth.

A modern reviewer of this historic episode noted that Durant came to the right conclusions but for the wrong reasons: he "failed to see that he had demonstrated that the effects of hypnotism depend on suggestion and imagination."[43]

The two academic authors who wrote the introduction to Durant's reprint cited this modern reference and concurred with his opinion. These reviewers failed to understand the unstated deception that took place in the experiments. Durant's demonstrations have nothing to do with suggestion and imagination— although compliance was likely present. Andros and his wife were cheating, just like Marcillet and Didier.

There is only one way that Mrs. Andros knew what was in those hats. She was cued in code by Mr. Andros. They colluded. Durant knew they were cheats when he held that handkerchief under her nose. Durant never tipped his conclusion of cheating to Mr. Andros, acknowledging only that Mrs. Andros was trying to please, and he never tells readers that the Andros duo was involved in deceit. He could not get mesmerizers to think and question, so he never told them specifically what his experiments demonstrated. Similarly, he wanted readers to come to their own conclusions. He decided that if people could not see the nose in front of their face, he was not going to describe it for them. Think and figure it out yourself. And he walked away to study seaweed.

18

MESMERISM

The nervous fluid, animal magnetism, is an imponderable. Chiefly eliminated from the blood by a glandular action of the brain and ganglionic centres, dispersed along the spine and the back of the lungs, liver, stomach and spleen. It can be sensibly felt by any one, and can be seen by the clairvoyant.

—Samuel Underhill, *Underhill on Mesmerism*, 1868

The thought concealed in the medullary neurine near the corpus callosum of John's cerebrum gets loose, travels down the hippocampus major till it gets well out of the locus quadratus, jabs a hole in the septum lucidum, crawls through the foramen of Monro, proper ganglion and then finds it comparatively easy work to run down the coraco brachialis to the pronator radii teres, and thence by means of the lumbricales and abductor metacarpi minimi digiti in John's hand, transfer itself to Washington's opponens pollicis, and finally gets through Washington's formamen magnum and into his brain. Any one of the five doctors who were on the committee Monday night will corroborate this simple statement.

—Burlingame, *How to Read People's Minds*, [1905].

We have now discussed the seven categories of Didier's performance repertoire and concluded that none required a mesmeric state. All his psychic effects could be accomplished by natural means, namely cheating. Did mesmerism even play a role in Didier's drama? We need to go back, now, to consider what mesmerism is and whether it exists as anything other than suggestion, expectation, and compliance.

The German physician Franz Anton Mesmer (1734–1815) identified animal magnetism as the active ingredient in his treatment. He said it was a pervasive fluid throughout the universe that followed meaningful laws. As historian Francis Yates would later comment, occult magicians have always assumed that laws and forces run through the universe waiting to be captured and exploited. Animal magnetism was not easy to capture, although it was easy to exploit.

A HIDDEN FORCE

Remember that Fontenelle's philosopher said the sages of the past were unable to understand how the heavenly bodies moved because the mysteries of nature were hidden. Now sages watched the strange influence of mesmerists but were stymied in explaining the force being applied. In 1784, the French Royal Commission found no evidence for the existence of animal magnetism. All it observed was suggestion, expectation, and compliance; animal magnetism was in the imagination of the subject and the operator (a commonly used term for the mesmerist). If something does not exist, the commission concluded, it cannot produce an effect. Thereafter, whatever mesmerists did to their subjects, the army of doubters felt obliged to declare that nothing had occurred except suggestion, expectation, and compliance.

Operators and subjects strongly disagreed because the mesmeric response was so obviously real for them that it was difficult to dismiss.[1] As E. Z. Campbell noted in 1853, skeptics and doubters suddenly became believers when they experienced the procedure, whether as operator or subject.[2] But those clutching their skeptical belief were convinced that the actors in this play followed scripted roles. They believed the French Commission got it right. No matter what the mesmerists demonstrated—clairvoyance or painless surgery—the skeptics had an answer. These two sides felt an obligation to cling to their belief and fight it out.

A few—very few—nearsighted individuals watched this conflict with curiosity. They tried to think of ways to test one possibility against the other, but there was always another variable to consider. It took a couple of centuries (and a couple of chapters here) to untangle this mess.

OCCULT ROOTS OF ANIMAL MAGNETISM

Mesmer wrote his doctoral dissertation on the pervasive influence of animal magnetism. To prop up evidence for this universal force, he resorted to metaphors instead of observations. He enumerated twenty-seven propositions that reveal how obviously he was immersed in the tradition of occult magicians who are entranced by metaphysical speculation.[3] He began with:

1. There exists a mutual influence between the Heavenly Bodies, the Earth and Animate Bodies.

And what is this influence? Proposition 2 answers:

2. A universally distributed and continuous fluid, which is quite without vacuum and of an incomparably rarefied nature, and which by its nature is capable of receiving, propagating and communicating all the impressions of movement, is the means of this influence.

This fluid was animal magnetism, and it was subject to astrological influence. Further propositions asserted that this wonderful juice, like Kool-Aid and luck, could be concentrated, stored, and transported. But even more importantly, he said this power could directly cure nervous disorders and other diseases. Once understood, the physician could reliably determine the origin, nature, and progress of the most complicated illnesses. Empowered by this certainty, Mesmer became nothing short of grandiose with his final proposition.

27. This doctrine will enable the physician to determine the state of each individual's health and safeguard him from the maladies to which he might otherwise be subject. The art of healing will thus reach its final state of perfection.

Mesmer's dissertation was little more than direct plagiarism from Richard Mead's *A Discourse Concerning the Action of the Sun and Moon on Animal Bodies*,[4] which suggested the existence of "animal gravitation." Mesmer turned Mead's animal *gravitation* into animal *magnetism*. Beyond that small modification, Mesmer presented no case examples, did not consult the books cited by Mead, and referenced almost no other authors. Nevertheless, Mesmer's exchange of gravitation for magnetism had some advantages. You cannot hold a bar of gravitation in your hand, but waving a magnet over patients offers its own symbolic attraction.

The magnet has always been esteemed for its power to cure, so Mesmer was not presenting anything new. Johannes van Helmont, a Renaissance physician and mystic, got in trouble for writing about the "natural" characteristics of magnets to heal wounds. The faculty of Reims censured his book, which then entangled him with the Spanish Inquisition for about eleven years, and he was thrown into a dungeon for practicing magic.[5] Similarly, Jacques Gaffarel, a chaplain of Cardinal Armand Jean du Plessis, duke of Richelieu, said the attractive influence of the magnet was natural, not diabolical or spiritual. For this and similar ideas, the Faculty of Theology in Paris forced Gaffarel to retract what he had written. The questions for the church were not *whether* the magnet cured but how it was used, who used it, and who got credit. Seeking a cure from a magnet was resorting to magic, an affront to the majesty of God.[6] None of this was of concern to Mesmer. He was a son of the Enlightenment, uninterested in religious and spiritual matters. His patients did not praise God for their healings. Mesmer could bask in the adoration of his subjects, and his presence became so powerful that he no longer needed to hold a magnet in his hands. He became the master of a mystery of nature: animal magnetism.

CRISIS IN PARIS

Mesmer caused a sensation when he arrived in Paris from Vienna in 1778. In his lilac silk coat, Mesmer waved his iron wand over the heads of his patients and his hands over their bodies. Young women were so gratified by the thrill of being transported into the mesmeric state (called "the crisis") that they begged to be his subjects. They were described as musical strings in perfect harmony. When one moved, the others similarly vibrated in harmony. Meanwhile, music from a glass harmonica created an ethereal milieu.[7]

Mesmer's success did not please the existing medical establishment, whose treatments were often harsh and frequently dangerous. Therefore, few physicians had any incentive to change their minds when two French commissions denounced mesmerism in 1784.[8] The first commission found no evidence of the existence of animal magnetism or a magnetic fluid. Shortly thereafter, the second commission (formed by the Royal Society of Medicine) issued a report concluding that the improvement of patients was not the result of magnetic influence but suggestion, compliance, and expectation.

The first committee, which included Ben Franklin, then issued a confidential report that outlined the moral dangers inherent in the practice. The committee asked one of Mesmer's assistants, "when a woman is magnetized and passing through the crisis, would it not be easy to outrage (molest) her?" He replied with something equivalent to "does the pope wear a beanie?"

These reports immediately started a war of pamphlets. Mesmer offered to conduct controlled trials to test his animal magnetism treatments, but that did not address the existence of animal magnetism.[9] Those engaged in the new mesmeric fad were eager to describe their successes while its critics condemned the practice. Alan Gauld was not exaggerating when he said that the battle was waged in a barbaric bog of medical ignorance and superstition.[10] Both sides blasted their opponents with uninformed opinions. No standards or models of hypnotic research existed to guide the curious mind about how to make sense of this confusion.

Mesmerism moved further down a metaphysical path when French noble-man Marquis de Puységur (1751–1825) began focusing on psychological factors. Puységur believed that mesmerism induced a magnetic sleep in which subjects maintained a special connection with the magnetizer. Although appearing to be asleep, subjects could reply to questions and convey information, just like Puységur's understanding of sleepwalking. This artificial state became known as "somnambulism," which enabled amazing things to happen. For example, Puységur believed the magnetizer was so intimately en rapport with his somnambulistic subjects that he could influence their will. Subjects could be mentally directed to pick up an object in the room if the mesmerist concentrated on that object and willed them to respond, even if the subject was blindfolded. The magnetic fluid acted like a lodestone, inexorably impelling a compass needle. Then, around 1784, Puységur began exploring whether his mesmer-ized subjects could become clairvoyant.[11] Mesmerism became so popular that one of Puységur's critics claimed that six thousand somnambules were now practicing their deceptions on the gullible savants of France, playing the game of magnetized subject.[12]

THE SAINTS COME MARCHING IN

Before all these secular somnambulistic wonders, the whole of Christendom celebrated miracles attributed to saints and mystics. Mesmerists looked back and considered that maybe beatific visions, for example, were induced by trance states and that other manifestations were earlier versions of mesmeric wonders.[13] Herbert Thurston devoted a whole chapter to saints who, like clair-voyants, transported themselves from one place to another by spiritual means.[14] But Montague Summers[15] is the one to consult to understand the full range of mystical manifestations: telekinesis, vision through opaque bodies, ecstasies, supernatural abstinence of food, illuminous irradiance, revelations, emission of flames, swoons, healings, supernatural fragrances, trance states, perception of apparitions, and power over nature and animals. The list goes on.

The pontificating of Summers only draws us deeper into controversy be-cause he informs us that all these states can be humanly imitated—or arise from the aid of diabolical powers.

> In the garden of the mystics alongside the lilies and roses of Paradise grow Deadly Nightshade, and the scarlet poisonous berry, so fair to the eyes, and the fascinat-ing exotery of brightly-blooming venomous flowers.[16]

Fear not. Summers claims that he can separate the sheep from the goats with certainty. But just when we get our hopes up that he will provide some rules for a differential diagnosis, he lets us down. It turns out that he alone can tell the dif-ference with certainty. Only he knows which modern clairvoyants are express-ing diabolical powers, personal pathology, simple paranormal phenomena, or autosuggestion. So we arrive at a dead end with Summers.

Modern historians understand that when Christianity was rapidly expand-ing into Europe, it had greater need for stories of miracles than truth. As one interesting example, Hildegard of Bingen described lifelong visions of flashing light, which she interpreted as messages from God. Recent speculation suggests she was experiencing "scintillating scotoma," which is associated with migraine headaches.[17] Hallucinations, delusions, and physical manifestations can result from many different natural causes and are interpreted according to the social context. During the last half of the eighteenth century, a whole shelf of books emerged to describe these phenomena, complete with endless examples arising from poisoning, diseases, head trauma, medications, narcotics, dementias, sleep disorders, fevers, and traditional mental maladies.[18]

So now we encounter the entanglement of religious raptures, mesmeric trance, and secret pretending. Sorting out these possibilities divided people like a civil war, with skeptics accusing believers of credulity and believers decrying the closed-mindedness of skeptics who denied facts demonstrated daily by trustworthy operators.

Everything changed when the French Revolution began. Mesmer fled, leaving his fortune behind, and Puységur spent two years in prison. However,

Puységur's publications spread throughout Europe, where practitioners explored his ideas of "higher powers." The mesmerism movement went underground then emerged into public awareness again about thirty years later. And it roared back: John Colquhoun said the somnambulistic state would, among other things, allow transference of sensation, a greater degree of intellectual power, insensibility of the organs, vision without the use of the eyes, superior knowledge of languages and topics previously unknown, the power of discovering hidden things, prevision, and prophecy.[19] He made these responses appear reasonable by describing an amazing spectrum of unlikely feats performed by people deeply asleep—sleep walkers, somnambulists. He fired the gun, and the race was on.

MESMERISM EVERYWHERE

Mesmerists were so convinced of their new discoveries that they rejected any interest in testing and were exasperated when questioned. They were satisfied—thrilled—by the drama of their patients' responses. A devoted follower of Puységur, Philippe Deleuse, said that thousands of respectable witnesses gave all the evidence needed; the only ones who opposed it had no experience.[20] He believed magnetism should be a subject of observation but never a subject for experiment. Similarly, Poyen said that none of the critics has been magnetized and disbelievers have scrutinized only experiments that were failures.[21]

When Galileo confronted different possibilities for the arrangement of the planets, he made a commitment to search for reality as it existed. He used his telescope not only to observe but to test different possibilities of planetary movement. He was determined to accommodate his understanding "to what nature has made" rather than to what matched his belief. He did not present his ideas as propositions supported by metaphors. Mesmer and his disciples, on the other hand, preferred the wondrous over the mundane. They were thrilled when somnambulists described the magnetic fluid as having a shiny, brilliant glow. Some detected a taste and smell. Evidence for its reality was everywhere, and testing was unnecessary.

In 1837, one of Mesmer's pupils, Baron Dupotet, went to London where he was permitted to treat hospital patients. Dr. John Elliotson, the supporter of Didier, was eager to apply this magnetic fluid as an invisible poultice to any and all disorders.[22] Elliotson was drawn to Dupotet's mesmerism because medical practice had stagnated. Few treatment options were available. The microscope was neglected for medical purposes; contagion and germ theory were still in the future; bloodletting and purging were commonly practiced.[23] As a result, Elliotson pushed boundaries by experimenting with massive doses of medications. Some said he was poisoning his patients, but physicians needed alternative treatment options. Elliotson was primed to try animal magnetism.

Young Dr. John Elliotson (1791–1868) was a rising star at the University College London, already president of the Medico-Churgical Society, and founder of the London Phrenological Society. He quickly became a close friend of Dr. Thomas Wakley, who made the brilliant discovery that medical students would rather pay sixpence a week to read good lectures than to pay larger fees to endure the drone of hospital instructors. Thus, Wakley started a new journal he called *The Lancet*. He intended to advance medicine by reporting controversies and new developments, but he also printed sensational issues, graphic descriptions, and scurrilous criticism of quacks and pretenders. As one historian concluded, *The Lancet*'s style was more like that of an American crime reporter than the temperate, influential weekly it has now become.[24] Elliotson's enthusiasm for better medical practice was exactly what Wakley wanted because students said of Elliotson "He did not lecture; he taught."

KEY FINDINGS

Elliotson gave demonstrations of mesmeric treatment on the wards of his hospital, but interest was so great that he soon filled more than two hundred seats of the medical lecture hall. His audience included poets and politicians. Charles Dickens attended and later became a mesmerist in his own circle of friends. Elliotson's usual subjects for demonstration were Jane and Elizabeth O'Key, young Irish sisters who had been hospitalized for hysteria and epilepsy. Upon first introduction, Elizabeth was a small and immature individual with no expression; she appeared unaware of the lecture Elliotson delivered to his audience. When he mesmerized her, however, she became lively, shrewd, and witty while interacting with the audience. Spectators were impressed, and they flocked to see what might happen next.

In one demonstration, Elliotson could not awaken Elizabeth until she informed him about the method he should employ. As sessions continued, Elizabeth became more difficult to control. In her normal state, she was a demure, lower-class servant girl. During her trance state, she sang suggestive songs, regaled the audience with jokes, and insulted Elliotson. Some observers concluded she was changing roles with him, especially when she began diagnosing his patients and recommending medications. Who was studying whom?

Hospital officials became concerned that Elliotson's lectures had devolved from medical demonstration to carnival entertainment.[25] Hospital politics, being what they always have been, caused administrators to wonder about the propriety of such displays. In 1838, Elliotson was pressured by the medical school council to end his lectures, but he refused. The controversy erupted in medical journals and public newspapers. Was Elliotson demonstrating something new or was this an impropriety devoid of scientific merit? Wakley was having his doubts, and he published some statements by mesmeric subjects who confessed they had feigned their trance effects. Wakley started to wonder if Elizabeth O'Key was a fraud.

The experiments of Elliotson began to look suspicious to Wakley. The O'Key girls could be put into a trance merely by drinking water that had been magnetized by means of Elliotson dipping his finger into the glass. Jane O'Key put herself in a trance by making passes while looking in a mirror, and Elizabeth read messages in sealed envelopes when held against her skin. The girls started making predictions of future events on the ward, and patients generally carried out those expectations. Instead of becoming doubtful like Wakley, Elliotson expanded his theories of the mesmeric force.

Wakley now took up the challenge of determining whether the sisters were acting in compliance. One August evening in 1837, Elliotson arrived at Wakley's home to confront the controversy.[26] Each side brought five supporters, but Elliotson was not aware that he was about to encounter the equivalent of a duel on the estate of his adversary. For his part, Elliotson attempted to show the effects of mesmerism, but Wakley's goal was to determine if Elizabeth was a fraud. One of Elliotson's presentations was to demonstrate that mesmeric fluid attached itself to nickel but not to base metals like lead. Thus, when Elizabeth was touched with a bar of nickel, she would suffer a convulsion and go into a trance, but lead would have no effect.

As the evening progressed, Wakley surreptitiously switched the bar of nickel with lead. As he walked toward Elizabeth, a confederate of Wakley whispered within her hearing that the nickel should not be applied too strongly. Elizabeth went into vigorous paroxysms with muscle contractions that produced striking rigidity. Elliotson demanded more trials, but the game was over. O'Key was labeled a fraud, and Elliotson was condemned as a gullible practitioner who had unwittingly signaled what he expected from his subjects. Wakley dismissed mesmerism in a style consistent with his tirades against quackery. He said the whole process was the equivalent of fortune-telling; he called it "mesmeric humbug." The O'Key sisters were discharged from the hospital. Elliotson offered his resignation.[27]

The O'Key girls were subjected to previous experiments that deserve some attention to complete our story. Seven different experiments were conducted in which Jane was presented with six glasses of water, one or more of which was magnetized in a way that had previously created a dramatic reaction. When she had no knowledge about their content, the effect was gone. Surely, one would think, believers would understand the implications. However, people with a strong belief system do not change their mind easily; instead, they change the interpretation of the study. Readers of *The Spiritual Messenger* were informed that these experiments were not trying to find the truth but to crush it underfoot. The only way Jane could know which glass was mesmerized was for them to *will* her to know.

> All experiments which are conducted in diametrical opposition to the laws of the science which they are presumed to illustrate, are utterly valueless; their effect is prejudicial to the well-being of mankind; and they are not only derogatory to the fame of those who conduct them, but are insulting to Him who has given to man his numerous qualifications to be used in showing forth His Divine power and goodness.[28]

Although Wakley dismissed Elizabeth O'Key as a fraud, her reaction to the water and bar of lead was certainly not so much intentional deception as a response of suggestion, expectation, and compliance.

Victorians loved controversy and scandal. Wakley and Elliotson publicly fought out their ever-diverging opinions, with no hint of their previous friendship.[29] Some say Elliotson was on the verge of a psychological breakthrough in his study of the unconscious mind. He was never able to let go of the animal magnetism nut he was gripping inside the knothole of Mesmer's tree. He hung on to the worthless metaphysical concepts the rest of his life because his belief system was stronger than common sense.

Elliotson developed his own private practice after leaving the University College Hospital in North London. He continued to rally his allies and rail against Wakley in *The Zoist*. This was not two blokes duking it out in the back alley; this was a cultural war with the medical establishment asserting itself as the referee. They were fighting over truth at a time when no one could establish who was a qualified medical practitioner and who was a quack. The goal was to establish "my truth" when there were no established research procedures or statistical tools to sort these controversies. As a result, Elliotson published articles that supported his belief at a time when there was little distinction between medical ignorance and popular journalism.

Other mesmeric journals of that day, which mostly published only a few issues, all begin with tirades against those who rejected this gift to the world. The *Spiritual Messenger*, quoted earlier, rails against "material philosophy" and derides materialistic medical schools that depend on drugs alone for cures.[30] *The Phreno-magnet* complains that its supporters have been subjected to scorn and ridicule.[31] Similarly, four of the first five issues of *The Mesmerist* (1848) open with articles aimed at critics of their practice. *Buchanan's Journal of Man* opens with a hope that the blind but powerful will cooperate with the wise but powerless inventor. With "loathing and horror" of disbelievers' past denials, the editor turned to the light of phrenology.[32] The *Zoo-magnetic Journal* (1839) devotes ten pages of its first issue to combat criticism that emerges from ignorance, prejudice, obstinance, and preconceptions. The editor described his critics as selfish and prideful, resorting to every maneuver, shift, stratagem, and evasion. And these are just a few invectives from the first paragraph.

As much as Elliotson's allies defended their mutual beliefs, additional developments over time damaged his credibility and status. The early enthusiasm for phrenology evaporated, and chemical anesthesia eliminated a primary value of the trance. Elliotson died penniless in 1868 in the home of a medical colleague. He was never able to differentiate legitimate responses of the trance from what was improbable. For that task, we can thank our old friend John Forbes, to whom we return shortly. But first we note that mesmerists did their best to make animal magnetism reasonable.

EXPLAINING TRANCE

Believers who refused to abandon animal magnetism provided shabby explanations for their belief. Adolphe Didier, the brother of Alexis, reflected a common view when he admitted that animal magnetism was not clearly proven. Nevertheless, Didier said we can presume that it exists because it sure looks like it does.[33] He described animal magnetism in grand terms. It was a spiritual force that acts in the physical realm because it is the principle of life. It is a vital fluid that cannot be called into question. It is "an imponderous fluid which can be neither felt nor measured." The effects are regular and constant yet incomprehensible. It is a mystery of nature. It is faster than lightning, instantaneously transmitting information from one extremity of the Earth to the other. It confounds reason but like Newton's gravity, its attraction has been established even though the cause is unknown.

Some mesmerists dropped the idea of animal magnetism and accepted the unlikely odic force as promoted by German scientist Baron Charles von Reichenbach. He spent two decades studying his force, which he said permeated all dead and living matter throughout the universe. In the preface to the second edition of his book, Reichenbach was furious that people compared his od to Mesmer's animal magnetism.[34] His experimental subjects could see and feel the heat of this force, establishing it "by the enumeration of a superabundant amount of incontestable individual cases." Oh, yes, maybe a couple of individuals gave an exaggerated response or two, but that was from misapprehension, not dishonest intention. Once Reichenbach caught on to the game perpetrated by some difficult subjects,

> it was impossible for anyone to continue to answer me falsely, even for a few minutes, without my at once detecting it. . . . [Now] all reasonable doubt must disappear before the evidence of truth.

For page after page Reichenbach railed against the lies and deceit of incompetent inquirers who did not know how to conduct scientific studies in order to obtain clear and instructive answers. He presumed that Michael Faraday had not encountered his odic research; otherwise he would not have ignored it. (Faraday was probably laughing too hard to write a review.)

All of this mesmeric and odic fascination had extended past fantasy into complete nonsense. Finally, Reichenbach's findings were challenged in 1846 by Scottish physician James Braid.[35] Braid was born and educated in Scotland but moved his medical practice to Manchester after a couple of years. Braid's criticism of Reichenbach extended to mesmerism as well, which was known in England almost exclusively from the publicity surrounding Elliotson and from the lectures by impostors and quacks. Mesmerism might have faced a dismal future in England if Braid had not become intrigued when he witnessed a mesmerized person unable to open his eyes when told he could not. Braid decided he could investigate the natural physiological responses associated with trance without accepting its prior trappings of magnetic fluids and mystical forces promoted by Mesmer and Reichenbach. Braid suggested the term "mesmerism" be dropped in preference to "hypnosis," which gave the whole enterprise of trance induction a new life. He said the trance state was a response to induction that rendered subjects susceptible to suggestion. It was a subjective response that created "a rapid exhaustion of the sensory and nervous system" so that the mind "slips out of gear."[36] Hypnotists changed their conceptualization of the induction process; they no longer manipulated magnetic fluids but verbally created a state of mental concentration in subjects.

What primarily distinguished Braid was the skepticism he carried into his study of the trance. He asked individuals to criticize his experiments in the most severe manner they could conceive. Braid wanted to protect himself against deception so that he could arrive at the truth. Elliotson ignored Braid in *The Zoist* for six years and never accepted his work, probably because it meant accepting challenges to his own articles. By ignoring Braid, he denied his readers the opportunity to consider alternative ways of conceptualizing a trance. Elliotson, like most practitioners, did not welcome criticism.

Braid's recommended term, hypnosis, began replacing mesmerism. Hereafter, I try to use the word as it reflects the belief of the operator and how the trance was conceptualized in the situation under discussion.

ART VS. SCIENCE

This brings us back to Dr. John Forbes, who offered a theory about how to separate the reasonable from fanciful in the mesmeric (or hypnotic) context. Forbes declared that animal magnetism or hypnosis was a new *art* but not a new *science*. Mesmerism was a new way of influencing or inducing physiological responses. No new force or agent was created by mesmerism. It was nothing more than a *method* of influencing the sense organs and consciousness. Mesmerism created no new effects to the human repertoire that did not already exist.[37] He noted, for example, that medical practitioners commonly see cases of hysteria in which symptoms are expressed as temporary loss of muscle function, like the inability to open one's eyes.[38] Hysteric conditions were mostly spontaneous and internally generated, whereas mesmerism evoked similar symptoms through external manipulation of expectation. Both created the same effects such as symptoms of catalepsy, seizures, and involuntary movement. All of these effects were the same whether observed in a doctor's office or at a mesmerist's performance.

Similarly, Forbes asserted there were known cases of insensibility to external pain without mesmerism, including the endurance of surgical procedures and rituals involving torture.[39] Some people tolerated these circumstances without complaints of pain, so again, mesmerism was dramatic in its ability to reduce pain without contributing a unique ability. Mesmerists carefully prepared their surgical subjects, just as shamans enabled boys to endure painful rituals that inducted them into manhood. In his view, the symptoms in naturally occurring situations and in mesmeric states are somewhat different, but nothing novel is introduced that would distinguish one from the other.

In contrast, Forbes noted that well-documented cases of spontaneous clairvoyance are not available. Patients with hysteria or other natural conditions do not see at a distance or into the past or future. Similarly, there are not spontaneous cases of transposition of sense organs as claimed for mesmeric trances, such as in seeing with the stomach or the back of the head. Therefore, Dr. Forbes concluded that mesmerism would likely not provide any mechanism to create such instances, and any presentation should be carefully examined to rule out other explanations. Physicians do not automatically accept what patients report or confirm what they believe. Anyone endorsing a clairvoyant should submit the subject to the same careful evaluation that a physician would conduct with any patient who described an atypical symptom.[40]

James Braid and John Forbes attempted to draw mesmerism from the miasma of occult belief. It was time to step back and be more thoughtful. Braid understood that something interesting happened when people were hypnotized, but he rejected the mystical forces as nonsense. Forbes observed trance responses and declared that some were probably variations of natural human behaviors, but other manifestations were likely unverifiable because they were outside the bounds of human ability. These two physicians placed some fencing around the mesmerism/hypnosis landscape. Now investigators had some guidelines for exploration.

But the powerful experience of trance induction maintained a grip on operators as well as on subjects. Operators brought their belief to the trance situation, and experience confirmed their opinions. Subjects were convinced they had experienced something uncanny. Everyone loved the new art, but no one was eager to explore the science. That involved searching for evidence that your theory is wrong.

MODERN HYPNOSIS

When all believe in the notion, the experiments will not contradict that notion.

—Etherology, 1850

In the mid-nineteenth century, mesmerism was promoted in the press, the pulpit, and the scientific community. Dr. John Forbes complained that emotions led people to conclusions before the intellect had an opportunity to make rational decisions about the mesmeric state. Operators accepted occult explanations and treatments based on magical thinking.

As noted, Dr. James Beard decided it was time to throw out the animal magnetism bathwater but keep the trance state baby. The new term "hypnotism" gave practitioners the advantage of treatment without the taint of occult contamination. Beard was able to continue his practice and conduct experiments, but most clinicians were not interested in such exploration. Most were content with their practice and could see no reason to change. Moreover, no standard methods of investigation were available to study hypnosis. As a result, mesmerists like Reverend Townshend, Professor Mayo, and Dr. Deleuse continued to encounter spiritual manifestations in their sessions. Avowed materialists like Dr. Elliotson did not.[1] Each saw what he expected. That's a problem.

SEEING WHAT YOU BELIEVE AND BELIEVING WHAT YOU SEE

Seeing what you believe has been an alluring temptation in mesmerism and hypnosis because the underlying mechanisms are hidden behind that curtain.[2] Thus, the operator can interpret whatever he sees as the product of whatever he believes. By way of example, consider the work of Jean-Martin Charcot, the famous nineteenth-century neurologist of the Salpêtrière Hospital.[3] The Société de Biologie asked Charcot to evaluate a new medical strategy promoted by mesmerist Victor Jean-Marie Burq. Serendipitously, Burq found that copper tended to irritate or burn patients sensitive to mesmeric influence. In contrast, gold and silver were pleasant and soothing. After further trials, Burq concluded that each patient possessed his or her own "metallic idiosyncrasy." Thus, he began treatments with external and internal applications of metals, including the ingestion of mineral salts and waters.[4] This metallotherapy became popular in France,[5] and Charcot's two-year evaluation resulted in an enthusiastic endorsement by the society.

While investigating metallotherapy, Charcot became acquainted with Braid's revision of mesmerism into hypnosis, and he followed the work of Charles Richet and other French physicians who were trying to rehabilitate mesmerism into hypnosis. Unfortunately, Charcot failed to incorporate Braid's skepticism and caution. Instead he brought the same disdain for safeguards that was evident in his study of metal sensitivity. Charcot's observational skills that served him so well in his exploration of structural neuropathology were insufficient for the study of human behavior. Some of the psychological findings are now so quaint as to be absurd. For example, he used metals to evoke emotions and magnets to draw out morbid delusions. He was convinced that patients emitted an effluvium that created an observable aura. Through suggestion, he transferred sensory anesthesia and catalepsy from one side of the body to the other or from one patient to another. With such sensational demonstrations, it is little wonder that his lectures were so popular.

Charcot believed he had taken sufficient precautions against deception, dismissing the possibility that poor, ignorant patients could simulate consistent responses. However, Charcot's ideas were attacked by Hippolyte Bernheim at the medical school in Nancy.[6] Bernheim accused Charcot of unwittingly training his subjects to produce the effects he expected from them, and he rejected Charcot's three stages of hypnosis. Instead of conducting better experiments, Charcot flew into a rage at the very mention of Professor Bernheim's name, and he was bitterly resentful when a Paris newspaper article denounced his public demonstrations of hypnosis as a ridiculous spectacle of no scientific value.

Charcot was famous for his Tuesday public lectures attended by politicians, journalists, artists, writers, actors, and actresses.[7] His meticulously prepared and dramatically delivered presentations were captivating because of his clinical skills and vast knowledge. As tourists travelled to Paris, so patients and students flocked to Charcot. Freud spent four months with him between 1885 and 1886. However, not everyone was pleased with his style. His sarcasm was ferocious, and he brooked no contradiction. Some described Charcot as resembling Napoleon in appearance and demeanor, and behind his back they called him the Napoleon of Neurosis or the Caesar of Salpêtrière.

Axel Munthe, a peripatetic Swedish physician, was highly critical of Charcot and his lecture demonstrations. Munthe, the author of an enduringly popular book, *The Story of San Michele*,[8] described Charcot's lecture amphitheater as filled with the morbidly curious. He was convinced that Charcot's hypnosis was problematic, especially after observing Bernheim's hypnotic treatments at Nancy. Charcot's act, he declared, was "an absurd farce, a hopeless muddle of truth and cheating." He noticed that Charcot's hypnotic subjects seemed delighted to perform their various tricks in public, acting out exactly the hysterics of what was expected. Munthe said many of these patients were in a state of semitrance all day long with their brains bewildered by absurd suggestions. Some would smell a bottle of ammonia with delight when told it was rosewater; others would eat a piece of charcoal when told it was chocolate. When Munthe tried to deprogram one of Charcot's Tuesday stars, Charcot expelled him.

In one of his lectures, Charcot wanted his hypnotized subject to display a particular symptom, which he believed was caused by a previous trauma and was now hidden in his subconscious mind. The reluctant patient gave in and produced his expected symptoms only after our Napoleon "insisted a little more." Professor Borch-Jacobsen described this lecture of Charcot as the birth of belief in dissociation amnesia: mentally blocking out an emotionally traumatic event.[9] Any symptom might now be attributed to a hidden trauma or "nervous shock." Thus, the therapeutic task was to uncover the memory of that trauma from the subconscious mind into consciousness so it could be analyzed and resolved.

Georges Guillain, Charcot's biographer, graciously said Charcot's hypnosis was a slight failing. Richard Webster said this claptrap theory was "one of the most significant misunderstandings in the entire history of medicine."[10] Before Charcot concocted his theory,

patients remembered quite clearly the psychic or mechanical shock that had triggered their hysterical paralysis and attacks. After, they would tend not to know the cause of their symptoms anymore.[11]

Charcot passed down to Freud this bogus idea that hidden knowledge was retained by the unconscious mind, which needed to be exhumed. If the patient does not remember past abuse, so goes the theory, then maybe the memory can be refreshed by hypnotic retrieval or recovered through a convulsive abreaction, which goes back to Mesmer's crisis. Freud and his horde of followers never checked to see if any of this was actually true.

In fact, we now know that hypnotic subjects can be easily influenced by suggestion, and therapists can implant false beliefs. In the 1880s, French physician Hippolyte Bernheim demonstrated that patients could be convinced of memories that came from therapists, not from actual events.[12] He suggested to his patient, Marie G., that she had fallen during the night and hurt her nose. Nothing could persuade her otherwise in the waking state. Following that success, he suggested that three months earlier she witnessed an old bachelor rape a young girl, and he provided her with explicit details. Three days later, Bernheim asked an attorney to question her in an official manner. She related the suggested event with conviction, saying she would swear the truth before God and man in court. Bernheim then tried to persuade her to doubt herself, but she was immovable. He had to rehypnotize her to remove the memory.

Was Marie acting or role-playing? Bernheim certainly did not think so, and the attorney who questioned her did not believe she was faking her belief.

STILL CRAZY AFTER ALL THESE YEARS

Charcot and Bernheim demonstrated that hypnotic subjects might be thoughtlessly complying with the suggestions of the hypnotist, playacting to please the therapist, or responding with convincing conviction. There must be a joke here in which a hypnotherapist says "My patients are cured, yours are playacting their success, and his are mindlessly pretending to be better." The point is that some meaningful research was needed to sort out what was happening to create such different responses. No one stepped up to investigate the questions of hypnosis better than Martin Orne.

Martin Orne, MD, PhD (1927–2000), emigrated to the United States from Austria with his family in 1938 to escape the Nazi invasion. After impressive educational achievements, he designed some diabolically clever studies to clear up some of the confusion about hypnosis.[13] His work confirmed what Forbes had suggested, namely that hypnosis did not create any new human ability but instead is a complex interaction between hypnotist and subject.

Orne showed that hypnotic subjects regressed back to a younger age, easily recalled earlier learned but now forgotten poetry, remembered details of their lives from the first grade, and drew pictures suggesting age regression.[14] For the most part, however, these presentations were "extremely deceiving" misrepresentations. Subjects' drawings showed adult motor coordination with mature and immature features intermingled. Subjects were not performing as they had at age six but were producing what they perceived they would have produced, and they behaved according to how they expected someone of that age to behave. One subject could not recall the name of his first-grade teacher when hypnotized, so he was asked to visualize her walking around the classroom. He suddenly identified her as "Miss Curtis." Despite his emphatic and positive assertion, he was incorrect. In his desire to comply with the expectation of the research study, he produced a compelling response. But he was wrong. Orne's research illustrated the importance of sorting things out, especially the need to verify past events, even when confidently asserted.[15] Subjects (and patients) give answers when expected, sometimes incorrect answers, often holding them with strong conviction.

Orne confirmed that hypnosis is not a magical tool for recovering memories of lost information or forgotten experiences of trauma. As Forbes would have predicted, hypnosis added no new memory retrieval mechanism to the human repertoire. On the contrary, Orne's studies, often cited by courts, showed that hypnotized subjects are highly suggestible, which stimulates confidence in their confabulated recollections.[16]

These findings created a serious challenge to the recovered memory movement that exploded in the late twentieth century. It should be no surprise that Orne was a founding member of the scientific advisory board of the False Memory Syndrome Foundation.[17]

With all this unexpected compliance, Orne wondered if psychological experiments (and analogous situations) might also be an unintended context for acting out expected roles, which he described as "demand characteristics." Again, subjects are not necessarily conscious of deciding to be compliant. Yet they comply while experiencing their responses as completely of their own volition. In multiple experiments, volunteer subjects responded to a wide range of cues (demands) based on their perception of the totality of the experimental situation.[18] Subjects cooperated with the experimenter in ways that validated what they perceived to be the experimental hypothesis. For example, if a researcher has a grant to study war trauma in combat soldiers, then interviewed veterans will relate stories of their suffering.[19]

Orne's study on demand characteristics was, for a time, one of the three most cited papers in American psychology, but these lessons now seem forgotten, just as the warnings of the French Commission were forgotten. Perhaps the lessons have slipped away from us because we do not see them operating. These are subtle but powerful and persistent influences, yet we easily ignore what is hidden from our direct experience. Bias can be managed in experiments by keeping all participants blind to the hypothesis being tested (as much as possible), but demand characteristics remain inherent in many situations.[20]

DUAL FALSE BELIEF SYSTEMS

We understand that the instructions of an experimenter or the suggestions of a hypnotist can subtly influence the response of a subject. Less apparent is the likelihood that the subject also influences the belief of the operator. Demand characteristics work in both directions in hypnosis or any dyadic relationship. In 2002, I wrote a tribute to Martin Orne in which I said his demand characteristics suggested the possibility of creating a "dual false belief system."[21] The model for this is analogous to the old joke about the city hall caretaker who set the town clock by the noon horn blast of a local factory. When he retired, he asked the factory manager how they determined the right time to blow the horn. He responded, "We look at the town hall clock."

In my tribute, I explained that I have examined thousands of pages of depositions, chart notes, and transcripts of recovered memory cases. From this outside perspective it seems so obvious how therapists and patients influence each other, yet both seem completely oblivious to the expectations of the situation. Patients gradually incorporate more and more of the therapist's belief, which confirms the therapist's confidence in a theory that can have devastating consequences for someone falsely accused of sexual abuse. These travelers were leading each other down a dangerous dead-end trail, ignoring all the warning signs.

As another example, Charcot created profound demand characteristics on his patients and on his students, who were known to coach patients to perform for their master. Without patient compliance, however, Charcot would never have been able to maintain belief in his fabulous theories about metals, magnets, hysteria, sensory transfer, and traumatic amnesia.

The concept of the dual false belief system is so robust that I decided I could analyze the details of a 1575 inquisition reviewed by historian Carlo Ginzburg.[22] A priest in a remote Italian village encountered customs that he did not understand. Further handicapped by not being familiar with the dialect, he stopped seeking information and started suggesting that their rituals were demonic. Weakened by the priest's pressure, the community slipped into accepting his obviously false interpretation of their ancient ceremonies. This in turn convinced the priest that his theology of demonology was correct.

Orne's research put many interactions under a powerful microscope, and we suddenly realize that some tiny pathogens can pollute the therapeutic waters. Hypnotherapists with good intentions have created some toxic results without any awareness that both they and their patients were confirming the other's belief while leaving reality behind.[23] Canadian psychologist Marilyn Bowman devoted a full chapter to considering additional reasons why clinicians often confirm traumatic stress and blame external circumstances when they should address the patient's current behavioral difficulties.[24]

CUTTING TO THE CHASE

What, then, can we conclude about the value of hypnosis? It does not provide any additional powers, but Orne suggested a way to determine if hypnosis might assist patients in certain situations. Select a task that could be enhanced by hypnotic suggestion and measure the outcome. Then compare similar but non-hypnotized subjects performing their best at the same task. If hypnotized subjects do better, that difference could be considered the "essence" of hypnosis.[25]

From the earliest days of mesmerism, pain insensitivity has been a thorn in the side of skeptics. Beginning in the 1820s, patients under a mesmeric spell tolerated with remarkable serenity the brutal slicing and dicing of primitive dental and surgical procedures. What was it that created tolerance for these operations? What was this "essence" that helped patients endure?

In England, the first surgery associated with mesmerism occurred in 1821.[26] In 1829, a woman had her cancerous breast removed during a mesmeric sleep, and in 1842 a leg was amputated after Dr. John Elliotson induced a mesmeric trance. In 1826, a tooth extraction was performed in Paris, and a similar extraction occurred in America in 1836. In 1837, a dentist removed the roots of four broken teeth without difficulty. American surgical operations with mesmerism began in 1843 with the removal of a nose polyp.[27]

The breakthrough in support of mesmeric surgery came with James Esdaile, a Scottish physician assigned to a hospital in India for his military service. He created a sensation in 1846 when he published the results of about seventy operations using only mesmerism for anesthesia. Even more impressive, the July 1851 issue of *The Zoist* reported Esdaile's surgical experiences removing 161 scrotal tumors that were endemic in India at that time, most weighing between forty and one hundred pounds.[28] Although 5 percent of Esdaile's subjects died, none was the direct result of the surgery. In contrast, hospitals in Calcutta suffered a 50 percent mortality rate with their two thousand yearly non-mesmeric operations on scrotal tumors. Moreover, Esdaile believed that he could not have performed most of his operations using chloroform, introduced as an anesthetic the year he published his book, because the physiological effects made recovery too precarious.

Elliotson predicted that mesmerism would replace chloroform and ether because of their harmful side effects, and he had good reason to express frustration with the medical establishment that ridiculed mesmerists. Elliotson and his supporters heaped scorn on their critics, asking if patients were denying pain as they were being savagely cut open and sawed off. Searching for reasons to maintain their doubts, critics pointed out that Esdaile had eliminated any candidate for surgery who coughed or displayed any nervousness.[29] This meticulous patient selection was considered a sufficient reason to reject Esdaile's conclusions.[30] No one acknowledged his success with several thousand operations, about three hundred being major procedures.[31] As the debate raged, improvements in the safety of chemical analgesics rapidly discouraged reliance on mesmerism.[32] Esdaile died in 1859 without any appreciation of his accomplishments, but you can believe there were some thankful souls back in India.

Until the time of Esdaile, mesmerism could be explained in terms of suggestion, compliance, and expectation. However, the success of invasive surgeries strongly supported the claim that something more was happening. What was it?

THE EFFICIENT BRAIN

Modern understanding of brain function suggests that our sophisticated brain operates efficiently in the unconscious area not because it has something to hide, as Freud popularized, but for efficiency. It's like having kindly mental butlers who know what we want and what is in our best interest. They anticipate our needs and take care of us without being asked.[33] Our brain processing is so efficient that it takes over much of our behavior automatically so that we do not need to think about or evaluate mundane or repetitive tasks.

If we can act without conscious awareness of brain function, then maybe our brain can place pain signals out of our awareness. There are some analogies to help us understand this. We often believe we have acted intentionally even when our behavior is autonomic. For example, you raise your arm to protect your head from a foul ball before you can process any intention to protect yourself. The hypnotist utilizes this disconnect by telling the subject, for example, that his arm will levitate without any apparent effort or intention. "Just notice how your arm feels light and wants to go up." And it happens. The subject stops attending to intention and just notices that the suggestion of the operator is actually occurring.[34] The brain makes decisions without the conscious mind involved.[35]

The experience is strange, which is why subjects have always insisted that induction of mesmerism/hypnosis creates something real, something more than simply agreeing to cooperate. The raised arm is not expected, and the subject has no sense of compliance with the request. Patients are sometimes so surprised that they laugh, breaking the trance state. Hypnosis is, therefore, more than a distraction or an ordinary response to suggestion. Hypnosis can reassign the neuropsychological pathways in a manner that creates confusion or disregard for the link between intention and action. Highly suggestible subjects probably have better ability to manipulate neural pathways, or they may have more neuropsychological plasticity.

For a moment, consider subjects who are told they are now six years old. They think and behave like a six-year-old child, and we can say that they are acting "as if" or "in the role" of a child. On the other hand, subjects tolerating a painful medical procedure are not simply playing the role of someone with insensitivity. Their situation is better understood by saying that they have turned off or diverted the typical neural pathways responsible for pain awareness. Some patients need to be told when their surgery has been completed.

ORNE'S EFFERVESCENT ESSENCE

The history of hypnotic pain control can be understood as the brain's diversion from attending to painful stimuli. Is that any different from the expectation of a placebo? Orne attempted to answer the question by comparing deeply hypnotized subjects with those given what they believed to be a painkilling drug. Results showed hypnosis produced an advantage, particularly for more susceptible subjects in a deep trance.[36] Hypnotized subjects experienced what the neuropsychologists predicted and what hypnotists long suggested, namely a perceptual distortion that helped manage a painful experience.

Still, every study raises more questions. Canadian psychologist Nicholas Spanos[37] told both hypnotized and nonhypnotized subjects to imagine their hand in a glove protecting them from painful ice water. Half of each group was told that pain reduction would occur naturally, whereas the instructions provided to the other half implied that their pain reduction would take some effort. Subjects in both groups (hypnotic and nonhypnotic) reported equal amounts of pain reduction, but those told that effort would be required (in both groups) said the pain reduction was more difficult. Thus, the sense that something is happening *involuntarily* during hypnosis is challenged by this study. As Orne might have predicted, subjects were highly sensitive to implied suggestion.

Although Orne's study suggests hypnotic analgesia is more than a placebo, a survey of studies (meta-analysis) found no particular advantage for hypnosis.[38] The general consensus emerging appears to be that the benefits of hypnosis begin to vanish as more variables are controlled, like better motivational instructions for subjects not hypnotized. That essence of hypnosis we thought we could capture appears to be fizzling out. Maybe the French Commission was right. But maybe placebo subjects are discovering and utilizing the protective mechanisms that hypnotized subjects are instructed to employ.

Some psychologists at Northeastern University in Boston designed some clever research to compare highly hypnotizable subjects with nonhypnotized subjects who were asked to pretend to be hypnotized. All subjects were told their arm would become too stiff to bend.[39] Then, attached to a lie-detection device, subjects were asked if they *really* experienced an inability to bend their arm. Hypnotized subjects really experienced a rigidity that most pretenders did not. From this experiment, it seems we must conclude that hypnotized subjects experience something different, which they have been telling us all along—at least that's what half of my mind believes.

Is there another way we can understand this complex and sometimes contradictory information?

THE BEST ILLUSION AWARD

While in the midst of puzzling how to understand this mess—and to explain it—I had the privilege of hearing a lecture on illusions by Susana Martinez-Conde and Stephen Macknik. They were instrumental in founding the Best Illusion of the Year contest.[40] As I listened to their definition and discussion of illusions, I realized that almost everything they said was applicable to hypnosis. It occurred to me that a trance could be conceptualized with a simple premise: hypnosis provides an opportunity to experience an illusion.

Illusion experts assert that all our sensory systems are vulnerable to manipulation, distortion, and misperception. We usually think of illusions as optical, but our emotions, thoughts, and memories are also susceptible, as well as all our sensory apparatus. Hypnotists have provided examples of illusions in all these areas.

An observer has no understanding of what subjects are seeing when they look at an optical illusion. For example, you have no way of knowing whether the person is seeing a rabbit or duck, the stairs going up or down. Similarly, the observer does not know if a hypnotic subject is experiencing a disconnect from reality—a perceptual illusion. Members of the French Commission observed mesmerized subjects, and they insisted it was all suggestion, expectation, and compliance. Mesmer's subjects, on the contrary, insisted otherwise because their reality was not one of intentional compliance. Similarly, those hypnotized subjects of the Boston study insisted they *really* could not bend their arms; nonhypnotized subjects admitted they were just playing the game of compliance. Pretenders had a different perception because they did not experience the illusion of those whose brains were operating in a different way.

Forbes told us that hypnosis does not provide anything new, but it can effectively transport some people into remarkable experiences. His friend William Carpenter called it a "dominant idea" that fills up the person's mind. Skeptics have difficulty accepting the subject's description of the situation. *Really? You were not able to bend your arm?* Subjects cannot bend their arms because they are controlled by an illusion, and that illusion is the person's reality. When subjects are told they will experience no pain, that is the illusion they experience. If it is suggested that the effect will take some effort, then the illusion is distorted to that extent. There is yet another illusion we must consider.

As we have already discussed, a hypnotist and subject can collude, which creates an illusion for observers. We need another peek behind that curtain.

YOUR CHEATIN' HEART

Esdaile did not bring his subjects into the operating theater, hypnotize them, and start cutting. He employed assistants who prepared his patients for their ordeal over varying periods of time. He did not always make clear that the process was much more complicated than most realized. When mesmerists brought this new analgesia to medical establishments and lecture halls, they found it inconvenient to search for volunteers who were sufficiently ready to illustrate pain insensitivity on the spot. Therefore, they resorted to recruiting subjects already prepared to manage their reflexive pain responses. From the earliest mesmeric lectures, operators used confederates, sometimes called "horses."[41] H. J. Burlingame admitted that all "professors" have ringers who fake the tricks, and they expect to be paid from $5 a week upward.[42] By the beginning of the nineteenth century, these horses were making $60 a month plus an additional $10 a day for those who feigned three days of sleep and $20 for each subsequent day.[43]

Groups of men were ready and available to demonstrate pain insensitivity and other endurance feats.[44] Some relied on previous hypnotic practice, self-hypnosis, or self-controlled stoicism.

> I have gone before a college of physicians and put myself into a cataleptic condition and allowed them to cut, torture and sew me up. Look at my arms; they are masses of scars, every one of them I have felt just as much as any other person would. What have I done this for? For nothing but fame and money, both of which I have made.[45]

Some managed tolerance through distraction or gritty determination. We need only consider that diabetics learn to ignore the inconvenience of daily injections. Joseph Ovette provided the following instructions for pain endurance:

> With the needle in your right-hand jab it through the skin. The only pain you will experience will be a slight sting. The needles can also be run through the cheek, from cheek to cheek, or through the loose skin under the neck.[46]

Perhaps William Gresham best understood the men who endured pain in sideshows. Early in his life, he worked an awful assembly-line job.[47] After watching a demonstration by a "human pincushion," he concluded that he would rather take that man's job than go back to life in the factory. His analysis: pain is simply a cost of doing business. Early hypnotic professors and sideshow performers jammed scarfpins and knitting needles into their bodies. Their secret was a previously punctured hole through the skin (often with the assistance of a physician) kept open with a thick gold wire until the perforation was thoroughly healed.[48]

In contrast, contemporary magician David Blaine endures painful events *without* using trickery, which everyone expects from him because he is a conjurer. A *Chicago Tribune* reporter asked him if he suffered posttraumatic stress symptoms after his shockingly hurtful stage shows. He admitted suffering in many ways until his body healed.[49]

The "cost of doing business" is less for individuals with congenital pain insensitivity.[50] Tommy Minnock often denied that he had a natural pain tolerance, preferring to say that he learned to endure distress through practice.[51] Minnock was less than forthcoming about many aspects of his life, however, and even his own mother once testified against him in court. In the 1890s, Minnock worked extensively with Professor Santanelli, and together they took bets from doctors that Minnock could tolerate excruciating procedures, like having his lips sewed together. They split their winnings. Santanelli's book on hypnosis does not mention Minnock, probably because they had a falling out.[52] The cover of Santanelli's book depicts poppies with the heads of girls resting peacefully inside each flower. Was he implying that his treatments were like opium or that he was getting by with a little help from his friends?

Narcotic abuse was a hidden part of nineteenth- and early-twentieth-century horses who tolerated painful demonstrations, staged burial stunts, and endured extended sleep. This darker side of hypnosis was admitted by an entertainer known as Parkyn. He exposed many secrets from his own experience and from confidential discussions with other hypnotists and their horses.[53] He said some operators gave their subjects "large doses of trional or chloral just before hypnotizing them," and the dose was repeated when the effect wore off. Trional, also known as methylsulfonal, is a sedative and hypnotic with fast-acting anesthetic properties, and chloral (chloral hydrate) is a commonly known sedative and hypnotic.

Herbert Flint, the travelling hypnotist who broke concrete on his daughter's outstretched body, got caught up in a narcotics problem.[54] Flint was entangled in many problems because of his abrasive personality. One correspondent snidely reported that Flint was touring out West, hypnotizing anyone who got in his way. On one occasion, a window sleeper travelling with him was arrested as a drug addict. In his defense, the man claimed his addiction resulted from Flint giving him drugs to keep him asleep during his stunt. This was a likely claim because we now know that many addictions begin with a prescription for pain medication.[55]

I have in my possession correspondence from the late 1920s between a young magician and Adolph Lonk, a chiropractor and hypnotic entertainer. Lonk offered this magician one long lesson for $10 that would "make you an excellent psychotherapeutist."[56] In this same letter and in his *Lessons in Hypnotism*, Lonk peddled the following drugs at bargain prices: cannabis, barbital, paraldehyde, amyl nitrite, and chloroform. Narcotics were not well regulated at that time, and Lonk said he furnished them only in small quantity, "as I do not wish to get you into trouble."[57] Chloroform was used for induction enhancement, secretly delivered through a tube hidden in the sleeve. The other drugs were purportedly to be used to enhance individual treatment sessions.

Almost all early performing hypnotists offered private treatments in addition to their entertainment and demonstration lectures.[58] The additional income was accepted without any competence of the individual to assess or treat the presenting complaint. Some of these hypnotherapists were predatory, which I discovered while serving on the ethics committee of the Oregon Psychological Association. At that time, every known unlicensed hypnotherapist in the state had been accused of sexual improprieties, and I offer my explanation.

Beginning in the early twentieth century, American hypnotists began offering training manuals and correspondence courses in hypnosis, and they produced books, brochures, and pamphlets that prominently displayed provocative scenes and suggested male conquest. The front cover of Herbert Flint's 1912 brochure shows a "Gibson girl" drawing of a woman wearing a dress with a plunging neckline. She is seated on a slightly reclining couch with her eyes closed, Flint dominantly standing nearby.[59] The 1917 book *Advance Course in Personal Magnetism* has a dustjacket showing a man holding a woman who looks captivated—or magnetized.[60] The training manuals of Harry Arons and Melvin Powers, popular promoters of midcentury hypnosis, show the hypnotist inducting attractive females into a trance state.[61] A pamphlet of lessons on hypnotism by L. E. Young features a man with electric beams or sparks coming from his eyes waving his fingers at a young lady.[62] A closer look shows what appears to be a third hand extended toward her covered breast. The content of this book was reprinted many times, each with its own variation of the picture.

These illustrations—all on the covers—lured men into the idea that they might be able to hypnotize women and say with conviction, "You are now under my control." A case can be made that individuals with "vulgar curiosity" and the need for dominance were attracted to the practice of hypnosis.[63]

I suspect that the drugs offered by Lonk were of additional assistance in facilitating sexual improprieties. This would not be a surprise to those who wrote that private report of the French Royal Commission in 1784: they knew what was happening in secret when they asked, "Would it not be easy to outrage [molest] her?"[64]

THE MYSTERIES OF
NATURE AND SECRETS
OF SPIRITUALISTS

20

RAP IF YOU
BELIEVE IN SPIRITS

Rapping had by this time become a flourishing trade, and me-
diums sprung up, as plentiful as mushrooms and as worthless
as toadstools.

—Vizetelly, *Spirit Rapping in England and America*, 1853

Modern spiritualism began in Hydesville, New York, at the home of John Fox, his wife, and their daughters, Maggie and Kate. Its humble beginning was at the end of March 1848, when the family was bothered by knocking noises or rappings at night. After the rappings responded to one of the girls snapping her fingers, Mrs. Fox asked questions that were answered by these sounds. The family concluded that the communication was with a peddler who had allegedly been killed in their house by a former resident. They summoned a neighbor to witness these manifestations, and she also asked questions that the spirit answered with raps. This neighbor, a Mrs. Redfield, reassured the girls when she noticed they appeared frightened. One responded: "We are innocent—how good it is to have a clear conscience."[1]

This quote was reported in a pamphlet written by family friend, D. M. Dewey. Harry Houdini owned a copy of this uncommon little item, and some alarm bells went off in his head when he read the girl's declaration of innocence. He made a mark in the margin and then recorded his thoughts on a sheet of stationery from the New York Commodore Hotel:

Why should a child say: "We are innocent," etc? It is not a child's expression. Just so, if you read this whole story analytically, you will find the same incongruities all through.

Houdini had reason to be suspicious of this expression of innocence. One of the sisters was denying guilt without being accused of anything, and she was also declaring innocence for her sister. I doubt that Houdini realized these girls were adolescents, not young children. Many spiritualists changed the ages of their founders in order to wrap them in a cloak of childlike purity.

PRESERVING INNOCENCE

In his *Encyclopedia of Occultism and Parapsychology*, Leslie Shepard says the Fox sisters were seven and ten at the time of these first spirit rappings.[2] He indicates that the first reports improperly identified them as eleven and fourteen, but he provided no source for his conclusion.

As I started looking, I was surprised to discover that no one mentioned specific birthdates of the Fox sisters. There is a likely explanation. The girls were born in a rural Canadian village where births were probably never recorded anywhere except in a family Bible.[3] This situation enabled believers like M. E. Cadwallader, publisher of a large spiritualist newspaper, to describe the Fox sisters in their "guileless innocence" as they encountered mysterious raps with "childish simplicity."[4] The birth of spiritualism was thereby established in this manger of innocence.

How old were they, really?

Those first raps were investigated by E. E. Lewis, an attorney working as a newsman. He obtained "statements" or "declarations" from twenty-one eyewitnesses that were signed and published a couple of weeks after the critical March 31, 1848, rappings. The testimony of Mrs. Fox, who best ought to know the ages of her children, said they were twelve and fifteen.[5]

> The girls, who slept in the other bed in the room, heard the noise, and tried to make a similar noise by snapping their fingers. The youngest girl is about twelve years old; she is the one who made her hand go. As fast as she made the noise with her hands or fingers, the sound was followed up in the room.[6]

Now if you reread this statement, you will notice that the structure is not particularly clear or organized, which is common in a spontaneous, unedited statement. As we see shortly, spiritualists took the liberty of revising the style and content of the mother's early statement. Believers preferred spirits talking to innocent little girls rather than to mischievous adolescents.

I reviewed nine books and pamphlets published between 1850 and 1856 that accepted the signed statement of Mrs. Fox.[7] After 1856, authors asserted younger ages for the girls. The first revisionist I found was Robert Dale Owen, a social reformer, member of Congress, and American foreign diplomat. Because of his impressive credentials, spiritualists regularly cited him as an authority for their cause. Owen mentions in a footnote that he possessed the "very scarce" Lewis pamphlet. Yet he clearly misrepresented it. "How old is my daughter Margaret?" Twelve strokes! "And Kate?" Nine![8]

In the original, Mrs. Fox asked the spirit to rap the ages of her children, but she never specified the spirit's response except to say the answers were correct. (Twelve and nine contradicts their own mother's statement that they were fifteen and "about twelve.") Moreover, Owen has Kate addressing the rapping spirit as "Mr. Splitfoot," a nickname for the devil, who is popularly depicted having cloven hoofs. Nothing like that appears in the original statements by Lewis. The raps were identified as the result of a spirit, not the devil. Owen must have made it up.

The issue of youth and innocence continued to flow from one author to another. Out of ignorance or intention, each revised spiritualism's beginning. For reasons discussed later, the sisters themselves were pleased that years had been removed from their age, and not just because of vanity.

RAP IF YOU BELIEVE IN SPIRITS

I began this chapter with one of the Fox sisters declaring their innocence. If they were not causing the raps, then who or what was responsible? Curious people tried to answer that question, some with better investigative skills than others. Mrs. Fox asked if the mysterious rapper was a spirit. Mother Fox and Mrs. Redfield, the friendly neighbor, quickly concluded the raps were "a revelation from the spirit world." Why did they come to this conclusion so effortlessly? We find a clue in the appendix of that 1850 pamphlet owned by Houdini.

Dewey provided his readers the "psychological facts" by reprinting extracts of various authors who supported the idea of spiritual communication with departed souls: the Seeress of Prevorst (clairvoyant), John Wesley (founder of Methodism), Emanuel Swedenborg (scientist-philosopher), Joseph Haddock (physician), A. J. Davis (clairvoyant-healer), and William Fishbough ("psychological writer"). Fontenelle fought the belief that demons were out there trying to harm us; now these six authors supported the idea that departed souls were trying to help us. Even before the Foxes, Dr. Edward Drury heard three mysterious raps on the bed board of his thirteen-year-old patient in 1841. That experience provided him a firm conviction "that we live in a world of spirits."[9] Moreover, raps were one of the defining characteristics of the poltergeist.[10] The cacophony of rapping tables, which might seem silly to us, was easily accepted as the communication of discarnate spirits.

Popular gothic novels described ghosts as individuals who had not been provided proper burials, and they were restless and intrusive until suitably laid to their eternal peace.[11] Mrs. Fox and neighbor Redfield defaulted directly to that belief. The news of spirit manifestations quickly caught up the Fox sisters in a whirlwind of publicity. Mr. E. W. Capron, a journalist, invited Kate to live with him and his family in nearby Auburn so that he could investigate the raps. Capron provided public support for Kate, which met with some initial opposition. Nevertheless, raps were soon prevalent in houses throughout Auburn, causing considerable social upheaval.[12] By the end of the following year, an estimated fifty to one hundred mediums were actively rapping.[13]

When Kate went to Auburn, Maggie went to Rochester to live with her older sister, Mrs. Leah Fish. Leah was married at fourteen and shortly thereafter deserted by her husband, although it was more expedient to say she had been widowed.[14] Leah recognized the opportunity for creating a new role for herself with Maggie, and they produced an immediate stir of interest. As in Auburn, houses in Rochester were soon beset with turmoil. A host of heavenly spirits suddenly decided it was necessary to rap on tables. The epidemic was spreading across the land.

RAP IF YOU BELIEVE IN EPIDEMICS

Most people accepted the new phenomena; they were convinced because they could think of no alternative explanation or because they just wanted to believe. Some rejected the idea, but no one had ever before considered how to investigate rappings on tables. As we shall see, more observant individuals noticed things that looked like cheating and were inconsistent with a supernatural explanation. However, no one could strike the death blow to stop the nonsense or, on the other hand, to prove the existence of spirits to skeptics.

Those able to contact spirits were called "mediums," and those who sat through their séances were called "sitters," which is the meaning of "séance" in French. Interaction with spirits evolved from simple knocks to having someone recite the alphabet until a knock indicated a selected letter. Letters formed words, and words created sentences, but the process was painfully slow. (In one situation, it took fifteen minutes to rap out a sentence of thirteen words composed of fifty-six letters.[15]) As an alternative to saying the alphabet, someone decided to point at letters printed on a card, one at a time, until a rap occurred. This process was faster but still unacceptably slow.

Some people wondered why spirits did not give more direct answers. Spirits responded by saying they were restricted by a higher power. And why do they make sounds only in the dark or under tables? Raps required spirits to assume a material form, so darkness was necessary or their presence "would frighten."[16] Some activities, like those mysterious touches of spirit hands, also required a tangible form, so it must be done in the dark because people were not prepared for a direct visitation.[17] Some spirits decided to reveal themselves with a luminous glow that looked much like phosphorescent paint, whereas others allowed their arms to be seen emerging from cabinets especially built to concentrate psychic energy. One clairvoyant was able to see spirits, which he described as transparent like gauze. The spirits told the clairvoyant that they produced raps by willing them to happen. No one seemed to care if his spirits were transparent, yours were invisible, and mine were materialized. Everyone was eager to hear those raps.

From this simple beginning, spirits flamed out across the land, and soon mediums introduced their rapping to England and Europe. All sorts of additional manifestations emerged. There was a spirit-contact race.

SPIRITS LEARN NEW SKILLS

With the next iteration of communication with spirits, a medium would drop her hand down over an alphabet card onto a specific letter as directed by a spirit. The medium's arm and hand were controlled by a process colorfully described by J. G. MacWalter.[18] He said that as the medium's hand dropped, the finger sensibly and irresistibly arrested at a certain letter to construct words and sentences. Readers may recognize this description as similar to the modern practice of "facilitated communication" that was developed to help autistic and other language-limited individuals express their unspoken thoughts. (Unfortunately, this modern work is at best a well-intended pseudoscience.)[19]

After the hand-drop era, mediums simply kept their hands on the alphabet card, moving their fingers from letter to letter. The first trials were characterized by jerks and convulsive movements, but most mediums soon settled into a smooth and skillful process. Cynics noted that the card method was in full vogue in Philadelphia in April 1851, but spirits in Springfield did not suggest this improved method until March 1852.

If a spirit could move the medium's hand to point at a letter, then it made sense that a spirit could move a hand to write. And so automatic writing emerged during the trance state, which allowed a spirit to guide the hand without the intention—or even awareness—of the medium. Automatic writing had a downside, however. The medium could not answer mental questions (asked silently in the mind) or gather clues from the sitter (watching expressions, etc.) as they did while producing letter after letter.[20] As a result, automatic writing was seldom used in séances but was generally reserved for transcribing longer messages. From the hand of Isaac Post came more than forty communications, mostly from famous individuals who had passed over to the other side.[21] At least six of them forgot how to spell their name.[22] Another automatic writer and friend of the Fox sisters, Charles Hammond, said he did not change a word from what spirits directed.[23] Maybe he should have. His essay, purportedly from Thomas Paine, was described as "slip-shod, ungrammatical, and confused," which was particularly ironic because Paine's *Age of Reason* was concise, direct, and full of wisdom.[24]

TRIAL BY COMMITTEE

In November 1849, a committee convened to evaluate the Fox sisters and the source of the sounds, but they "failed to discover any means by which it could be done."[25] Dr. Langworthy relied on his stethoscope to conclude that the sound was not coming from their lungs nor were the sisters using any sort of ventriloquism. A committee of ladies found nothing in the girls' clothing that might have produced the sound. Raps were evident while they were standing on pillows with a handkerchief tied around the bottom of their dresses to expose their feet. However, some of Rochester's leading citizens complained that the committee had been stacked with sympathizers.

Feelings were running high. When the sisters began making considerable amounts of money, some thought it time to observe more critically.[26] Meanwhile, three physicians at the University of Buffalo discovered a young lady in Rochester who was able to produce séance sounds by "cracking" her knee joints. Kate and Maggie responded by offering to be personally examined again. This time, a different group of physicians placed the heels of the girls on a cushion with their legs extended so they could not tense their ligaments as necessary for cracking. In this position, no sounds were detected during the next half hour. The sisters admitted it was useless to continue, but the sounds resumed once they were freed from that constraint. The physicians then held their knees through their dresses so that any dislocation could be perceived by touch, and again no raps occurred.

Following these investigations, letters appeared accusing the committee of bias while others cited Bible verses suggesting inherent dangers in these satanic knockings. Adding to the confusion, a Professor Loomis explained in an 1850 issue of the *American Journal of Science* that water flowing over local dams was transmitting vibrations through the ground to local dwellings. These vibrations, he said, caused unexpected rappings that people were reporting in their homes.

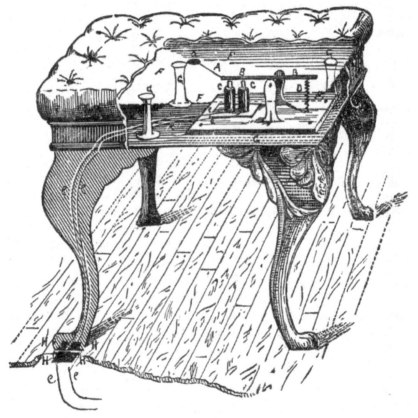

An electromagnetic rapping device offered for sale to spiritualists by Harry Hermon. Harry Hermon, *Hellerism: Second-sight Mystery*. Boston: Lee and Shepard, 1884. *From the author's collection*

Around the same time, a Mr. John Hurn asserted what few wanted to hear. He said that all manifestations of spiritualism were the most miserable imposition ever palmed off on a civilized community.[27] Hurn admitted that he provided the Fox sisters an ink that mysteriously appeared on a wall shortly after a message was written. He also claimed that he touched sitters in séances to make them believe in the presence of spirits. Hurn's comments were promptly denied.

In the winter of 1850, a Reverend Potts delivered a lecture in Rochester in which he proposed that the raps were caused by the cracking of toes, which he demonstrated to the delight of his audience. Potts was the first to publicly exhibit this theory, which was then widely publicized by Chauncey Burr, a New York investigative reporter. In response, a physician presented an affidavit describing raps by a medium in which the cracking of any body part was restricted so as to be impossible.[28] Burr had an answer. He had by this time investigated more than fifty mediums and discovered seventeen different methods for creating raps.[29] Spiritualists were discovering what magicians already knew, that they could produce any effect by multiple methods.

As one example, a carpenter confessed that he had manufactured a table with a rapping device carefully concealed inside. He said, "If people will give a dollar a piece to hear a little hammer strike inside a table, and run around crazy about it, it is not my fault."[30] Several of these table-rapping devices were discovered,[31] and a manufacturer of magnets confessed that he sold his products to mediums who constructed electromagnetic rapping mechanisms.[32] Some rapping devices, however, were as simple as a child's toy cricket, a metal device that crackles when pressed,[33] and some mediums merely gave a sideways tap of their shoe against the leg of a table. French magician Robert-Houdin, our friend of Didier, baffled everyone when he produced mysterious raps from within an empty box hung from the ceiling by a single wire.[34] Startled witnesses to this miracle did not know that the wire was attached to a hook in the ceiling that was struck by the plunger of an electromagnetic device controlled by a switch under Robert-Houdin's foot. The impact of the plunger was transmitted down the wire to the box, which acted as a sounding board.

The Fox sisters, especially in those early days, often used a solid object to project or enhance the sounds they produced. The editor of the New York *Commercial Advisor* said the clumsiness of the sisters' imposture was so great that one literally had to divert one's attention or shut one's eyes to avoid seeing it. The editor also said their motive was equally easy to see, namely that they were making from $50 to $100 a day, and their supporter, Mr. Capron, was a printer selling pamphlets as fast as they could be supplied. Capron immediately denied ever setting a line of type in his life, although he failed to mention that he was a journalist successfully selling his booklets promoting the sisters and spiritualism.

If all this rapping nonsense was so obvious, why were there so many believers? People found something satisfying in spiritualism, which proposed the idea that everyone happily entered the next world still carrying their earthly quirks and flaws. Up there, issues that had not been finished on Earth were resolved at an unhurried pace.

Spiritualism was without demanding clergy or restrictive dogma. Mediums developed routines to make their séances interesting, providing surprises and entertainment instead of boring sermons of condemnation. For example, the Fox sisters opened some of their sessions by requesting that each sitter write down the names of three individuals, one being deceased. Then the cause of death was written, as well as two other possible causes of death. Finally, they were to write three locations, one being the place of the chosen person's death. When the sitter pointed at the names, the sisters tried to identify the selected person by rapping, which was supposedly to indicate his presence. Similarly, raps signaled the cause of death and location.[35]

The editor of the *Commercial Advisor* said it was easy to see what they were doing, and others had also detected problems with their presentations. One man, however, stepped up to conduct a systematic investigation.

A PAGE TURNER

That man was Charles Grafton Page, a chemist, inventor, and patent officer who admitted his prejudice against any rapping by spirits. When Page attended his first séances with the Fox sisters, sometime during 1853, raps were heard on the underside of the table.[36] The strike was loud and obviously created by something hard. His colleague in the investigation feigned surprise and alarm, bending down to look beneath. The raps ceased immediately. Pretending seriousness, Page informed the gathered believers that spirits were invisible so there was nothing down there to see. The girls backed away from the table, and the rapping then commenced on the floor near their feet. However, both girls were wearing long dresses so that nothing could be seen.

Page asked the girls if sounds ever came from places other than near them. "Yes," answered Leah, who typically managed the séances for her two younger sisters. "Sometimes the sounds have been made in that wardrobe." To his surprise, one of the girls got into the wardrobe and asked the spirits to rap in there. Raps came from the floor, and then she requested the raps be made on the sides and back of the wardrobe, which occurred after she adjusted her position. The girl then pointed to various parts of the wardrobe, pretending the sound came from that area. Page noted that the sound was exactly the same no matter where she pointed. He monitored these sounds carefully because he knew that the sense of hearing is highly dependent on expectation and suggestion.[37] One of the sisters also made sounds along the boards of the wardrobe, but again Page knew that sounds are transmitted through solid objects in ways that confuse our perception about the origin of the noise.

At a second séance, a colleague of Page asked the girls if they could rap while in a swing or standing on a pillow. "Yes, all that had been demonstrated previously," Leah affirmed. One girl went upstairs to get a pillow, but Page realized that a common pillow would be hidden by her long dress. Thus, he made a cushion of their cloaks covering an area about three-and-a-half feet in diameter. It was sufficiently thick that a sound could not be produced on the solid surface of the floor. As he expected, no rapping sounds occurred. Realizing her predicament, Kate stood partially on the cushion so that her dress hung over the edge, but Page objected on the pretext that the dress would conduct the electricity away. No raps came after repeatedly beseeching the spirits, so Kate asked her sister to join her, apparently hoping she would get a chance to rap. Page prevented Maggie's contact with the bare floor, and no raps occurred.

Page noted that Leah often misspoke by saying that the "girls" could rap in particular situations, when she should have said the "spirits" could rap. It should also be noted that Leah listed herself in the 1851 Rochester City Directory as "mysterious knocker."[38]

Page said that raps occurring under the table sounded only when the girl's feet were tucked out of sight. When raps occurred when the girls were more in the open, the floor was never fully visible at the critical time. He detected a small flinch in the lower abdominal muscles concurrent with the rapping. When one of the girls noticed Page observing her efforts, she immediately pulled a curtain to darken the room, and she covered this area of her body with her shawl. Similarly, another observer noted the raps were associated with evidence of effort on the face of the girls.[39] Moreover, the raps ceased when the girls were not near a solid surface, and the sound of raps was obviously associated with the surface they were standing on. Why would spirits create raps that were always near the girls, under their skirts, or where the impact could not be observed? Page concluded that the overwhelming evidence showed that the Fox girls were willful agents in producing the sounds.

Page was convinced that the Fox sisters used some mechanism besides cracking of toes, and he experimented with multiple possibilities. He produced the closest approximation of the sounds with a lead weight shaped like a dumbbell attached around his big toe with one part on each side. He believed this accounted for the "double-thump" sound that was peculiar to the Fox sisters' rapping. Another investigator independently duplicated this double sound, which he believed was caused by one end of a lead bar hitting the floor a fraction of a second before the other side.[40] Kate was noticeably embarrassed when she got up to walk with the mechanism still attached to her toe: it sounded like she was walking with a peg leg.

But how did they know when to rap if the question was asked silently, only in the mind of the sitter? The sisters were obviously getting clues from the sitter's reactions in the situation. Page and his companions decided to avoid any flash of facial cues or hesitations that might signal correct responses, and the sisters failed. In contrast, a believer obtained a satisfactory interaction with his deceased wife, which gave Page an opportunity for comparing the two conditions.

CHARGES AND DENIALS

Capron claimed the report by Page was "full of sound and fury, signifying nothing." He ridiculed the possibility that raps were made with a lead weight attached to the toe. But Capron faced a continual flow of accusations against the sisters. In April 1851, Mrs. Norman Culver, a sister-in-law of the girls, made a declaration of her involvement in the conspiracy of spirit manifestations.[41] Mrs. Culver, who was about the age of oldest sister Leah, became suspicious about the spirit manifestations and decided she wanted to learn their secrets. Mrs. Culver waited until Maggie was away and then told Kate, the youngest girl, that she could provide information about a particular sitter in exchange for learning how to make raps. Kate requested that Culver touch her arm when someone pointed at the correct letter on the alphabet card, and Culver learned how to observe the faces of sitters (and gather other clues) so that she could pass information to Kate. After Culver assisted in séances for a couple of weeks, Kate reciprocated by teaching her how to produce raps with her toes. Heating her feet in warm water made the process easier, and eventually she could produce 150 raps in succession using all of her toes. Maggie also accepted Culver as a confederate and told her she had made raps with her knees and ankles when tested by the committee in Rochester. They also used leveraged floorboards to create raps, and in some situations a Dutch servant girl tapped under their floor from the cellar when she heard them call on the spirits.

Capron responded by claiming that Mrs. Culver was a bitter enemy of the Fox side of the family. Culver's deposition, he alleged, was arranged by Chauncey Burr, that journalist who had engaged in unscrupulous trickery. No leveraged floorboard was ever found, he said, and the Fox family never had a Dutch servant girl. Culver's confession, he said, was a fabrication made up under the mesmeric spell of Burr, who was determined to confirm his theories. However, Capron failed to mention that one home where the sisters were tested employed a Dutch servant who was a spiritualist.[42]

Critical observers noted that the sisters used hand signals to communicate with each other, and Charles Elliott of New Haven was of the opinion that the girls were such poor actors that they often had difficulty controlling their emotional responses.[43] He said they were shallow and simple cheats who were encouraged by credulous people who wanted to believe. Because of their repeated exposure, he concluded that they were not imposing on the public, because the public was well informed about their trickery. Instead, the public was imposing on them. Anyone with an observant eye had little difficulty identifying problems in their presentations, but believers helped the sisters become successful. The whole process looked like it had devolved into something like suggestion, expectation, and compliance.

RAISING KANE

Believers helped spiritualism become successful even as the sisters fell deeper into their own personal problems. For Maggie, it was the turmoil and trials of love. In 1852, Dr. Elisha Kane fell in love with nineteen-year-old Maggie Fox when she was in Philadelphia with her mother. Kane was a medical officer in the U.S. Navy, already famous for his polar expeditions. His love letters, later published, reveal his passion—and his ambivalence.[44] He feared being disinherited by his family, and he dreaded the potential damage to his popularity in the press if he was known to be associated with her. In February 1857, Kane unexpectedly died. Maggie claimed they had been united in a common-law marriage, but she could offer no proof. Kane had set aside $5,000 in trust for her with his brother, but his brother was appalled by the differences in their social classes and eventually stopped paying her the pittance he had been dribbling out.[45] As she sank deeper into depression and alcoholism, a friend of hers published their love letters. Maggie was depicted as a young, innocent girl rejected by the cruel Kane family. "Poor girl! With her simplicity, ingenuousness, and timidity, she could not, had she been so inclined, have practiced the slightest deception with any chance of success."[46]

The exposé implied that Maggie was only thirteen years old when she met Kane.[47] She was nineteen. Despite repeated characterizations of Maggie as child-like in her demeanor, the letters show that both Maggie and Kane understood her spiritualism act was fraudulent. Maggie returned to rapping due to financial stress.

MISSING LINK

During the early years of their act, Kate and Maggie's older sister Leah managed their séances, but the family broke apart as the years passed. As the younger girls struggled, Leah entered her third marriage with a wealthy New York banker. She published a book introducing her family as the founders of spiritualism, the missing link between this life and the spirit world. It was a beautiful 477-page publication with steel-engraved portraits and illustrations. The cover featured a dramatic gilt illustration of two spirits placing the final link in a chain, representing the connection between heaven and Earth. The content of the book was figuratively gilded. Leah claimed that the sisters' psychic power was inherited from their father's side, and she gave herself credit for most of their accomplishments. Leah also hung out some of her sisters' dirty laundry to enhance her own image, which created even more conflict within this already-troubled family. She distorted the facts of history, as well. For example, pretending to quote from that early Lewis deposition described earlier, she said:

> The children, who slept in the other bed in the room, heard the rapping, and tried to make similar sounds by snapping their fingers. My youngest child (Cathie) said: "Mr. Splitfoot, do as I do," clapping her hands. The sound instantly followed her with the same number of raps. . . . Then Cathie said, in her childish simplicity: "O mother, I know what it is; to-morrow is April-fool day, and it's somebody trying to fool us."[48]

Notice that Leah cleaned up her mother's awkward wording and added words that were not in her mother's signed statement: "Mr. Splitfoot," "childish simplicity," and "April-fool day." She omitted Kate's stated age of "about twelve."

Leah's book attempted to stake the family claim as founders of spiritualism because history was passing them by. In the early days, spiritualism diverged in a couple of directions but the sisters never kept up with either side. One branch became associated with women's rights, prison reform, healing, abolition of slavery, utopian experiments, and other social justice movements.[49] In the role of medium, women were empowered to express new ideas, including some that might be otherwise unacceptable. Their theology had no hell, no angry god, and no dogmatism; they envisioned a golden future on Earth as it was in heaven. They asked little while still promising mansions in heaven. However, the tolerance they permitted led to indifference. Reform-minded mediums elicited no passion in sitters for the causes they promoted. By the end of the nineteenth century, reform movement groups abandoned spiritualism.

The other branch of spiritualists developed more exciting encounters with spirits, preferring theatrics and miracles. Listening for knocks became antiquated and was surpassed by more entertaining manifestations: slate writing, spirit drawings, luminous phenomena, visual and tactile stimulation, clairvoyance, escape from bonds, and an embarrassing plethora of other manifestations.[50] Many sitters did nothing more than watch mediums as they developed increasingly innovative theater. There was nothing to join, no public commitment or institutional loyalty. Talking about spirits was sufficient; God and one's soul were generally ignored.[51]

In terms of the first divergence of spiritualism, the Fox family had no attachment to any particular social movement, religious theology, or moral belief. Their séances were not about spiritual enlightenment. In terms of the second divergence, no one ever described their performances as particularly entertaining. The closest they got was probably their attempts to figure out who died, where, and of what disease. This interaction was like the modern game of Clue: who killed Colonel Mustard with a candlestick in the library? This might have been diverting but not compelling. Although their presentation was simplistic, a Dr. Schlessenger transformed the basic format into a profound mystery that fooled the most intelligent people. Magician David Abbott declared that Schlessenger, who travelled the country at the end of the nineteenth century, could accomplish miracles when people wrote things down that he could not see.[52] He divided a sheet of paper into six parts, instructing a sitter to write the name of a deceased person in one section and to fill the others with names of living friends. He appeared to pay no attention while this writing occurred, and the sitter then cut out the sections and folded them into billets. These were placed in a hat held under a table. The sitter removed one at a time, throwing each on top of the table. Three loud raps erupted when one billet hit the table, the one containing the name of the deceased person. Schlessenger then went into a trance and stated the name, which proved correct when the slip was opened.

Abbott noted that Schlessenger was correct about nine times out of ten, which created great excitement wherever he performed. His method was similar to that of the Fox sisters, but his strategy was professionally subtle and utterly confusing. Without reviewing all the specific details, his incessant talking and subtle suggestions caused people to write the name of the deceased person in the third spot half of the time, with the remaining half usually written in the fourth space. His distracting verbalizations increased the time it took for individuals to write the names of the living persons, which he pretended not to notice. In fact, he perceived cues too subtle for the most observant skeptics to detect. While feigning poor eyesight, he removed his glasses and appeared inattentive to the whole procedure, all the while gaining information that sitters would later swear that he could not know.

SEYBERT COMMISSION

In 1884, Maggie was interviewed and tested by members of the prestigious Seybert Commission at the University of Pennsylvania. After some typical examples of knocking, all raps completely disappeared once controls were in place. In response to a question asked more personally, Maggie shocked everyone by declaring that she never said the raps came from spirits. Further, she declared that she did not know if the raps were entirely independent of herself. Then she asked for paper and wrote a communication from the supposed spirit of the deceased Mr. Seybert, who declared his intention of helping the committee believe. The message ended with a Latin phrase, but one member of the commission confided that Seybert knew no Latin whatsoever.

Maggie acknowledged that the séances had been unsuccessful and said her present state of health was such that she doubted another meeting would be more successful. The committee concluded that "the so-called raps are confined wholly to her person."[53]

Time was not kind to the Fox sisters. Both developed serious alcohol problems that disrupted their séances.[54] Early in 1888, Joseph Rinn attended a slate-writing séance conducted by Kate, but she was so drunk the séance was a complete failure.[55] She was probably using opium as well.[56]

Then Maggie announced she would make a formal confession and renunciation.

THE "DEATH BLOW" WAS A WILD SWING AND A MISS

In a highly publicized event on October 21, 1888, Maggie confessed that spiritualism started with them as childhood pranks. She said she had now matured, and she denounced the new religion as an absolute falsehood from beginning to end. She removed her shoe and cracked her toes on a small table that acted as a sounding board. A committee from the audience consisting of three physicians agreed that the raps were made by action of the large toe. Critics cheered and spiritualists booed.

Maggie did not reveal all the methods they had used to produce raps. Admitting their other tricks could have made them appear less innocent and more intentionally devious. A professional writer and critic of spiritualism, Reuben Davenport, quickly published a book after the confession. He portrayed Maggie and Kate[57] as victims of childhood exploitation who had been carried away by the sensations they created: "From so little a plant has grown a gigantic wood of deceit, corruption and fraud, nurtured upon the fattening lust of money, and of the flesh."[58]

The sisters were now grateful that they had been portrayed as children when those first raps occurred, too young to know they were doing anything wrong. Indeed, in some ways, they were just innocent, giddy girls who enjoyed frightening their cousin with strange sounds when they were all sleeping in the same room. They did not realize that spiritualism would become so popular—all based on what was false and disreputable. After her confession, Maggie went out on the lecture circuit to speak against spiritualism, but people were more interested in considering the possibility of a mystery than learning the secrets of cheats. So, she went back to conducting séances. Perhaps we should not be surprised that the lives of the sisters were compromised by alcoholism, drug addiction, mental disorders, child neglect, and marital conflicts.

Maggie's confession had no lasting effect, and Davenport's wild swing at the heart of spiritualism hit nothing. The movement had already developed a life independent of the Fox sisters. Spiritualism's contemporary historian, Emma Hardinge, believed the *truth*, not the excitement, propelled the movement:

> The ball once set rolling in New York City, sped on with an impetus which soon transcended the power of the press, pulpit or public to arrest, despite of every force that was brought to bear against it.[59]

The Fox sisters started this ball rolling, but the impetus was powered by other mediums who created new manifestations and championed social causes.

As times changed, mediums became more sophisticated. They were now carrying messages from departed spirits that could not be known by watching facial expressions. One sitter accepted that the toes of the medium might be causing the raps, but she wondered how those toe joints knew all about her deceased friend.

21

THE TURING TEST

It is easy for men of acute intellect, apart from experiments and practice, to slip and err.

—William Gilbert, *On the Magnet*, 1958

If spirits up there somewhere can contact mortals down here, would they be eager to prove their existence to honest inquirers, or even to doubters? Arthur Conan Doyle, perhaps more than anyone, passionately wanted the spirits to convince Houdini of their existence. Moreover, Houdini desperately wanted to talk to his departed mother, with whom he had a relationship that bordered on obsessional. Yet time after time, no medium could contact her. Despite his desire for connection, the blatant deceptions of mediums were always evident to him.[1]

In contrast, books on spiritualism from that period give hundreds— thousands—of testimonials from people who were convinced they interacted with a spirit. We know, of course, that many were deceived by clever impostors preying on the expectations of the bereaved. How would you know for certain if someone was in contact with the other side or was flimflamming the sitter?

Cambridge mathematician and cryptologist Alan Turing, the one who broke the Enigma code during World War II, speculated about whether computers could ever be said to have consciousness. He suggested staging a conversation between an interrogator and unknown respondents, one of which would be the Turing machine. If a human cannot reliably tell the difference between the computer and humans, the computer passes the Turing test.[2] If the machine passes for intelligent, it *is* intelligent. Would it be possible to create a spirit Turing test?[3]

The computer is a winner if it cannot be distinguished from a human, but a spirit would need to do more than be indistinguishable from a medium's pretense. What would transcend the manifestations that a medium could contribute? What would provide convincing evidence of communication with the other side?

I suggest we appoint an independent panel of thoughtful people to determine if a medium has brought us something that might qualify as evidence of a spirit. After a séance, our committee could evaluate the process and the content. If they are not convinced, they can recommend precautions and additional questions for further evaluation.

THE CLEVER PHRENOLOGIST

In the previous chapter, the Fox sisters engaged in a task that would qualify as a simple spirit Turing test. They identified names of deceased persons, with their causes and locations of death. But when the sitter maintained a stoic demeanor, the spirit Turing test was always a failure. After other séances, Charles Page concluded that the Fox sisters were not producing any information that one would expect from sapient spirits in the next world.[4] He rhetorically asked how some people can disregard the incongruities and failures or pretend not to notice them.

Page gave some useful guidance to our hypothetical Turing committee. He said that mediums were like the shrewd itinerant phrenologists who were at that time giving performances across the country. Phrenologists measured and fumbled over someone's head, asked a few key questions, and then portrayed the life and character of the individual with a wonderful degree of accuracy. Page was describing what would become known as "cold reading," the ability of the phrenologist or psychic entertainer to meet someone cold (for the first time) and appear to know all about them.[5] Someone sitting for a phrenologist was not interested in whether the underlying science was valid; that person wanted to hear something meaningful about his or her own life. Page said spirit rappers (like phrenologists or remote travelers) were keen observers, and even skeptics were fooled by experienced mediums. Sometimes not much skill was needed, because most sitters quickly betrayed their emotions and revealed unsolicited information, usually without awareness. Page said that spirit rappers can "become as successful as the Greek oracles of Delos and Lesbos."

Some people have difficulty controlling their communication. Remember that patients of Dr. George Beard unintentionally and unconsciously told him their symptoms at every stage of the interview when he pretended to be a clairvoyant doctor. On the other hand, strict precautions did not protect people from being fooled by Dr. Schlessenger, the medium who perfected the game of spiritual Clue that the Fox sisters bumbled through.

A chemist at the University of Pennsylvania had a great idea to prevent mediums from exploiting unintended signals.

ROBERT HARE AND HIS SPIRITOSCOPE

Robert Hare (1781–1858) was a scientist who made prolific contributions over his lifetime, mostly in chemistry. When he was seventy-two years old, he wrote an editorial in support of Faraday's claim that electricity could not be the source of power that made tables mysteriously rotate. (More on table tipping and table turning later.) In response to Hare's rejection of table turning, he received letters inviting him to attend a séance, which he accepted. The medium received a message from a spirit that was addressed to an attorney who accompanied Hare to the seance. Life was about to change for Hare.

One of the "spiritoscopes" constructed by Robert Hare. Robert Hare, *Experimental Investigation of the Spirit Manifestations, Demonstrating the Existence of Spirits and Their Communion with Mortals* (New York: Partridge & Brittan, 1855). *From the author's collection*

Light is dawning on the mind of your friend; soon he will speak trumpet-tongued to the scientific world, and add a new link to that chain of evidence on which our hope of man's salvation is founded.[6]

At another séance, Hare was startled to hear raps that he could not attribute to anyone at the table. Nor could Hare explain the table tipping he observed. The muscular response explanation he had endorsed, as proposed by Faraday, was not an adequate solution. Perhaps some of the people at the séance were capable of deception, but Hare was confident that no one present was using trickery. These were all worthy people whom he confidently proclaimed would not waste their time with legerdemain.

Hare had never been a religious person, and he rejected the idea of satanic interventions. He reasoned that if a devil existed, it would have been created by God, which would make God complicit in evil. But he was open to "spirits," and they were apparently communicating through auditory raps on tables. Hare considered ways he could investigate this mystery without exposing any cues to the medium when sitters pointed at letters. His brilliant solution was his construction of a device he called a "spiritoscope."[7]

Hare placed letters around the edge of a disk that was slightly larger than a foot in diameter. An arrow at the center of the disk, like the hour hand of a clock, swiveled so that letters could be selected one at a time by the medium. This pointer was connected to a table through a system of cords, weights, and pulleys. Finally, he used a zinc sheet to prevent the medium from seeing the letters on the dial. With this mechanism, he reasoned, the medium could not give an organized response unless she was communicating with spirits or spirits were controlling its movement. That got the attention of our spirit Turing committee.

At what was apparently his first séance with his spiritoscope, Hare asked if any spirits were present. The arrow indicated "yes." Hare asked the spirit to identify itself by providing the initials of its name. The index pointed to R and then to H. Hare asked, "My honored father?" The index pointed to "yes." Hare asked his father, Robert Senior, to spell out the name "Washington," which was successfully accomplished as were other tasks of a similar nature, all without the medium seeing the letters on the dial. The small company of sitters "now urged that I could no longer refuse to come over to their belief." Adding further pressure for his conversion, the revolving disk spelled out, "Oh, my son, listen to reason!"

Hare was convinced he was in contact with his father, so our panel had to consider whether he was listening to reason or not. Was this a positive spirit Turing test? A careful review of his book reveals some issues to consider. Hare constructed several spiritoscopes, but his writing is so disorganized that it is unclear which one he used at any given time, or even if at all. One device shows a rolling mechanism on top of the table that moved the pointer. This was something like a Ouija board with a dial. One of his instruments was not a table but a levered beam that the medium could press to move the dial. Another illustration shows the medium looking at the back of the disk, thus providing opportunity to identify specific letters by noting imperfections on its back. No zinc sheet is evident on any of these.

Thus, our Turing committee recommends that the zinc sheet always be properly situated. The letters on the dial of the spiritoscope should be scrambled instead of arranged alphabetically. As an added precaution, the committee suggests that a dial with different letter locations be surreptitiously inserted during a session break by an independent observer so that Hare and his medium are both unaware of the change and, therefore, uninformed about letter location. No matter how worthy his spiritualist friends might be, someone could be signaling the medium, whether aware or not, for example, by tilting the head forward as the wheel nears the correct letter. Thus, only the independent person observes the dial, recording the letters silently so that the medium does not obtain clues about what words she spells. Credit must be given only for correct responses, not simply for being near the correct response. Get back to us after you repeat the experiments with these improved controls.

Some mediums refused to use the spiritoscope. Others tried but failed to obtain meaningful results. Hare excused failure by saying his father was not present during those times. At one séance, Hare sat for more than an hour on an excessively warm evening. He received no meaningful communication through this medium because, "as I supposed, of the effect of the heat upon her organism."[8]

Some mediums managed to produce messages, but the content was unsatisfactory to Hare. In one séance, Hare asked his father about life in the next world. The information he received from the medium was disappointing because it was "too much blended with her own prepossessions." As a result, he discarded many pages that we cannot now examine. On the other hand, a Mrs. Gourlay provided information that was acceptable to him, so he had no difficulty attributing the results to his father rather than prepossessions of the medium. Meanwhile, our Turing committee is wondering why Robert Hare, Senior, can work only with certain mediums.

After some brief success on the spiritoscope, Mrs. Gourlay convinced Hare that spirits could communicate faster if they wrote their responses. With his permission, Mrs. Gourlay enlisted the assistance of a friend whose hand was seized by an invisible and intelligent power when she put her pen on paper. The results are reported by Hare with some quotations, but we do not know how much has been modified or interpreted by Hare himself.

Nevertheless, does the spirit Turing test committee believe the automatic writing is from the mind of the medium or Hare's father? The results speak for themselves. We are informed that the spirit world is located between 60 and 120 miles above the surface of the Earth. It consists of seven concentric regions called "spheres." These spheres are not projections of the mind or ethereal vapors but physical locations that have their own atmosphere of spiritual air that support a diversified landscape of vegetation on lofty mountains, all corresponding to the best features of Earth. Each sphere has increasing beauty and grandeur. Although the spheres revolve on an axis similar to the Earth, they depend on a spiritual light that baffles description.

After receiving this information, Hare wondered how astronomers—or anyone else looking up—had failed to see these regions. Then he thought of polarized light, and that concept seemed to satisfy him. Our panel is shaking their heads, but Hare later returned to this conundrum. His solution was to assert that Christians have always believed heaven is "up there," so spiritualism proposed nothing new. His heaven was "up there" as well. Panel members are still shaking their heads in disbelief.

Hare's father—or the medium—populated this newly identified world with ministering angels who assisted the inhabitants. The demand for wealth was always equal to the supply, and moral disagreements were always managed peacefully. Denizens had different social and familial relationships, but marriage was born of God so that all found their eternal counterparts. Spirits up there communicated with our world through raps produced by voluntary discharges of the vitalized spiritual electricity as it came in contact with the animal electricity emanating from the medium.

Hare discovered additional information about the solar system that later appeared in his *Christian Spiritual Bible*.[9] In 1901, the son and daughter of Hare published this massive 379-page volume that was "given through the angel Robert by his intelligences." I have been able to locate only two copies: one at the University of Pennsylvania, where Hare taught chemistry, and one in my private collection.

Hare had never been satisfied with the Christian Bible because he believed an omnipotent God would never have given a document that created so much distress in the world. Moreover, Genesis had misrepresented how the world was created because it was now known to be more than a few thousand years old. These were different times, and Hare wanted something better. His version of the Bible never caused any wars, but it did not stand the scientific test of time any better than the creation story in Genesis.

From Hare's bible, we learn that 260,000 years ago our moon exploded, throwing a considerable portion of its mass into space, destroying all plant and animal life. Human spirits living on the moon were destroyed, although it is possible the explosion brought some of them here to populate the Earth. The embarrassing content of his bible goes on page after disorganized page. His family should be pleased that more copies have not survived.[10] I'll vote for Genesis.

HARE'S FAILURE TO TRANSFER HIS CRITICAL SKILLS

The arena of science relies on criticism to stimulate research that distinguishes truth from wishful thinking. Science requires an independent Turing panel (or peer review) that makes recommendations to prevent the acceptance of faulty conclusions. Hare's work as a chemist received only praise because he knew the field and his conclusions were valid. Thus, vigorous criticism was new to him, and he was irritated by the negative comments about his research in spiritualism. His response should have been to follow the advice of his critics and our spirit Turing panel by replicating his experiments with better controls, which is standard procedure in science. Instead, he lashed out, reminding readers of all the distinguished people who had joined the spiritualist movement. He also reproduced messages received from the spiritual realm that praised his efforts. For example, his deceased sister told him that she watched over him as he disseminated truth to the ignorant. She assured him that the two of them would soon revel in delight "when all nations shall become as a band of brothers." She failed to anticipate the impending Civil War, which would soon tear families—and the nation—apart. She also sent him poetry about the beauty of flowers on the other side. These responses did not satisfy our panel, although one member admitted that she was negatively influenced by the truly insipid poetry of his sister.

Although his sister presumed to know the fate of diverse human civilizations, none of the spirits contributed anything helpful to the emerging sciences, like chemistry. They told Hare they did not like to communicate about certain topics, and spirits were themselves conducting scientific experiments. They were apparently no more advanced in their findings than anyone here on Earth. One supposes that Hare would have given anything to obtain some secrets of electricity or magnetism to convince Faraday that spirits are available to assist us.[11] We have, instead, moral platitudes. Tenacious readers of Hare endure these for page after page.

Surprisingly, Hare recorded the criticism of a distinguished physician, a Dr. Bell, who concluded that nothing was ever communicated by mediums that was not previously in the mind of mortals.[12] Bell conducted his own spirit Turing test with mediums by devising an experiment in which he asked a series of questions. Interspersed were queries "involving replies unknown to the interrogator." When the questioner did not know the answers to his inquiries, spirit responses were a "set of perfectly wild and blundering errors, the responses often being obviously formed out of the phraseology of the question, as a schoolboy guesses out a reply!" Bell's conclusion was that "what the questioner does not know, the spirits are entirely ignorant of." Another failed spirit Turing test. The panel was impressed that Dr. Bell had written out all replies he received because people can be easily impressed by glib but meaningless answers in the excitement of a stimulating situation. In the cold light of day, the nonsense was obvious.

Hare's response to Bell was an invitation to hear the messages of his father and other intellectuals who became believers. Moreover, Hare said, "My experience does not tend to establish that there is less folly or more wisdom in the inhabitants of the spirit world than in this." In other words, a spirit Turing test was irrelevant to Hare.

The spirit Turing panel noted that Hare was not disturbed when different spirits gave contradictory information. But the committee was disturbed. Hare conducted experiments to confirm his belief and ignored information that contradicted his expectations. One panel member made a motion to disband the committee. There was no discussion. They all voted "aye," got up, and walked out. One was heard mumbling something about wasted time.

There is one final footnote on Robert Hare; his gullibility resulted in the spread of spiritualism to England. Hare credited Mrs. Maria Hayden with teaching him about spiritualism. The truth, however, is that Hare taught Hayden the skills she needed for success across the Atlantic.

MRS. HAYDEN CONQUERS ENGLAND DESPITE HER BLUNDERS

Mrs. Maria Hayden was the wife of a successful Boston newspaper editor who became interested in spiritualism after witnessing a séance of D. D. Home, the most famous male spiritualist of all time. Then, observing how the Fox sisters read the expressions of their sitters, Mrs. Hayden had the notion that she could solve Hare's spiritoscope. After extensive practice with Hare, Hayden confidently sailed to England in 1852.

The advertisements of Mrs. Hayden's ability to contact departed spirits caught the eye of Charles Dickens.[13] He commissioned two gentlemen, Brown and Thompson, as his Turing testers to check out the claims. They probably considered their payment of a guinea each for a sitting something akin to the entrance fee to examine P. T. Barnum's bogus mermaid. And they did it all with a straight face, literally.

Mrs. Hayden declared she was in contact with the mother of Thompson, whom he would have gladly greeted because she died when he was a small child. As he pointed at letters of the alphabet, his mother rapped out a message: "Dear son, I am well pleased to see you. I watch over you, and God bless you."

This was not a very poignant response from someone so untimely taken from her child. Thompson asked for his spirit mother to identify her given name. The letters identified were wrong, wrong, and wrong. Mrs. Hayden asked for more light so she could better see the alphabet card, but that was an excuse to cover the fact that Thompson gave her no facial responses or other cues to provide an answer.[14]

Mrs. Hayden then introduced the spirit of Brown's mother, who was alive and well at the time. In contrast to Thompson, Brown purposefully allowed his pencil to pause over certain letters on the alphabet card so that "Mary" was spelled out for his mother's name, which was wrong. In response to a silent question (asked only in the mind), Brown was told that he would have 136 children.

The conclusions of Brown and Thompson were reported in a popular English magazine, *Household Words*. Then Hayden got some mild support in a letter to the editor of a popular newspaper, the *Leader*. The editor, a Mr. Lewes, respectfully disagreed with the endorsement he printed in his paper, so he decided in fairness he needed to test her himself. Lewes tested Hayden by asking if she could summon the spirit of someone unnamed whom he wanted to contact. He had in mind a mythical Greek goddess. Eager to oblige, Hayden announced her contact with this individual, indicating that her soul had passed over in 1847 and that she was the mother of six children.[15] At the end of the séance, Lewes could not resist one final question: "Is Mrs. Hayden an impostor?" The spirits tapped him their answer: "Yes." Our spirit Turing panel would have smiled.

Lewes concluded that seekers were giving her clues, just as with clairvoyants, but he had not played into that role. Here is how he explained it to his readers: "You tell all, and fancy you are told. You do not tell it in so many words, but unconsciously you are made to communicate to the very thing you believe is communicated to you."[16]

Before visiting Mrs. Hayden, Lewes informed his colleagues of his hypothesis, the proposed tests, and expected results. Everything emerged as he predicted. Unfortunately, Lewes lamented, few people follow these simple steps to ensure they are in contact with a real spirit.

One would think that Mrs. Hayden's days in London would have been numbered. However, by this time, Mrs. Hayden had attracted a small group of followers who attempted to protect her reputation. One spiritualist dismissed all criticism, accusing the *Household Words* reporters of buffoonery intended only to "collect materials for a funny article."[17] Someone else said she failed because "having a delicate and sensitive mind, any insults directed against her will tend to interrupt her powers." As in America, exposures did not stop the progress of spiritualism.

THE AFTERMATH

Mrs. Hayden returned to America after a year in England, leaving a few converts to perpetuate spiritualism. One was Professor Augustus de Morgan, a famous but eccentric British mathematician, and his wife.[18] He provided an extensive introduction to a book that his wife wrote in praise of spiritualism.[19] Their involvement with this new religion was a strange admixture of encounters that now appear overly naive and excessively revealing. On one hand, they acknowledged failures and misinformation by mediums, which they summarily dismissed, while on the other hand, they accepted some fantastic events without question.

Mrs. de Morgan admitted that her first communication with spirits, as presented by Mrs. Hayden, contained information that was wrong. The language and style she presented was "wholly unlike" anything that her relative would have used in life, and the spelling was that of the medium, not her deceased relative. Mrs. de Morgan set aside all the bad poetry and uninspired messages as the intrusions of spirits with a lower mental and moral character. Inexplicably, she remained committed to the extent of declaring that she never heard a word that could shake her conviction of Mrs. Hayden's honesty. Similarly, she discovered that her own personal servant, Jane, had abilities as a medium. Mrs. de Morgan declared complete trust in Jane, saying, "she had not skill to deceive."

It is worthy of noting here that the British were of the opinion that someone of higher social standing would never deceive a peer and that lower-class individuals and children were not sufficiently ingenious to fool them. For example, Sir William Crookes said of D. D. Home: "I thoroughly believe in his uprightness and honour; I consider him incapable of practicing deception or meanness."[20] Similarly, Crookes was convinced of the "perfect truth and honesty" of his prized medium, Florence Cook. Such misplaced trust damaged the value of much research of that period.

Professor de Morgan argued that he did not want to discourage suspicion, but skeptics refused to examine the truth offered by the testimony of rational people. He said that a biologist might not believe in the possibility of a duckbilled mammal, even to the extent of considering that it might be a fake when placed in his hands. However, he should not ridicule the creature because everyone will soon be acquainted with the platypus. No one would doubt a sea captain who discovers an island in the Pacific.[21] Just because you have not seen the territory does not negate its existence. The spiritualist, he said, appeals to evidence: he may not have enough, but he relies on what he has seen and heard. Newton did not have all the facts about gravity because he provided no explanation for how it worked. No one ridicules gravity as a "pully-hauley crankum." Facts that he collected, he declared, were incontestable. Those facts came from his direct observations of honest Mrs. Hayden.

Before his days as a spiritualist, de Morgan once declared that the advancement of science depended on learning from the mistakes of the past, not just the successes.[22] He left that advice behind when he needed it most. The Turing committee understood this problem all too well.

WHIRLED PEAS

The de Morgans were caught up in the excitement of a new world here on Earth as well as the paradise awaiting them. Like Hare's expectation of a new brotherhood of nations, spirits told the de Morgans that they were riding the crest of a wave that would soon break into an era of world harmony. Another convert of Mrs. Hayden, social reformer Robert Dale Owen,[23] said that spirits told him:

> God now commands all nations, through the new manifestations of Spirits from superior Spheres, to prepare for universal peace, that man may commence on earth a new existence.[24]

Finally, recall our old friend Dr. John Ashburner, whom we encountered as a supporter of Elliotson's mesmerism and Didier's clairvoyance. He was now converted to spiritualism by Mrs. Hayden and acknowledged that it would only be his own "weakness or folly to suspect her of any fraud or trickery."[25] This was a natural transition. He previously dismissed all the superficial thinkers who had treated mesmerism and clairvoyance with disdain. Now he defended spiritualism with the same fervor. He was ready to fight his "enemies, persecutors, and slanderers," those enemies of science who were denying the reality of mysterious forces that were turning tables in the homes of Christian people.

"Modern demoniacs." Daniels (1856) said this was the natural consequences of spiritualism. All hell breaks loose during a table-tipping session. J. W. Daniels, *Spiritualism versus Christianity; or, Spiritualism thoroughly Exposed* (New York: Miller, Orton & Mulligan, 1856). *From the author's collection*

22

TABLE TURNING

What philosopher would ever dream that his own preferred
option was affecting the result of his experiment?

—Campbell, *The Spiritual Telegraphic*
Opposition Line, 1853

At 8:00 p.m. on Thursday evening, June 16, 1853, the Reverend Nathaniel Godfrey, his wife, and his pastoral assistant placed their hands on a small, round mahogany table.[1] The top was about twelve inches in diameter, and it stood on three legs. After three-quarters of an hour—these were patient people—the table began to move in a circular motion. Godfrey first commanded the table to move to the right, then to the left, toward the door, then the window.

Similar séances were occurring across England, but Godfrey was so startled that he called his two female servants and a nearby schoolmaster to witness the event despite the late hour. At first, Godfrey wondered if electricity caused the table to move. At that time, electricity was mostly an unknown force, but people attributed all sorts of vital powers to it. Then he realized that some intelligent force was involved because of the table's ability to follow commands.

Godfrey instructed the table, "if you move by electricity stop." The table immediately stood still. Once it began to move again, he asked it to stop if an evil spirit was responsible. It continued to move.

Godfrey was not one to be tricked by a devious table or by an evil spirit hiding inside. He whispered to the schoolmaster to fetch a small Bible and softly slip it on the table at his signal. The table instantly stopped. Everyone in the room was "horror-struck." Carefully testing alternative possibilities, they discovered other books had no effect.

"If there be a hell, I command you to knock on the floor twice with this leg." The table was motionless. "If there be a devil, knock twice." No answer. "If there be not a devil knock twice." The leg slowly rose and knocked twice. This was clearly a lie.

Godfrey then said, "Tell me the day of the month." The table lifted its leg and tapped sixteen times. Then it rose slowly and fell the seventeenth time. Godfrey declared the table was wrong, wondering if they had miscounted. His pastoral assistant looked at his watch. It was three minutes past midnight, the 17th of June.

Godfrey said he did not know the hour or the day that the Savior would return, but this manifestation was obviously a sign of the times that would usher in the Antichrist.

TABLE TIPPING CRAZE

Table tipping is also known as table turning or moving; table lifting implies that a table rises in the air without its legs touching the floor. Table talking refers more specifically to situations in which the table answers a question by tapping a leg in some designated way. The table craze blazed into popularity early in the spiritualism movement then rapidly declined, perhaps because of the analysis of Michael Faraday, of whom more is said later. The Fox sisters most likely started the practice as a variation of their table rapping.[2] The fad soon arrived in Germany through mail correspondence. In Hamburg, a professor from Cologne gave a concert featuring "the galvanic electro-magnetic dance of table, with grand musical accompaniments."[3] Meanwhile, the phenomenon in France was "engaging the voracious appetite of the lovers of the unaccountable." Reverend George Sandby said that all of Paris was in excitement at the dancing of tables. They experienced associated electric shocks, mysterious knocks, strange sounds, and visions.[4]

About one week after Reverend Godfrey's experience with table turning, England's most famous scientist of the day, Michael Faraday, published his explanation for this mysterious occurrence. Although it seems unlikely, the science behind Faraday's investigation begins with understanding the movement of the humble pendulum. We are not concerned here with Galileo's mathematical principles of a swinging weight. Instead, something entirely different happens when the supporting string of a pendulum is held by the human hand.

GETTING INTO THE SWING OF THINGS

For centuries, the cause of the swinging of a handheld pendulum was considered a mystical force and a method of divination. Roman priests attached a string to a golden ring, waiting for it to swing toward letters arranged in a circle.[5] In 1677, John White said "the working of your pulse will make the ring to move striking upon the sides of the glass the hour of the day or night, and then the ring will stand still again."[6]

In 1798, a group of German scientists utilized a handheld pendulum to gather information about various materials, providing information otherwise unknown. A book written by the group reached Michel-Eugene Chevreul (1786–1889), a popular twenty-six-year-old French savant. Chevreul was highly suspicious of the German work, despite the scientists' hundreds of experiments. He believed they had only contributed "learned error."[7] He was surprised, therefore, when he obtained the same results they predicted when he held his pendulum over various metals and living creatures. All movement stopped, however, when glass was placed between the pendulum and mercury. That gave him the idea to instruct someone to insert the glass while he was blindfolded. Now the glass had no effect on the pendulum's movement. Therefore, he concluded that the muscular movement of his own arm must have been influenced by his belief about the glass despite having no awareness of it.

Galileo's pendulum perfectly followed mathematical predictions, but all bets are off once a human hand touches it. Suddenly we are in the territory of deception—or more confusing still—self-deception. Chevreul tested this mystery by resting his arm against a support, which reduced the swing, and the swinging stopped altogether when he stabilized his fingers. Once he understood the underlying principle, he no longer reproduced any of his previous results.

Chevreul's experiments led to an important understanding of how individuals can be unaware of their own influence on an otherwise well-designed study. Readers note that I have previously discussed this with regard to George Beard's muscle reading, the social expectations of John Elliotson's belief about the O'Key sisters' response to metals, and the hypnosis experiments of Martin Orne. Belief has a powerful influence that is often hidden from awareness. Chevreul demonstrated how his own expectations magnified the swing of the pendulum, even though he detected no physical movements of his fingers and despite his intention to remain perfectly still.

One might think that people would have lost interest in the pendulum following Chevreul's findings, but his conclusions were not widely known. Dr. Herbert Mayo, our supporter of Didier and professor of anatomy and surgery at the Royal College of Physicians, conducted a series of experiments with a pendulum. He attached a gold ring to a silk thread, which he held over various substances. Each created a different swing pattern, which gave him confidence that he could separate organic from inorganic materials, life from death, and determine the electrochemical properties of any substance. All this was possible, he believed, because of the odic force described by Baron von Reichenbach. He named this modest little instrument an "odometer."[8] Unfortunately, he was caught in the same trap as those earlier German scientists.

Being the wise professor, Mayo had fleeting concerns that his results might reflect some self-deception. Not knowing about Chevreul's findings, Mayo quickly dismissed his apprehension because he knew that he was an honest, unbiased person. As evidence of his good faith, he noted that the ring periodically swung in a direction that he did not expect. Moreover, the ring moved exactly as predicted over a piece of silver hidden under a handkerchief.

Substances he tested variously created transverse, longitudinal, oblique, or rotatory oscillations. He was confident that he had established the laws of odic force and resolutely defended his conclusions and summarily dismissed contradictory information. You may remember that Mayo also excused Didier's mistakes, and he likewise excused any unusual results of his odometer by claiming the weather was cold and wet or that he was suffering from a flare of rheumatism.[9]

As eventually happens in science, other investigators took up his experiments to confirm his findings and expand the body of knowledge. However, someone discovered that success vanished when experimenters did not know what was being tested or had their eyes diverted. Instead of providing evidence of odic force, Mayo demonstrated that "expectant attention" of the experimenter can result in muscular reaction undetected by the most honest individual. Why did he get the correct response with the silver under the handkerchief? He could have known at some level what was most likely there, the swing could have been correct by chance, or dozens of other factors might have influenced the result. The clear message is that ideas are not confirmed until vetted by independent investigators. That is a basic principle in science. And make certain that something really exists before investigating it.

EXPLOITING THE PENDULUM

By the middle of the nineteenth century, the force causing the swing of the pendulum was well described as involuntary muscle action. However, the swinging was not any less mysterious, which allowed for continued exploitation for fun and profit. In the 1920s, an advertisement for a "sex detector" appeared in a magic magazine. It was promoted as a scientific triumph that would baffle and entertain the whole world by indicating the sex of any human, animal, or plant—dead or alive. But wait. If you act now, the seller will throw in a secret explanation for locating any selected playing card. For the considerable sum of $2, the purchaser obtained a little red ball on a string and a stupid card-force trick. Those who knew the simple secret instigated a flurry of complaints. The same sex detector sold for ten cents from the famous Johnson Smith catalog, but the company advertised it only as "an amusing and entertaining novelty rather than for any scientific value it may or may not possess."[10]

These sex detectors started appearing for sale everywhere, but Claude Conlin was particularly successful at exploiting them. Conlin was a travelling mentalist who promoted himself as "Alexander, the Man Who Knows." He claimed that he obtained the secret of the pendulum from an old Indian chief but later changed its provenance to Asia, calling it the "Arnola," the miracle of ancient India. The Man Who Knows was especially knowledgeable about separating money from his audience.[11]

In 1967, Milton Bradley revived interest in the pendulum with its sale of "Kreskin's ESP" game box. Kreskin enhanced sales by his appearances on national television, using the pendulum to locate a mentally selected card. It was, of course, that same old $2 card-force trick. His suckers got a plastic plumb bob attached to a cheap eight-inch chain.[12] No explanation of the card trick was included.

All this handheld pendulum motion was confusing, but in the latter part of the nineteenth century, a spirit medium created a mystery so deeply hidden that it was not solved for many years.[13] Several pendulums were hung from a rigid frame, with each weight dangling into a different vessel like a bottle, crystal vase, wine glass, and a cup. Someone then asked the medium a question and designated a specific pendulum to provide the answer. The medium lightly placed his fingers on the edge of the table and invoked a spirit to swing the selected pendulum. After a brief interval, the chosen weight began to swing, striking the side of the vessel a specific number of times to indicate the answer.

The secret is that each string was of a different length so that slightly different but undetectable movement of the hands on the table caused a different response. For a variation, ordinary individuals can place their fingers on the table with instruction that everyone focus on a specific pendulum. Eventually, the selected pendulum will swing while the others remain still.[14] In 1915, *The Boy Mechanic* passed along this secret so that teenagers with a little knowledge of physics could exhibit the trick.[15]

Whenever spiritualists devised a mystery, magicians were likely to improve the presentation. Indeed, conjurors have created ways of swinging a pendulum without anyone touching the table.[16]

FROM PENDULUM TO TABLE TIPPING

Because Chevreul had solved the pendulum question, the French Academy of Sciences called on him when the table-turning craze invaded Paris. Surprising as it may seem, the same principle that swings the pendulum explained the movement of big tables. Chevreul said that a single principle was involved: unconscious (involuntary) muscle movements initiated by autosuggestion. Around the same time, Englishman William Carpenter and American Charles Grafton Page came to the same conclusion.

On Friday, March 12, 1852, William Carpenter gave a formal presentation at the weekly meeting of the Royal Institution of Great Britain.[17] He addressed the question of how he, a physiologist, understood the current excitement about the phenomena called "electro-biology." He said this fad was one more attempt to label what had been animal magnetism and mesmerism, then artificial somnambulism, and later hypnosis. Carpenter provided a simple but compelling example of what happens when someone is "biologized." He said that an operator so fills the mind of a subject with an idea that thoughts exert a powerful influence on the body. For example, an operator might suggest a subject cannot get up from his chair, and that belief becomes so fixed that the act becomes impossible. Similarly, if told a glass of water is milk or wine, that idea can make a stronger impression than his personal experience received through his sense of sight, smell, and taste. The subject becomes controlled by a "dominant idea."

Carpenter then took this concept a step further. He said that an idea can become the source of muscular movement.[18] We are obviously aware of the connection between our thoughts and any resulting action; however, it is possible that an idea can be reflexive or hidden outside of awareness. He recommended the term "ideo-motor" to describe the response when a subject had "expectant attention." It was a general condition in which expectation created involuntary muscle action, even sometimes in opposition to the will. For example, physical movements of the hand could occur even while attempting to keep the hand perfectly still. This would not be surprising, he stated, to anyone who knows how difficult it is to prevent tremors while holding a long telescope. Then he reviewed an array of studies to demonstrate its reality. This "force" was not humbug or imagination. Instead, the ideo-motor response needed to be considered before appealing to the existence of an occult explanation. Individuals making reports about the movement of a divining rod or handheld pendulum were not influenced by the devil or responding to an odic force. They were not cheating or lying about their actions. Instead, their expectant attention, however minimal, was creating an ideo-motor response outside of their own personal awareness.[19] He did not mention muscle reading, but the same principle is involved.

TIPPING IN AMERICA

About the same time that Carpenter and Faraday were studying table tipping in England, Charles Grafton Page was testing the cause of table tipping in America.[20] Page judged that any report of table movement without the natural force of hands was false and should be given no credence. Yet he was repeatedly surprised by how many intelligent people reported that they observed table tipping and movement without visible agency.[21] They told him that they did not understand how such things could happen, but they could not discredit the compelling testimony of "Mr. A., Dr. B., Professor C., Reverend D., Judge E., Honorable Mr. F., etc." ("Judge E" was an obvious reference to Judge John W. Edmonds,[22] who eventually resigned his position on the Supreme Court of the State of New York because of his endorsements of spiritualistic nonsense.) Page said that if anyone offered the idea that they had witnessed gravity suspended, he would reject that opinion.[23]

To test his theory, Page had subjects place each hand on a piece of paper on a table in the same position that had successfully produced tipping. In this manner, the paper slipped on the table when the fingers applied lateral pressure, preventing any table movement.[24] Page had another clever idea. After sitters successfully tilted a table, he turned their chairs around so that they now sat straddling the seat, leaning forward against the chair's backrest. In this position, they were stretched forward as far as possible in a way that prevented their hands from creating any forward motion. As a result, the table no longer turned. Merely having hands on the table was insufficient to turn it, even if their fingertips were in the exact same place that once fostered success. He concluded that the rules of friction and gravity remained intact and were not overcome by an unknown force. He predicted that new methods for moving tables would be forthcoming, but they would all be trickery and cheating.

He was right. Magicians and spiritualists rapidly developed an astounding array of methods for tilting and lifting tables.[25] However, Page probably underestimated the ignorance and self-deception that contributed to the continued flood of correspondence in support of table tipping.

As an aside, Page noted that whenever he challenged his friends about their convictions regarding the illusion of table turning, they did not take umbrage at being told their senses had been deceived. Faraday was not so fortunate.

FARADAY'S ANALYSIS

Many people attributed the movement of tables in séances to magnetism or electrical forces, so people naturally wanted to know the opinion of Michael Faraday. Faraday was perhaps the most popular and influential scientist of his day.[26] His discoveries regarding electricity were closely followed by common people and scientists alike. His success came despite his humble birth in a slum and lack of formal education in a society rigidly dominated by social class and titles. But Faraday understood electricity and magnetism in ways that changed the world. His innovations allowed the development of electric motor technology, and it was largely due to his efforts that electricity came into practical use for daily life.

Faraday responded to an inundation of letters on table turning by diverting his scientific research to study what many believed was a new force.[27] Faraday's excursion into table turning appears, on the surface, to have some flaws. He did not pretend to have an open mind about the cause of table movement: fingers had to be the cause. He appeared to lack the characteristic impartiality we expect of the scientist. The difference, however, is that Faraday looked for information to contradict his belief, and he conducted experiments conceived by his cunning insight that provided compelling information.

Once Faraday ruled out the presence of an electrical or magnetic field, he quickly moved into an area completely outside his familiar research. Nevertheless, his lynx eyes knew exactly where to look in order to gather necessary information. His experiments illustrate the mind of a gifted scientist, aristocratic background or not.

Although deeply religious, Faraday had no kind words for those who asserted spiritual explanations. He said that people should first inform themselves about known forces before claiming diabolical or supernatural agency because it was not scientifically proper to make up forces for which no experimental evidence existed. Just giving something a name did not make it real. He was even more appalled that some people believed tables moved because of the rotation of the Earth, "as if the earth revolved around the leg of the table."

Almost everyone who wrote about table turning claimed they had conducted "experiments." But "experiment" means different things to different people. Historian Thomas Hankins points out that *demonstration* experiments exhibit what one wants to show or teach, and *research* experiments search for information about the unknown.[28] Table turners were not searching to find out what

Michael Faraday (1791–1867), English physicist and popular lecturer at the Royal Institution of Great Britain. *Photo by Michael Nicholson/Corbis via Getty Images*

they did not know. They demonstrated their fascination with an unknown force, and they elicited what they believed they would find.

Looking back on this history, we see "research" reports on table turning that focus on everything imaginable with complete disregard for what now appears to be reasonable controls and realistic variables. Believers studied the effects of weather, the influence of insulators, the characteristics of table construction, and the direction of movement. In considering the influence of sitters around the table, investigators evaluated gender effects, personality types, current mood, and mode of dress. The variables were endless—and ridiculous. Visetelly said that men of science daily forwarded the results of their research to medical journals, all declaring their open-mindedness. Faraday, in contrast, went straight to the obvious cause of movement—the fingertips.

On June 25, 1853, three months after Carpenter described ideo-motor action, Michael Faraday presented the results of his research in the *Times of London*, and a somewhat expanded version appeared in the *Athenæum* on July 2.[29] His hypothesis was that table movers did not intend to move the table, but they had an expectation that created a "quasi-involuntary action." He began by attempting to determine if any mechanical pressure was inadvertently exerted by the people at the table. Any downward pressure, which happened when fingers were placed on the table, would not cause turning. In order for the table to turn, some horizontal pressure was required, so he created a way to determine if such pressure was activated. Thus, he prepared a sticky mixture of wax and turpentine, which he placed between pieces of cardboard, and the bottom was firmly attached to the table. The goop between the sheets of cardboard was sufficiently sticky to offer considerable resistance against sliding yet weak enough to give way to continued horizontal force. The result was exactly as he expected. When hands of sitters pushed the top card to the left, the table moved in that direction. Therefore, the table had not moved on its own as experienced by sitters, because then the top board would have drifted in the other direction (i.e., the table would be pulling their fingers). In trials that resulted in no table movement, the cards showed no movement.

Faraday tested his theory using other strategies. When the cardboard pieces had small glass rollers between them, friction was insufficient to allow the table to move. In another test, Faraday placed a simple straw through a hole in the top piece of cardboard and attached the bottom end to the table. As a result, table turners had a visual clue when they started to push sideways. As soon as they perceived they were exerting sideways pressure, they were no longer able to deceive themselves, and all table movement stopped. When that straw started waving and smiling at them, it ruined their day.[30]

These experiments, Faraday said, "may be useful to many who really wish to know the truth of nature, and would prefer that truth to a mistaken conclusion." He said over and over that people believed and experienced one thing while unaware of the invisible motion they exerted once their hands experienced a little fatigue. He ended by saying he did not expect to convince everyone, but his test apparatus was available for examination, and he provided the address to come try it out. He said he had earnestly sought cases that would contradict his findings but found none. Therefore, he considered the case closed and would not engage in further debate by answering every question posted to him. He wanted to return to his research that had benefits for humanity. Meanwhile, he said, someone ought to consider why it was necessary for him to set aside time in this present age in this part of the world to dispel the question of whether an occult power had been newly discovered. "The system of education must have been greatly deficient in some very important principle." Ouch. This guy from the ghetto really knows how to throw a punch.

RESPONSES TO FARADAY

Faraday was correct in anticipating that not everyone would believe him. So now we leave him undisturbed in his laboratory but take time to consider the opposition that refused to accept his conclusions. Believers in occult forces had different theories, but none of them went to Faraday's experimental table to show fingertip pressure was not involved. The informal rules of science mandate that the person who makes a claim is responsible for providing evidence; he cannot demand others prove his idea is wrong. Believers in occult forces did not want to play that game. The Galileo problem was here in full force: *I don't care what your telescope says: here's how the heavens are arranged.*

Faraday was inundated with letters inviting him to séances, which he ignored just as he said he would.[31] One anonymous author decided that Faraday's learned theories were not needed as much as common sense.[32] Common sense told him that a subtle matter pervades all nature, which we know little about, but it is sometimes called electricity. Because electricity in the form of lightning is sufficient to shatter the timbers of a church steeple, common sense told him that this same electricity emitted from a hand would be sufficient to excite the movement of a table. Frank Podmore opined that this "common sense" pamphlet was a typical "apt illustration of the arrogance of sheer ignorance."[33]

Reverend Chauncey Townshend said that Professor Faraday fell below his reputation because he provided an insufficient explanation of table turning and proved nothing.[34] G. W. Samson was convinced that electricity was the interme-diary between the physical and the spiritual worlds.[35] He declared that healthy males were positive electrically, whereas delicate females were negative, which is why the stronger may control the weaker. Around a table, an enormous nervous energy builds up among the participants until, like a Leyden jar, the battery is discharged. That electricity makes the table tilt and move. George Sandby said it was doubtlessly a monstrous act of presumption to call in question the infal-libility of Professor Faraday. Yet, "I have lately been an eye-witness of a fact, in which I perceive no opening for mistake."[36] (Yes, profoundly presumptuous.)

Alfred Russel Wallace experienced gyrations he said could not have been produced by anyone around the table.[37] Wallace persistently declared that many of his experiences could not possibly be explained through normal laws of sci-ence. In page after page of his book on miracles and spiritualism he described experiences that he insisted had no possible natural explanations.

Faraday had said the failure to understand table turning showed people's igno-rance of their ignorance. The editor of the *Spiritual Magazine* said Faraday was the one who was ignorant of his ignorance. "Let him undertake a serious investi-gation and he will find there is a law of spirit that controls the law of Newton."[38]

Perhaps we should not be too critical of all who rejected the Faraday ex-planation. Some sitters may have experienced manifestations not explained by ideo-motor action because someone was cheating. For example, Charles Davies described a situation in which all sitters had only their fingertips on the table, yet "it forthwith rose up perpendicularly and came down with a crash that com-pletely shivered it in pieces."[39] He had not the slightest idea how this happened. A psychological explanation was as mystifying as a diabolical one. Yet he just shrugged. Davies had witnessed many unaccountable spiritual manifestations, but he also observed magicians perform even more startling illusions on a lighted stage. Davies did not need an explanation for every mystery, and he had no inclination to attribute an inscrutable effect to an unknown force.

GASPARIN AND HIS MYSTERIOUS FLUID

The first Frenchman to conduct formal studies on table turning was Count Agénor de Gasparin,[40] who was temporarily living in Switzerland. He was a gentleman of unbounded energy for liberal causes, but in May 1853, he turned his interest to the mystery of table tipping. Gasparin conducted "experiments" on tables with his wife, children, friends, and household domestics.[41]

On their first try, Gasparin and his companions sat with their fingers on a table for more than an hour before it finally began to rotate. The tabletop was about thirty inches in diameter, supported by a heavy pillar that terminated with three feet about twenty-four inches apart. The table tipped and rapped out the ages of persons in the room, a number merely whispered, and even one held only in the mind. Those who participated in the experiments with Gasparin were convinced their own motions could not cause the action of the table movement. Moreover, muscular action exerted was insufficient for the weight lifted, and the table could rap out numbers corresponding to those only in the mind. Gasparin's gang also declared that the table rotated three or four turns with their hands held an inch above the table. His results got better and better because of the increase in "the fluid" within his sitters.

After nearly one hundred pages of trivial discussion, Gasparin got to the question of what that fluid might be. He did not know. However, he affirmed that there was some "agent" and that agent was not supernatural. It was a physical agent communicated to physical objects as determined by the will; it was an admixture of moral and physical forces. His experiments confirmed this because the hands needed to be in certain positions in order to create the best results.

Gasparin defended his hypothesis of "the fluid" with a rage against critics. He told his reader to admit that "you do not know everything, that moral nature and physical nature, each and both have obscurities for you."[42] Gasparin was referring to critics like Jacques Babinet, a French physicist who put powder on the table so that fingers could not stick to the surface and create a lateral force. Gasparin shrugged, saying: You see our experiments and reject them by asking for another. You refuse our experiments unless clothed in your fashionable costume. Do not lay down the laws of nature against me when you do not know all of those laws. The magnet moves objects at a distance, and so does our fluid.

Gasparin insisted that men of science have not provided answers, only ridicule. He wanted others to accept his eyewitness testimony, and he was furious and sarcastic about comments that challenged his belief. He perceived science as the enemy of spiritualism, and thus the title of his massive two-volume work: *Science vs. Modern Spiritualism*. Gasparin had promoted good social programs, but his spiritualism was a jumble of poorly thought-out metaphysical expectations supported by self-deception.[43] He could have saved himself much embarrassment and grief if he had only attempted to see if that straw stood up straight when his table moved.

A BATTLE OF PAMPHLETS

We return now to the story that started this chapter, the table turning of Reverend Nathaniel Godfrey. He immediately published a pamphlet about his experience, which quickly sold out. He issued another printing, then another. While preparing his seventh edition, someone brought to his attention the explanation of table turning by Professor Michael Faraday. Godfrey added a brief paragraph stating that Faraday's opinion did not have the remotest bearing on his experiments. Those around Godfrey's table imparted no motion to cause a response. Faraday's table did not answer questions, which his did, and it responded in ways that were contrary to his own belief. Faraday's experiments, he declared, had been disproved by his experiments. When they placed the Bible on the table, for example, their own emotions were of curiosity, and their own wills were suspended. Their bias was against the table stopping. And their muscles could not have caused the table to tap out seventeen days when they believed it was the sixteenth. But Godfrey's conflict was not with Faraday. His battle was spiritual warfare.

Godfrey published another pamphlet in which he answered those who ridiculed him. He forcefully insisted that his belief in demons and devils was based on observations of his table and on the authority of the Bible.[44] At one séance, Godfrey learned that the table was being moved by a twenty-two-year-old man who had died five years earlier. The man said he could move the table without Godfrey's hands touching it but then claimed he was constrained by the devil when put to the test. This lost soul, Alfred Brown of Wortley, said that he had lived a sinful life and was now suffering the eternal consequences. He confirmed the literal existence of hell, complete with fire and brimstone.

Tables moved better than other pieces of furniture because the devil was mocking the table of the Lord's Supper. Table turning was a spiritual manifestation beyond the boundaries of science that only an unprejudiced mind could study and comprehend. Everything made sense to Godfrey because he knew the scriptures.

Christianity was in a crisis in the middle of the nineteenth century because of rising concern about punitive dogma that Christians had spread to the nation's colonies.[45] The emergence of science and biblical criticism challenged the literal interpretation of scriptures. People were perplexed and spiritually disoriented. Clergy viewed the world around them as falling into materialism and atheism. They needed evidence of an afterworld, but they were horrified when the visions of clairvoyants portrayed everyone living happily in a secular heaven.

The Reverend Edward Gillson and members of his Bath congregation confirmed Godfrey's findings, quite contrary to their expectations.[46] Thus, Gillson agreed with Godfrey that Professor Faraday had never witnessed anything like what they experienced. How could such violent reactions be caused by the slight pressure of their hands? Gillson ended by saying, "Let us gird our loins, trim our lamps, and be ready for the Lord." Bath resident R. C. Morgan was also certain the end-times were near because the scriptures declared that chariots would be running with flaming torches, which anyone could understand after seeing a railroad train at night.[47]

A newspaper correspondent said that if the devil were in these table movements, then the devil was a greater fool than he expected. Gillson replied that such self-admiring and self-complacent faultfinders had better examine their own wisdom. As the Bible predicted, the great enemy was deceiving the whole world. Ministers of the gospel—implying ministers like Godfrey and himself— were assigned by the church as "watchmen." It was time for watchmen to warn others of the coming satanic invasion.[48]

Reverend R. W. Dibdin said that table moving was as well established as any fact in history or science.[49] The table responded correctly even when no one in the room knew the answer and was always correct when the facts were checked, although the devil sometimes lied when answering. Faraday was simply wrong. Anyway, only men taught by the Holy Spirit were competent to give a trustworthy opinion. Dibdin sneered at the idea that the table turner should not believe his own senses but instead fall for the dictum of those wise and learned persons who knew all those wonderful laws of science. They had made a priori conclusions and ignored the power of Satan. Dibdin said it was impossible to reason with those who would not believe their own senses or accept the testimony of honest people.

Religious leaders with alternative views soon fired their pamphlets into the fray. Reverend Magee, one of the ministers of the chapel in Bath, was concerned that both rich and poor of the city were caught up in the mania of table talking.[50] Asking questions of tables was unnecessary, degrading, useless, and probably a grievous sin. How Magee's rant changed his relationship with fellow Bath clergyman Gillson is unknown, but it was clear that not all watchmen were seeing the same dangers lurking about.

Without a pope to impose meaning on the scriptures, the Anglican Church has a more diverse doctrinal foundation; rectors vary on their interpretation of their ancient texts. Unfortunately, scriptures referring to Satan and his hoodlums are particularly confusing if not downright contradictory. The bishop of London, in a terse letter, informed Godfrey that he could not prevent Godfrey from lecturing about his table-talking experiences, but he hoped Godfrey would cancel his lecture scheduled in London. If he did speak, the bishop said, Godfrey would subsequently be prohibited from officiating in any church or chapel of the London diocese.

Godfrey responded by reminding everyone that spiritual forces were now active, and the Antichrist was bringing delusions of hope. For example, the German Seeress of Prevorst[51]and the French Celestial Telegraph[52] both described a heaven where everyone is happy "but none in punishment."[53] This contradicted the Bible.

The pamphlets kept coming, each slamming the ball back over the net to his opponents. Reverend William Vincent of Islington issued a thundering diatribe against people turning tables.[54] Reverend Francis Close, rector of Cheltenham, called out Godfrey, Gillson, Dibdin, and Vincent by name for their promotion of satanic agency. Turning of tables was as innocent as the spinning of a top.[55] Gillson responded by attempting to educate Close about the proper interpretation of scriptures as they related to satanic influence and about prophetic declarations of the end-times.[56]

Reverend Henry Parr, curate of Tunbridge, warned that the devil was out propagating disease and causing storms. This fearmongering was too much for an anonymous Anglican author who reminded people that they were living in the nineteenth century.[57] He said that clergy should stop promoting superstition that causes fear among the lower classes of uneducated people, kindly not mentioning the fear in superstitious, educated fellow clergy. Did we not get rid of superstition, he asked, when we broke from the Catholic Church?

Another commentator said Godfrey's gullibility to illusion was proportional to his ardor for religious hopes and fears.[58] Reverend Glazebrook of Blackburn suggested Godfrey test himself on Faraday's table instead of lecturing him.[59] Yet another author noticed that no table turned without a human hand pressing on it.[60] Perhaps the demon was not in the table but in the hearts of men.

One investigator discovered he could move tables as well as objects like candlesticks, plates, and chairs.[61] When he and his young daughter held a doll, they soon felt it struggling to move. This reminded him of pilgrims in Tramutola, Italy, who carried a statue of the Virgin Mary through the city during a time of crisis. As they moved, the statue began to show signs of pushing against the robust men as they proceeded through the streets. These mysteries could not be investigated by science because they were of a different order of truth. Science was forced to stop at the border of the natural, leaving investigations of the supernatural in the hands of those without prejudice.[62]

SCIENCE: GOOD, BAD, AND BOGUS

When I reread the writings of Godfrey, his errors in processing his experiences jumped out at me more obviously, just as they did when I reread reports on Didier. For example, Godfrey forgave any mistake that was a near miss. When he asked the table for a man's address, the table indicated "11" when the correct answer was "1." Godfrey considered this a success. In another situation, he asked how many coins were under a piece of paper. The answer was wrong, but he noted that the response was correct for the number of *copper* coins under the paper. He drew the bull's-eye around the arrow after it struck the barn.

One summer evening Godfrey noted: "I was there two hours, but scarcely anything was done; we could get no answers to any thing." In reality, he got answers to his questions, just not the answers he wanted. He mentioned, as an aside, that not all the regular sitters attended this session, but he never considered the possibility that missing individuals might have influenced previous results.

Many who experienced table turning insisted the cause was not the result of Faraday's unconscious, lateral finger pressure. All of them declared they were honest seekers of truth. Most went out of their way to indicate that they considered alternative explanations and that their experiments contradicted their expectations. They cited others who agreed with their findings. These men were not crackpots but persons of social and educational standing.

At first glance, these individuals followed the expectations of scientific practice. However, a superficial knowledge about the strategies of science is not sufficient protection against nonsense. Table turners trusted their own experience, and they were infuriated when challenged. They promoted a nebulous theory about the force behind the table movement, like a mysterious fluid, satanic agency, or an electrical force. Think about it. Why would the universal laws of physics be suspended only to be replaced by an unknown agency when my friends and I place our fingers on a table?

One gets the impression that a spirit Turing committee would not have influenced them because they had already rejected Faraday's encouragement to try experiments on his tables, which provided ways of measuring the force of human hands.[63] They hurled invectives at Faraday, but they stayed away from his table as if it was swarming with demons. No one had the courage to put their fingers on such a dangerous piece of furniture.

23

PALLADINO: YOU JUST HAD TO BE THERE

Small fibs are useless. A lie obtains credence in proportion to its enormity; for, though the statements you make are difficult to believe, it is still more difficult to conceive a woman audacious enough to invent them.

—Sweird, *Spirit Rapping Made Easy*, 1923

Eusapia Palladino baffled more psychic investigators than any other medium. Palladino promoted no spiritual path or social cause, but she produced psychic phenomena like no other. While carefully monitored, sitters experienced mysterious raps, the levitation of their séance table, and the touches of spirit hands in the darkened room. Sometimes she was able to free one hand or foot when supposedly held by someone, but she admitted her cheating when caught. Then she went right on to baffle her sitters again. Thus, Palladino presents one of the most fascinating studies of mysteries and secrets in the history of psychic investigation.

Eusapia Palladino was born in 1854 in a rural Italian village, and her mother died shortly thereafter. Palladino said she was hit in the head with a pan when she was about a year old. "People have told me that a current of cold air rushes from that dent. It may be so, I do not know."

After her mother died, her father placed her on a farm, and he was subsequently killed by brigands when she was about ten years old. She then lived with a family in Naples but was unhappy and uncontrollable. At about age thirteen she became a household servant for a wealthy family that was interested in spiritualism. After she attended a séance, she said they told her she lifted the séance table without touching it, mysteriously moved books, and floated a wine decanter in the air. People began placing gold coins in her pocket after séances.

Curious men and journalists began visiting her until she grew tired of séances. She married and began a domestic life, helping her husband in his shop.[1] But demand for her séances became so great that she was taken to cities across Europe for study. Men of science remained mystified, unable to explain her manifestations, which were described by the editor of the *British Medical Journal* as silly games that might amuse a child: the plucking of guitar strings, loud sounds on the séance table, touches by an unseen hand, untying knots, air currents from her head, movement of her cabinet curtain, and movement of objects inside the cabinet.[2]

CRIMINOLOGIST MEETS CHEAT

Palladino got her first taste of fame from Italian psychiatrist and criminologist, Cesare Lombroso (1835–1909).[3] He examined her in Naples in 1891 and later participated in a series of seventeen séances with her in Milan. Although Lombroso was an atheist, he became attracted to spiritualism later in his life, much to the dismay of his friends. It appealed to him, he said, because it broke down "the grand monism which is one of the most precious fruits of our culture." Because spiritualism (or spiritism, as it was called in many parts of Europe) violated the known laws of science, Lombroso got a certain iconoclastic pleasure in promoting it. As an example, Lombroso began his book (the only one he wrote on the occult) with the introduction of a rebellious fourteen-year-old girl who displayed an amazing array of unlikely symptoms described as hysteria. Lombroso reported that when her eyes were blindfolded, her sight was transferred to the tip of her nose and left earlobe. This conclusion should have been sufficient warning that Lombroso's assessment skills had either fallen off the turnip truck or he was transporting an overload of credulity. The same mistakes were also evident in his psychiatric work, which was an embarrassment to his profession by the first part of the next century.[4] For example, he concluded that physical features like large ears and tattoos were the *causes* of criminality. Such unfortunate thinking illustrates how poorly research was understood at that time, even for someone with academic standing.

Lombroso was unaware that his book gives some obvious clues about how Palladino levitated a table. One photograph shows all four legs of a table about six inches off the floor.[5] Palladino sits at one end with two sitters on each side; everyone looks very serious with their fingers touching the tabletop. All very impressive—until you look closely and think. Palladino's two shoes are clearly visible from under her billowing skirt, but her skirt is touching the table leg; in fact, a small fold of the cloth is *under* the left leg of the table. This is even more evident in a later figure. The most obvious explanation is that Palladino removed her left foot from her shoe, placed a toe under the table leg (and under her skirt), clamped the table directly above with her left hand, and lifted it. (Removing a foot from its shoe was a common trick with mediums, one that Palladino was known to use.[6]) Lombroso even admitted, right there on page 45, that she was unable to elevate a table unless her skirt was touching the table leg just below her hand, nor would she attempt to lift the table while standing or when seated at the long side. Elsewhere, Lombroso said,

> it is now certain that supernumerary spectral limbs are superimposed on her true limbs and act as their substitutes. . . . Add to this the fact that no movement at séances takes place except in the immediate vicinity of the psychic, and especially in contact with her skirts, which makes some suspect artifice.[7]

Yes, I admit that I suspect some artifice here. But then, I also suspected artifice from the girl reading with the tip of her nose. Please, Dr. Lombroso, if you suspect cheating when her skirt touches the table, then eliminate that possibility before proposing "supernumerary spectral limbs." But then, one psychic performer said that in his whole career he had never met such a gullible sucker as Lombroso.[8] He was hopeless.

Similarly, astronomer Camille Flammarion printed a picture of Palladino lifting a table, which he offered as evidence of levitation. The fourth leg of the table, near Palladino's left leg, is not even visible.[9]

On the other hand, Lombroso was astute enough to notice one time when Palladino freed her hand from control during a séance, which enabled her to reach out and startle others with touches in the dark. Palladino had endless tricks: she was caught stretching a hair from her head across the balance of a scale while attempting to influence it by psychic power. Another time, she was observed picking flowers that were later introduced as mysterious apports[10] materialized from another world, and she was known to use her two hands and a handkerchief to create the illusion of a face. Once she told Lombroso that someone had secretly ordered her to cheat, and she felt compelled to obey. He acknowledged that Palladino had many artful wiles both in the trance state and out of it, and he believed she had learned "certain special tricks" from her mountebank husband. Nevertheless, he still affirmed her supernatural powers, and others continued to clamor for séances even as she began demanding more and more gold coins for her pocket.

EUROPEAN SÉANCES

In 1892, a distinguished committee of scientists and professors, including Professor Lombroso, held a series of séances with Palladino in Milan. They issued strong statements in support of her, although Charles Richet, a French physiologist, later issued his own more cautious statement. The committee said she was able to diminish her weight by at least 8 kilograms (17½ pounds). However, the scale was later considered unreliable when someone noticed variations related to where one stood on its base. Thus, a better scale was introduced, and her change of weight was then only trifling. Incidentally, the trifling difference occurred when her skirt was touching the floor, which the committee dismissed as insignificant. No change at all was observed when her skirt was kept off the floor. The committee concluded that nothing she did could possibly account for her weight variation, ignoring the absence of effect after taking the presumed "unnecessary precautions" of keeping that skirt off the floor.

Richet admitted that "the results degenerated as the conditions were made more stringent." Nevertheless, the committee concluded that "doubt was no longer possible." Even more surprising, the English somehow decided that Palladino deserved more attention because of the high scientific standing of the witnesses and the care taken during their observations.[11]

In 1894, Charles Richet invited Palladino to his summer home on the Mediterranean along with English spiritualists Oliver Lodge, physicist, and Frederic Myers, an English poet and one of the founders of the Society for Psychical Research. A few weeks later, Henry and Eleanor Sidgwick joined them, but they did not witness the more impressive phenomena observed by the others. Eleanor was devilishly clever in her ability to identify fraud,[12] and Palladino was smart enough to know when she had met her match.

Based on belief in her power, Myers invited Palladino to his home in Cambridge during the summer of 1895. From the beginning, these sessions did not go well because Palladino was providing only manifestations that involved freeing her hands and feet from restrictions, which she could easily accomplish. The Cambridge sitters finally gave up pleading with her to stop the trickery; they allowed her to set the conditions of control in the séances. Not surprisingly, they concluded that all her methods depended on fraud. The almost constant motion of her hands and feet made it impossible to evaluate her. Moreover, her overt sexuality, so appalling to women, seriously interfered with the men supposedly sitting in judgment of her. She spoke the truth when she said the cooperation of strong men gave her added power.[13] But even those who hated her were unable to explain a few of her manifestations that might not have been trickery.

The mixed English report was criticized by the French.[14] Richet complained that the English were "disdainful and haughty" when Palladino needed cooperation and rest. As a further insult, in Richet's mind, the Cambridge investigators had the audacity of consulting Maskelyne, a famous London illusionist who had lectured against spiritualistic phenomena. He exposed some of her manifestations.[15]

Even before the Cambridge group finished, Dr. Ernest Hart, the editor of the *British Medical Journal,* said these investigators were "waiting for the stirring of the muddy waters of deception, which they eagerly drink in as manifestations of a 'new psychic force.'"[16] Being touched by an invisible hand or seeing a table float in a darkened room was not sufficient reason to become excited when stage magicians were floating ladies and sawing them in half on a fully lighted stage. Hart had given up chasing psychic phenomena.[17] He was directly acquainted with the O'Key sisters, who inappropriately responded to nickel when led to believe it was iron. Additionally, Hart had travelled to France to evaluate the amazing claims of Dr. Jules Bernard Luys, chief physician at La Charité. Hart was astounded that Luys used actresses as his research subjects, one of whom admitted the whole performance was fraudulent and that she enlisted confederates.[18] Hart discovered many French "researchers" were promoting spiritualism, and they avoided even the most obviously necessary controls in their investigations. The point here is that English folks had reasons to be cautious with French endorsements of psychic forces—and of Palladino.

Myers, who had invited Palladino to England, suddenly changed his mind after Palladino left for France. Now he concluded that not a single phenomenon was plainly genuine and nearly everything was tainted by fraud. Eleanor Sidgwick[19] admitted she was confused by some of Palladino's manifestations, partly because adequate controls were never established. Meanwhile, the negative publicity from England did not diminish her fame, and she continued to collect large sums of money for her appearances. As Liberace famously responded to criticism, "I cried all the way to the bank."

Palladino returned to France where she provided forty-three séances for the Institut General Psychologique, a French psychical think tank.[20] On more than one occasion when Palladino was tied to a bed for a test, the face of a woman appeared for fifteen to twenty seconds. And other human forms similarly appeared from behind a curtain despite the restraints. Mysterious touches were felt even when she permitted enough light for her hands to be seen, and she allowed her wrists and ankles held. The investigators could not conceptualize a way that trickery might be employed to create the effects.

Still, not everyone in France was excited by Palladino's performances. One examiner said, "Eusapia is a subject but a very bad subject. We are obliged to take her as she is." Palladino produced unexplained phenomena mostly when the lights were darkened, and she stymied everyone by her unwillingness to allow appropriate controls. She continually diverted the attention of those trying to observe her. The institute ruefully admitted that its focus on her pulse, urine, electrical aura, and other physiological measures was a waste of time. It had nothing to show for its 25,000 francs fee.[21] Palladino, on the other hand, was even more rich and famous.

SEEKING ANSWERS IN NAPLES

The contradictory French messages stirred the Society for Psychical Research into disregarding its policy of not investigating mediums who had once been detected in trickery.[22] Eleanor Sidgwick grudgingly admitted that Palladino had been endorsed by a larger number of scientific men than any other medium, and some of her phenomena had not been explained by *detected* methods of trickery. Therefore, the Society decided to send Hereward Carrington, Everard Feilding, and W. W. Baggally to Naples to conduct a series of séances. Feilding and Baggally were both familiar with the methods of investigating mediums, and Baggally and Carrington were further experienced as amateur conjurors. Carrington and Feilding went to Naples in November 1908 to secure rooms in the Hotel Victoria, where they stayed and held séances.[23] They engaged a secretary who could take shorthand dictations and translate for them, and Baggally joined them later. The séance room arrangement and the table's construction were all according to Palladino's request.

The first couple of séances were not particularly convincing. Nevertheless, the three investigators were gradually persuaded that Palladino produced phenomena that were not the result of fraud. Carrington's concluding note said that his experience with Palladino left him with a conviction that could be understood only by being there. In session six, for example, Carrington reached his hand toward the cabinet while he and Baggally were controlling Palladino's hands, yet he felt touches of a hand that could not be Palladino. He arrived at an intellectual and emotional conviction that Palladino could not have physically touched him. You just had to be there.

After that sixth session, Carrington said, "I felt there was nothing more to be said." This troubling conclusion suggests a finality of belief as opposed to what might have been a springboard into further investigation. When someone with an enquiring mind encounters an enigma, the expected response is eagerness to explore further possibilities and determine limitations. Carrington's decision to quit may have been in large part due to the resistance of Palladino to allow controls appropriate to the situation.

The overall experience with Palladino was more than they could fathom, and Feilding said it clearly in his final note:

> My colleagues, then, having come to the deliberate opinion that a large proportion of the manifestations of which we were the witnesses in Naples were clearly beyond the possibilities of any conceivable form of conjuring, entertain no difficulty in saying so in precise terms, and so far as my own position as a layman entitles me to it, I associate myself entirely with their conclusions without apology for our seeming lack of scientific caution.[24]

Most likely, all three went to Naples believing they could unravel Palladino's deceptions because of their experience in spiritualism and conjuring. When they were stymied, they felt so overwhelmed that they concluded the only solution was an unknown force. Their endorsement was not very satisfying, however, because their conclusions were so subjective. They provided nothing that supported their claim or shut the door on doubt, and Feilding became exasperated by armchair skeptics who pointed this out to him.[25]

THE DISCOMFORT OF HAVING NO EXPLANATION

What happens when you experience something impossible? Modern magician Simon Aronson says:

> The magician must affirmatively raise and destroy any hypothetical solution which the spectator might be likely to consider. The spectator must be actively engaged, so that his own mind and senses together eliminate even the possibility that—let alone any explanation of how—the effect could have taken place. There is a world of difference between a spectator's not knowing how something's done versus his knowing that it *can't* be done.[26]

Expressed more succinctly, Juan Tamariz said, "[The spectator] should not believe that any possible solution is the right one. All solutions should be rejected and regarded as impossible."[27]

I have experienced the intellectual and emotional response of being confronted with magic tricks that seemingly cannot be done.[28] In some cases, I have subsequently slapped my head at the simplicity of the method after it was revealed to me. "Impossible" effects do not necessarily require elaborate methods. Theodore Annemann (1907–1942) recommended and used methods so bold that other magicians avoided them for fear of detection. He emphasized the direct and elegant methods that allow the performer to concentrate on the presentation, which is the main determinant for creating an illusion in the spectator's mind. For example, Annemann studied *Popular Card Tricks*[29] by Walter Gibson, the prolific writer and creator of *The Shadow*. Gibson's paperback magic book was available for 25 cents at every newsstand in the country. Armed with common card tricks that no respectable magician would touch, Annemann exploited these tricks for mental effects, convincing audiences he was a gifted psychic.[30] People never considered the possibility that anyone would cheat so obviously. As another example, Penn and Teller were baffled on their television show, *Fool Us*, by John Lovick. In his act, Lovick used what might literally be the oldest method in magic as an alternative to the standard procedure for the common torn-and-restored paper trick. The choreography and timing of the critical move prompted Teller to later admit that it was "a pleasure to be so elegantly fooled."[31]

Perhaps more pertinent to Palladino, I have stood behind magician Jerry Andrus (with his permission) and watched him make amazing sleight-of-hand moves that his audience never sees and thus never appreciates. Andrus never made any fast moves to distract his spectators' attention, which leaves no explanation for his miracles.[32] Similarly, Palladino could secretly free her hand or foot, leaving her "controllers" no understanding for the miracles they witnessed.

As previously discussed, people who have no explanation often invoke psychic explanations. Ray Hyman suggests that skeptics, in particular, are likely to experience an emotional wallop when confounded with an anomalous experience.[33] They are overwhelmed with a sense of *knowing* something anomalous has occurred, which diverts any interest in outside evaluation or further questioning. Our trio of men who examined Palladino were handpicked by the Society as experts in detecting fraud. They were walloped. The giveaway is Carrington's statement that you just had to be there. He was baffled by an experience that could not be explained by anything in their vast knowledge of explanations. But there is another possibility.

Carrington would ultimately become the manager of psychic tricksters, including Palladino herself. Baggally, despite his skepticism and training in magic, wrote a book in support of telepathy, and he was completely fooled by the Zancigs, a pair of clairvoyant entertainers.[34] Feilding's objectivity in spiritualism was probably compromised after his brother died in a drowning accident.

Additionally, he was duped by Rudi Schneider, a medium who was able to free his controlled arm to simulate telekinesis.[35] This investigative trio was not as good as advertised at detecting fraud, and maybe Carrington was already counting the money he expected to make on the book he was planning to write about Palladino. Moreover, Richard Wiseman has made a remarkable case that Palladino had an accomplice (probably her husband) who obtained access to the séance room during their Italian investigations.[36]

Massimo Polidoro recently made a compelling case that Palladino was born with particularly flexible hip joints that allowed her to use her left foot in ways that might seem impossible for someone her age and weight.[37] He said the Naples committee failed to adequately control her lower extremities, citing evidence from their transcript and the observations of Hugo Munsterberg. Thus, Polidoro believes many of her seemingly impossible manifestations were accomplished by her left foot being released from her shoe. Moreover, Polidoro documented how poorly the Naples trio controlled the séances despite the accurate claim that their investigation utilized the most rigorous controls of Palladino. The records also indicate that Palladino repeatedly threw temper tantrums when she was controlled in ways that restricted her manifestations. The trio humored her in order to appear cooperative and to experience successful séances. Even more telling, Palladino requested that her controllers inform her when they believed she was in a position capable of cheating. As a result, she was controlling them. She knew exactly when she was free to use her moves without being caught.

The individuals controlling Palladino told her when she could cheat.

PALLADINO: CHEATING IN AMERICA

Nor did these impostors have much difficulty in imposing on him, because he met them above half way.

—Andrew Oehler, *The Sufferings of Andrew Oehler*, 1811

In a 1909 *Collier's* magazine article, Hereward Carrington laid out to the American people his case for a new force exhibited by Eusapia Palladino that could not be explained by science.[1] *Collier's* was noted for its investigative journalism such as the exposé by Upton Sinclair[2] of Chicago's meat-packing scandal and the patent medicine frauds reported by Samuel Hopkins Adams.[3] Now *Collier's* was introducing America to a woman who had convinced the scientists of Europe that her strange powers were beyond science. Who better to introduce her than Hereward Carrington, a man who had blown the whistle on the scams at Lily Dale, the New York hotbed of spiritualism, and had written the definitive book on the deceptions within this national religious movement?[4]

Carrington brought Palladino to America with money advanced by S. S. McClure of *McClure's Magazine*, and he selected a committee of scientists and professors to evaluate her at Columbia University.[5]

Like Didier, Palladino produced manifestations that baffled everyone. Unlike Didier, she was frequently exposed for her cheating. Those instances of deceit were not carefully obscured in reports but clearly asserted. Surprisingly, Palladino usually confessed when confronted, stating she cheated because she had not been properly controlled to prevent trickery. Why then were scientists so quick to conclude her manifestations were the result of psychic powers unknown to science? Perhaps Carrington was unaware of the irony in the title of his *Collier's* article: "Eusapia Palladino: The Despair of Science." Scientists despaired of evaluating Palladino properly, and the story of their bungled attempts is as interesting as her amazing deceptions.

Palladino demonstrating a "complete levitation." Cesare Lombroso, *After Death What? Spiritistic Phenomena and Their Interpretation* (Boston: Small, Maynard, 1909). *From the author's collection*

Detail of the photograph showing her skirt under that table leg.

STEPPING INTO A NEW WORLD

Palladino arrived in New York on the German liner *Prinzess Irene* on November 10, 1909.[6] Her first response was to sell her story to *Cosmopolitan* magazine.[7] Although the article was written in first-person voice and Palladino received the byline, she could barely spell her own name according to those who knew her, and her English was primitive.

What's striking is that Palladino denied any role or responsibility for the unfolding of her life.[8] She presented her travel to distant cities as if she had been transported as an object to be studied. All her startling effects were reported as if she knew about them only because of what she had been told. She said that during the trance state she could feel a force but as for an explanation, "I have none." All she could say was that she needed the cooperation of strong men to give her added power. A cabinet was useful for concentrating the force, and darkness was better than a disturbing light.

Palladino ended her story to *Cosmopolitan* by declaring her desire to help American scientists. With deep sincerity, she wrote "this is why I am willing to submit to any test—in the darkness or light, with a cabinet or without it. I will go anywhere."

Reporters gathered in Carrington's apartment the next day to learn that he was charging $250 to $300 for a single seat at a séance. Palladino said she liked New York because it was big like Paris, London, and St. Petersburg. She also told reporters that she was an agnostic who had no use for clergy. In fact, she confessed, she once beat up a priest who annoyed her.

Six reporters gathered for a séance late the following evening.[9] A special table had been constructed for her, which the reporters watched levitate about two feet from the ground. This occurred with sufficient light to see both of her hands and feet, convincing them that the laws of gravity had been violated. When the lights were dimmed, the black curtain behind her was pushed out by an alabaster arm with a phosphorescent glow. Palladino ordered the hand to touch those in the circle. The hand, now invisible, touched several, including one man who was quite horrified as it repeatedly brushed him. Raps were heard as well as sounds from instruments in the cabinet behind her. The table levitated several more times, and one reporter decided to resist it by pressing down on the table with his forearms. Palladino complained that someone was pushing against her, not realizing that she was admitting her role in the levitation. Nevertheless, everyone concluded that Palladino was highly successful in demonstrating psychic phenomena.

S. W. REILLY CO. COLUMBUS, OHIO

No. 505 Luminous Ghostlite Paint

The Aristocrat of Luminous Paints and a favorite of professionals for its Quality Service and Satisfaction.

Goes Farther . . . Lasts Longer . . . Looks Better . . . Safe To Use . . . Easy to Apply . . . Costs Less.

Luminous Ghostline Paint . . . Is the finest quality made for making ghosts and other objects visible in the dark . . . Covers and adheres to any clean surface and will not peel, rub off or crack . . . Is everlasting, as it will permanently retain the quality of shining in the dark after exposure to light . . . Gives off a mysterious vaporish light the same as you would expect to see if you came face to face with a "real ghost" . . . Does not contain anything poisonous and is not harmed by soap or water . . . Brushed or sprayed with the ease of ordinary paint and dries over-night . . . The net result is the Best Luminous Paint Money Can Buy, at the lowest selling price consistent with its very high quality.

GLASS VIAL $1.00 1 OZ. JAR $2.50 1 LB. CAN $25.00

"WE MAKE EVERYTHING THAT SHINES IN THE DARK"

CLOSE YOUR ACT WITH REILLY'S GHOSTS

Luminous paint available from a supply catalog for mediums, 1901. Ralph E. Sylvestre, *Gambols with the Ghosts* (Chicago: Sylvestre, 1901). *From the author's collection*

Things were not as successful after this. The seeds that Carrington scattered in his *Collier's* magazine article landed on rocky soil. A *New York Times* writer said that Carrington was "raising the stock of human credulity."[10] Carrington's confidence in someone who admitted cheating was a confession of his "unsoundness of judgment." As further evidence that he had spent too much time in the dark, the author noted that Carrington's recent book promoted a quack theory that food does not supply energy for the body.[11] Both Carrington and Palladino were taking hits, but both were making money.

One of Palladino's early séances was held specifically for S. S. McClure, who had advanced money to Carrington so that he could bring Palladino to America.[12] At that session, McClure was struck hard in his chest by a fist, which obviously startled him. It was probably the same fist that hit that bothersome priest.

Nine days after Palladino arrived in New York, Carrington was caught buying luminous paint from Martinka's famous New York magic supply company.[13] When Martinka was asked if he believed in Palladino's powers, he simply laughed. He was selling the secrets for tricks that she was using to fool the uninformed. The luminous paint incident ruined the mystery of the alabaster arm observed a few days earlier,[14] and the disasters continued for Carrington and Palladino.

James Hyslop, professor of ethics and logic at Columbia University, pontificated with a stinging criticism of Carrington in the *Times* but not because he was an unbeliever. Proclaiming his authority as a psychical researcher, Hyslop complained that Carrington resorted to vaudeville publicity rather than taking the proper scientific approach.[15] He declared that the whole affair had degenerated into a money-making scheme.[16] But Hyslop was interested in money as well; he suggested that the American Society for Psychical Research should have an endowment as big as the Carnegie Institute, which would result in "unimagined benefits" to humanity. His funding request from Carnegie was rebuffed.[17]

Carrington exploded at the flood of bad publicity. He denied making any profit from Palladino and claimed his only interest was science. Hereafter, he said, newspapers would receive no further information from him, and no one examining her would be permitted to give information to the press because negative publicity was resulting in a lack of scientific interest in her. Her séances were misrepresented, he said; henceforth, all would be held in seclusion.

But boycotting the press presented a problem for Carrington. After returning from Italy, he had rushed out a lengthy book supporting the claims of Palladino's psychic manifestations.[18] He was advertising the book in the same papers that were criticizing him, which did not help his sales.

Carrington's next problem was W. S. Davis, an expert in conjuring and a reformed medium.[19] Davis wrote an essay that began by praising Palladino for her courage in opening her first séances to the more skeptical newsmen. Then he laid it down, saying that when she gets arrested for obtaining money under false pretenses, he would be on the pier handing her roses as she made her escape back to Italy. He declared that almost any medium could fool the best experts if they were not permitted to move out of their seats, which was standard practice for Palladino's séances.

Then Davis threw down the big one: he said that those who have a mercenary interest in Palladino should be kept out of the séance room because they could secretly assist her. Carrington must have been steaming.

Davis went on to explain the methods Palladino had used in England. Palladino could easily free her hands and feet despite the avowed certainty that they were under control. Similarly, in contrast to many opinions, she (or her dress) was always in touch with the table she levitated. Palladino objected, sometimes viciously, when researchers suggested a screen or obstruction between her and the table. Davis said her séances were like a Punch and Judy show: you can't see what's hidden down below that creates the action up above. If the power emanated from her body, then it was capable of penetrating her thick black clothing. However, she refused to sit enclosed within a mosquito net, claiming that it would block the passage of psychic fluid.

Davis outlined multiple ways that Palladino could levitate a table: black thread loops, hidden clamps, wires under her dress, clips up her sleeves, suction cups, and so forth. If Palladino was using deceptive methods, scientists were out of their league. They needed the help of those who understood the methods of trickery. Carrington knew these methods, yet he failed to protect himself against deception before endorsing the assistance of spirits. Instead, he gave uninformed readers the false impression that spirit agency was the only explanation. Davis outlined multiple ways in which Carrington actually facilitated Palladino rather than honestly evaluating her, as was his declared intent.

The article by Davis was brutal. He provided revealing visual illustrations to explain Palladino's methods of cheating, so it was not even necessary to read the whole article. The professors and scientists assigned to assess her took note of the article and recruited Davis as a consultant.

Meanwhile, Hugo Munsterberg, professor of psychology at Harvard, wrote an exposé of Palladino in the February 1910 issue of *Metropolitan Magazine*. He admitted that he was unqualified to test her, being unacquainted with fraud, so during a séance, he concealed someone in the cabinet who caught Palladino's foot as it extended back to strum a guitar. Palladino let out a wild scream. As Davis had suggested, it takes trickery to catch a trickster.

Munsterberg was criticized by two individuals, but for radically different reasons. Stanley Krebs, an educational psychologist, complained that Munsterberg was right about Palladino's cheating but wrong about some of her methods. Krebs reviewed all the phenomena she produced in his presence and provided his take on her ploys and tactics, about which I will say more later.

Munsterberg's findings were also attacked by James Hyslop, who was still angry about not getting his Carnegie endowment. He said that Munsterberg was either lacking in intelligence or dishonest. Why? Because she might have provided "valuable scientific results if investigated properly."[20] Hyslop was furious that Munsterberg had hidden a man on the floor of Palladino's cabinet to catch her foot. Palladino's value in America, he said, was frivolously dissipated. He had anticipated that she would provide evidence of life after death; now all had been diverted into doubt and dispute. No one needed to say it, but someone commented that Hyslop was ignoring the cold, hard facts that Palladino was looking more and more like a flexible-hipped cheat.[21]

Finally, the committee was set to test Palladino at Columbia. Their findings did not provide any comfort to Hyslop, Carrington, or Palladino.

FORMAL TESTING

The formal testing of Palladino took place over ten sessions with different combinations of the committee members participating. Her ninth séance on April 17, 1910, started like all the others except that Palladino was asked to try her powers of influence on an electroscope, as she had attempted in Paris.[22] The committee's excitement about her efforts was actually a ruse to bring two additional observers into the room. A Columbia University student and Joseph Rinn, a lifelong friend of Houdini, were both dressed in black outfits. They secretly positioned themselves under the séance table while Palladino was distracted.[23] At the table, the committee strategically placed W. S. Davis, the previously mentioned expert in conjuring, to control her left side. Controlling her right side was James L. Kellogg. Both were close friends of Joseph Rinn, and for years the three of them duplicated the tricks of mediums. All three were acquainted with the distractions that provided opportunities for escape from control that Palladino commonly used. She played right into their trap. Rinn had watched Palladino reach her foot back to kick the cabinet curtains: "This was done several times so daringly that under the chairs where I lay it seemed impossible that people above the table could not have observed it." Palladino left the session pleased with her results, failing to perceive that she had been outwitted.

However, some of the committee members witnessed no trickery during the séance, only miracles above the table. Committee member Joseph Jastrow, a popular writer and early president of the American Psychological Association, suggested they re-create the séance with Joseph Rinn playing the role of Palladino.[24]

Rinn's demonstration was just as convincing as Palladino, leaving committee members completely baffled. I pause here to note that Palladino's baffled evaluators were not blinded by a belief system. Their eyes were wide open, and they knew many of her methods, yet their scrutiny was insufficient to catch her trickery. After Rinn demonstrated her hidden methods, the committee conceded that most of her manifestations were amazingly simplistic. Once her trickery was known, she was finished.

Here's how her trickery was revealed.

The primary individuals controlling Palladino at the table (Davis and Kellogg) gave no indication that they were suspicious of her maneuvers. They gave her the leeway required without any indication that they knew what she was doing. They ignored the ruses she employed to free her hands and feet, pretending they did not notice. Then she produced all the phenomena that had made her famous, even when experts swore she had been under control.

But did she cheat because she had an opportunity, as she and her supporters bravely declared? The team wondered how they could refute that excuse.

The following week, a team composed of Kellogg, Davis, John W. Sargent, and Dixon S. Miller met to determine if they could counter that argument.[25] They concluded that Palladino would reject them as controls if they asserted any restrictions to her critical moves. Therefore, they reviewed ways to gradually eliminate the possibility of trickery, being extremely gentle to keep the conditions as favorable to her as possible. For a time, Palladino ignored the efforts to check her movements, which resulted in a marked deterioration of the manifestations. Eventually, she could no longer continue, embarrassed by her poor performance. Suddenly she realized that she had been trapped into giving away her methods. Her humiliation erupted, and she threw up her arms, wailing and screaming in Italian so loudly that the noise was heard in the street.

Carrington knew nothing of the strict controls implemented during the investigation until he read about them in the *Times* on May 12. Palladino was labeled "a fraud, and a conscious, clever, and aggressively determined fraud."[26] Her tricks had made her rich while men of learning looked like fools. Despite the warnings, they had paid the princely sum of $12,000 for ten sittings, and she relied only on common human gullibility.

A couple of days later, Joseph Jastrow laid bare the details of what occurred during the sessions. The surgical precision of his analysis resulted in the cancellation of McClure's contract with Carrington. Carrington was nothing more than a dupe of Palladino, if not a conspirator.[27]

PALLADINO RESPONDS

When interviewed by a reporter, Palladino was at her defiant best.[28] She told him to inform all the gentlemen of the *Times* that she was ready to give them a demonstration to prove that the committee was wrong about its charges of fraud. All she asked was to be treated like a decent woman. She complained that the tests had been unscientific.[29] When shown pictures of her reaching back with a free hand, she said it was a fake picture. She said that when people expect tricks, that is what she gives them, and she asked for a committee that would not come expecting to catch her in tricks, "and I promise you there will be none."

Carrington remained defiant as well, asserting his belief that she could produce genuine phenomena. "Let a further series of tests be undertaken, in which fraud is rendered impossible, and genuine phenomena will result."[30] Joseph Rinn immediately responded with a challenge: put Palladino in a sack up to her neck and sit her in the corner of the séance room. If she can produce anything before an unbiased committee, he would give her $1,000.[31] Carrington was up to his neck in trouble and over his head with this challenge, which he could not refuse but only delay.

Palladino did not help their plight. She said that she had been placed in a sack in Venice and still produced phenomena, but it nearly killed her. "After they took me out, I fainted and had hysterics." (I found no evidence for this unlikely occurrence—the test, not the hysterics.) She admitted that when netting was placed between her and a table, nothing happened; conditions had to be right for her to get results. However, she agreed to have her wrists bound to the controls sitting next to her if she could have three or four inches of slack.

This started a war of words about acceptable controls and what constituted a genuine phenomenon. Carrington claimed these disagreements showed that Rinn wanted out of his bad bargain.[32] But even Carrington was annoyed by her unreasonable demands, such as when Palladino insisted she should collect the money if she produced a single rap.[33] Nevertheless, Palladino signed the *Times* proposal for a séance on May 17, which Rinn also signed. Then, Palladino changed her mind. Rinn offered her an additional crisp $1,000 bill if she could float a pencil a few inches above the table. If she, or her spirit helper, could lift a table, why not hoist a small object like a pencil?

A news reporter who caught Palladino on the street offered her some sympathy. Maybe she was having bad luck because the Earth was going through the tail of Halley's comet, which was being daily reported in the news. Plaintively, Palladino asked why no one believed her, especially since she was a good Catholic, apparently forgetting her boast about being an agnostic who beat up a priest. Then she seized the reporter's hand, looked deeply into his eyes, and said,

> All these persons who do not believe in my genuine phenomena have no faith, no religion—nothing. They are atheists. They have turned all beautiful things into nothing and now they want to turn God into nothing.[34]

Palladino failed to appear for her scheduled testing at the *Times* on Tuesday. Someone eventually located Palladino and visited her, begging her to keep her promised seance. She responded with an array of "Neapolitan expletives" described as eloquent and vociferous. With hand on heart, she pledged that she would appear at the *Times* the next Sunday and promised the results of her seance would be "extraordinary, marvelous, wonderful."

Rinn simply laughed at Palladino's failure to appear, saying she could not demonstrate a psychic power because it did not exist.

DEPARTURE

Palladino returned to Italy in June 1910. I wonder if Davis was on the pier honoring her with the roses he had promised. Although she left America, her specter remained. A couple of months later, Carrington gave a presentation at the Berkeley Theatre on West 44th Street.[35] He demonstrated a wide variety of psychic phenomena, such as lifting tables, as Palladino had. He exposed how these tricks were accomplished, but he still affirmed his belief in Palladino's supernatural ability. We can only speculate about his motives. Carrington was described as "half-assed" and a "godawful magician."[36] His doctoral degree, which he had purchased from a diploma mill, did not prepare him for a career in anything. Carrington, like Palladino, was simply earning a living by following the money.

Then Carrington did what any red-blooded American would do: he sued the *Times* for libel because its exposure of Palladino belittled him. The suit was soon dropped, however, because he realized that "public sentiment is all on your side."[37]

Once Palladino was back in Italy, it was said that her powers waned. One wonders, however, if that was similar to Plutarch's claim about the decline of the oracles. Palladino's old tricks were now so commonly understood that she was powerless. Nevertheless, when she died May 16, 1918, she had fooled more scientists than any medium in history.[38]

ANALYSIS OF THE AMERICAN SÉANCES

Most of the committee selected to investigate Palladino signed a brief statement concluding that the sittings were "unfavorable to the view that any supernormal power in this case exists."[39] A couple of individuals indicated that they could not determine how some objects beyond her reach moved, but they, too, said that test conditions were never adequate.

However, the most insightful commentary on Palladino was issued by someone who was not part of the assigned committee and, thus, it is mostly overlooked today.[40] That person was educational psychologist Stanley Krebs, who discovered how Henry Slade secretly wrote on slates and who also exposed the Bangs Sisters, Chicago mediums famous for their "spirit portraits," which sitters could watch being painted by an invisible hand.[41] Krebs is not mentioned in the *Encyclopedia of Occultism and Parapsychology*, nor is his name known to many historians of spiritualism. Yet Krebs had the eyes of a lynx that could see into the soul of Palladino's tricks.

Like Galileo, Krebs observed what everyone else witnessed, but he detected small clues that were oblivious to others. He tested his suspicions and made certain that he could replicate Palladino's effects in order to confirm his findings. During his two séances with Palladino, "all seemed straight, impressive, and wonderful." But Krebs noticed the unusual size of Palladino's table (18 × 36 inches), which he correctly surmised was specifically designed and constructed

The National Hymn of Siam

As Sung by Sir Konan Toyle and Mr. Karrington at
Seances for the Exudation, Exteriorization
and Ectoplasmic Materialization of
Super-Hypnagogic Hallucinations.

A very remarkable phenomenon in connection with this song was revealed by an Indian Spirit called "Whaterdub" through a magnetized Ouija Board, and it is certain mortals could not have discovered it without celestial aid. ☞ If the Hymn is sung to the tune of "My Country 'tis of Thee," the foreign words will gradually take the form of English words and you will quickly know how well qualified Sir Konan and Mr. Karrington are to sing it.

NATIONAL HYMN OF SIAM

Ova tannas Siam
Geeva tannas Siam
Ova tannas.
Sucha tamas Siam
Inn ocen tas Siam
Osucha nas Siam.
Osucha nas !

AN APPEAL TO THE PUBLIC

If you know of any Fakir who apparently exhales air from a scar in the forehead; or if you know of any Magician who can stretch his arm a foot beyond its normal length, or move a table without visible contact, or pluck hot coals out of a fire with his fingers, or give other evidence of a future life will you kindly communicate with the Carwood Herrington Mythomania Laboratory?

Mr. Karrington, who conducts the investigations, is exceptionally qualified, as he is one of the loveliest cigarette smokers in the spirit profession, and he will appreciate donations, contributions and endowments. But do not ask for demonstrations, as they can only be had in the presence of the non compos mentis. A sane skeptic at a seance creates an inharmonious influence and the medium is apt to have a fit.

Hereward Carrington found himself the object of ridicule. *Flyer from the author's collection* according to her specifications. Palladino sat at one of the short sides with controllers close to her at the longer edges. Two other sitters were next to each controller on the long side, and sometimes another observer was seated at the short side of the table opposite Palladino.

This narrow table brought her two controllers, sitting across from each other, very close together. As a result, the controllers' hands were nearly touching, and their feet were parallel below and out of sight. This allowed Palladino to angle her foot, placing the toe of her shoe on the toe of one controller and her heel on the toe of the other controller, leaving one foot free. Similarly, under the cover of darkness, she could touch both of the controllers' hands with only one of her own hands. The person on the left believed he was controlling Palladino's left hand and foot while the controller on the right thought he was restraining her right hand and foot. However, she had a hand and a foot free. To test this possibility, Krebs gradually moved his hand toward his body. Palladino immediately responded by pressing her thumb down more firmly on his fingers to prevent this movement, all under the cover of darkness. If he withdrew his hand, she was unable to reach the other controller's hand. The pressure of her thumb, to keep his hand in place, convinced him that her other hand was free.

Palladino needed her hands to be on top of the controllers' hands, rather than being held, which prevented her chicanery, so she pretended that her hands were hypersensitive. As noted by Count Solovovo: "She constantly objected to my holding even her thumb or her little finger. She constantly complained of my squeezing her hand too much."[42]

Krebs noticed that Palladino complained more about the sensitivity of her hands when it was dark but did not respond when he touched her hand when she was distracted. Krebs tested his assumption without her awareness.

As another example of his surreptitious investigation, Krebs observed that Palladino cried out as if in pain when someone attempted to explore her cabinet. Therefore, he tried to determine whether she was experiencing real physical distress or complaining to avoid detection. At the next séance, he brought a black glove that matched his black coat jacket and quietly put his hand and arm inside the cabinet a dozen times. Not once did she cry out.

Krebs also noticed that the lighting of the séance room was adjustable so that Palladino could request various levels of light. He concluded that each manifestation was always performed with the same illumination, which he confirmed by reviewing past records of her séances. Similarly, Palladino needed to know the locations of all the observers in the séance room, their movements, and how carefully each monitored her. He determined that specific manifestations corresponded with how closely Palladino was being observed. Her séances were choreographed as carefully rehearsed stage performances.

Palladino succeeded for two reasons: she dismissed or replaced anyone who complicated her routine and she employed various techniques that baffled the most astute individuals. Additionally, she made it difficult for anyone to focus on was happening because she engaged in constant, meaningless movement.[43] She further distracted her controllers by asking everyone to talk or sing. Sitters were constrained by her demand to keep the circle of touching intact in order to maintain the circuit of her power, and she would not allow anyone to pass a hand between her leg and the leg of a table she levitated. No one could take unannounced photographs because it might hurt her and also because good results occurred only with sympathetic participants. Carrington enabled this demand by providing each sitter with printed instructions for her séance sessions. Item four stated:

> It is very important to remember that, whatever the attitude of the sitter toward the medium may be, no suspicion be openly manifested at the sittings, as this is liable to spoil the phenomena.

In other words, Palladino provided mysterious manifestations if no one explored with a free hand, demanded sufficient light, took a picture, or stood in the wrong place. Most importantly, no one should exhibit any suspicion of trickery. Investigators ostensibly evaluated Palladino, but in fact Palladino and Carrington controlled them. When her controllers said that her hands were secure, she had their hands secured, using only one of hers to immobilize two of theirs. Under these conditions, Palladino could concentrate on her performance.

Palladino's table was a stage prop. Because it was so narrow, she simply spread her knees and raised her heels in order to lift it a few inches. To raise it higher, she needed a free foot. She rocked the table onto her toe, clamped it with her hand above, and raised it. It only weighed twelve pounds. She also had alternative methods for table lifting.

Early during Kreb's first séance, someone accused Palladino of using a string to move a flower stand in the cabinet behind her. The séance stopped, all lights were turned on, and Palladino was examined for the presence of a string. No string or hook was discovered. However, during the search, Krebs observed Palladino's shoes, which were usually hidden by her long black skirt. The outside heel of the soles protruded about a quarter of an inch. (Those lynx eyes!) He concluded it was a little shelf for raising the leg of the table, which had square legs with sharp edges on the floor.

By constructing a similar shoe and table, Krebs discovered he could produce levitations similar to those of Palladino by clamping one hand on the table directly above his foot. It was easy to get the heel under the table leg and lift it by raising his heel. For Palladino, anyone looking under the table saw only the billows of her skirt touching the leg of the table, as pictured in Lombroso's book.[44] No one saw the table perched on that shelf of her shoe, and she could lift her heel even if someone was controlling the toe of her shoe. Palladino distracted sitters from the critical point of the lift by dramatically raising her right hand above the table, making movements as directing a psychic force for the levitation. Now we know why Lombroso said that Palladino would not stand up or sit at the long side of her table to produce a levitation.

Krebs devised a way to corroborate this heel-and-hand clamp theory. He noticed, for example, that while the table levitated, the lowest point of the table was always the corner where her hand rested, whereas the diagonal corner opposite was always highest. He obtained the same results when he experimented with levitations. Also, Krebs observed that Palladino's knee rose when the table levitated. Moreover, the clamp method required the hand and foot to be directly aligned, which is what he always observed. These insights brought convergent confirmation of his theories.

Krebs reconceptualized Palladino's long dress as a costume. The long black sleeves allowed her to move her arms in dim light without being seen, except when she pulled them up to expose the glow of that luminous paint. What looked like the quaint dress of an Italian peasant woman was actually a necessary part of her performance.

Palladino designed and orchestrated a masterful stage play: her dress, the luminous paint, the table, her constant movement, her personal sensitivities, the situational expectations, her required restrictions, the lighting. Everything came together to produce majestic deceptions while researchers believed she was under their control. After years of experience, she became adept at improvising something to baffle the most skeptical, and anything unexplained was interpreted as evidence of an occult force.

Krebs said that Palladino liked flattery, and he was sincere in praising her as unique in her specialty—one of the greatest sight jugglers and sense manipulators ever.

WHO'S CONTROLLING WHOM?

Palladino was remarkable in her ability to avoid controls appropriate to a situation. One European investigator said, "She constantly manages to render impossible any permanent control."[45] Sometimes she responded with vitriolic oaths when restricted from producing the phenomenon requested. Years earlier, Carrington had said that mediums can be completely controlled, but they would never allow it because manifestations would then be impossible. For example, mediums would permit themselves to be tied with rope but never with a simple piece of white thread, which cannot be easily untied: "Instead of binding the medium with ropes . . . employ a simple piece of thread, and see how quickly your offer is rejected."[46] Joseph Jastrow ruefully noted that if Carrington had followed his own advice and used thread, the investigation of Palladino would have been over in a matter of minutes, not years. If she refused appropriate restrictions, then it was useless to test her.

If someone caught Palladino cheating, she blamed the sitter for not controlling her, which is like an embezzler blaming his boss for failing to prevent his pilfering. Caught rapping with the toe of her boot, Palladino once responded by saying, "It is queer, anyhow, something is pushing my foot toward the table."[47] She asked that her foot be fastened to that of her accuser, then she knocked with her heel.

Carrington was right when he called Palladino "the despair of science." Scientists thought they could control her, but they were not prepared for her methods of deception and distraction. They did not know where or how to observe her. As Martin Gardner later said, gerbils and electrons don't cheat. It took individuals with extraordinary perception and knowledge of conjuring skills to expose her. In a grim admission, Dickson Miller said that the scientific literature on Palladino is "a monument to a groveling imbecility of judgment."[48]

LAST MAN STANDING

As Palladino sailed back to Italy, she left few believers behind. Carrington, of course, was giving lectures on Palladino, but those $300 seats at séances were gone. At one time he might have thought she employed some unknown psychic force. Now he was in an awkward position of explaining why he bought luminescent paint in a magic shop and why he told evaluators they could bring no suspicion to her séances.

Some individuals stayed faithful, even after overwhelming evidence of deception. Baron von Schrenck-Notzing became indignant when shown that a medium was not emanating ectoplasm but was holding images clipped from newspapers.[49] A medium confessed to hoaxing physiologist Charles Richet, but he still refused to accept what everyone else considered obvious. Physicist Sir Oliver Lodge's book about his deceased son was criticized for its many blunders and contradictions. His son, so brilliant in life, failed to communicate a single message of serious importance or utility in death. Nevertheless, Lodge responded by claiming he was a scientist obliged to be ruled by the evidence he had received.[50] These men, and many more, were often cited as scientists who believed in spiritualism. But they gave up their science once they entered the unexplored territory of psychical research.

Prominent scientists capable of identifying fraud declared their open-mindedness and asserted their dedication to truth. Behind their stated goals, however, many of these men grieved untimely personal losses, and they yearned for confidence in the hereafter. Most did not have any religion, but their belief was fervent. They possessed a deep wish for a vital force or some irrationality that would upset what they perceived as a cold, mechanistic world. They were not satisfied with the memory of human love that persists beyond death. They wanted more and pretended to have captured it through their interactions with spiritualistic mediums who were using tricks of the conjuring arts.

Although seldom acknowledged, these late-night encounters were sometimes inflamed by excessive alcohol and narcotics, which were easily available and commonly in use among the intellectuals and upper class. Mediums were frequently offered stimulants because they were thought to enhance manifestations.[51]

Camille Flammarion said you cannot demand a photographer develop his plates in the light. Similarly, he declared that investigators had no right to impose conditions on Palladino. They had to accept her constraints in order to get favorable results. He failed to acknowledge that her restrictions allowed opportunity for cheating. Flammarion got favorable results in more than one sense. Palladino openly bragged about her sexual conquests with her sitters. She made otherwise conscientious men forget their responsibility to science when she put her bare foot on their crotch during seances. Historian of spiritualism Ruth Brandon was surprisingly candid in describing the extensive intimate involvement of Palladino with Flammarion and Richet.[52] She does not mention Carrington, but his affair with Boston medium Margery Crandon suggests he was a likely participant, as well.[53] Frederic Myers, noted for his excessive sexual predations, likely invited Palladino to his residence in Cambridge with mixed motives.[54]

The dark séances provided an unusual opportunity for Victorians to express their repressed sexual urges.[55] Schrenck-Notzing photographed Eva C. in the nude but retouched those pictures he printed for the public.[56] The medium Florence Cook was a beautiful, seductive young woman who needed confirmation of her powers by Sir William Crookes to maintain the generous financial support of a patron. She caught him when he was at a vulnerable point in his career and personal life. His "research" with her was the context for an affair.[57]

Ah, the sweet mysteries of nature—and the secrets of men.

25

SLATE WRITING

It has been proved again and again in wartime that debriefing must be done immediately to be accurate. If there is a delay, the memory stops recollecting what happened and begins to rationalize how it happened.

—Wright, *Spy Catcher*, 1989

In his extensive review of spiritualism, Frank Podmore concluded that no physical manifestation ever won such universal recognition as slate writing.[1] Slates are now curious items of the past, but they were common when my mother carried hers every day to a rural Oregon school in 1912. Writing on a paper tablet would have been an unimaginable expense for her pioneer immigrant parents.

Spiritual mediums called on spirits to write on one of the inner surfaces of two slates held tightly together or on the inner side of a slate tightly clamped against the underside of a table. That straightforward manifestation, as Podmore said, was convincing.

SLADE'S SLATES

Spirit slate writing was popularized by Henry Slade, one the world's most celebrated spiritualists.[2] He started his career as a medium around 1857 and practiced his craft for about twenty years before venturing from New York to London.[3] He made more money every two weeks than most workers earned in a year, but his career was briefly interrupted when a Professor Lankester snatched a slate from his hands. The writing was on the slate before any spirit had been called.[4] Slade was convicted of fraud in a highly publicized trial but was soon released on a minor technicality of the law. His popularity soared despite the exposure of some slate-writing methods in court by popular stage magician John Nevil Maskelyne.[5]

Slade dashed to Europe, where his fortune was further enhanced by appearances before royalty. In Leipzig he convinced a professor of astronomy, Johann Zollner, that spirit manifestations proved the existence of a fourth dimension.[6] For example, Slade tied knots in a rope while Zollner held both ends. (Ray Hyman, an Oregon psychologist, showed how this can be accomplished using some ingenious rope magic.[7])

After Slade left England, William Eglinton quickly established himself as an even more accomplished slate writer. If he was using trickery, not even magicians could catch him. Famous stage conjuror, Harry Kellar, decided to expose him, but after observing Eglinton's séances, Kellar concluded:

> I still remain a sceptic as regards to Spiritualism, but I repeat my inability to explain or account for what must have been an intelligent force that produced the writing on the slate, which, if my senses are to be relied on, was in no way the result of trickery or sleight of hand.[8]

Eglinton's slate-writing accomplishments were overwhelming. Commonly, Eglinton placed a piece of chalk between two slates provided by the sitter. These were held, tightly clamped, under a table by both Eglinton and the sitter. After the sound of writing was heard, perhaps twenty lines appeared on the inner surface of the slate. Handwriting on slates often matched that of the deceased person called to respond to a question. On more than one occasion, Eglinton had someone randomly choose a book and select a page. The very words of that page were subsequently discovered on a previously blank slate, words presumably never seen by Eglinton's eyes.

Over a period of three weeks, a high-level British ambassador, Mr. Cholmondeley-Pennell, and several of his aristocratic friends attended séances with Eglinton, who was living in the Hyde Park district of London. The ambassador had previously investigated spiritualism but always considered the evidence inconclusive. But William Eglinton was not an ordinary medium. Cholmondeley-Pennell decided that these séances "conclusively established the existence of some objective, intelligent force, capable of acting externally to the medium, and in contravention of the recognized laws of matter."[9]

Spirits wrote messages on slates despite eighteen specific precautions to prevent trickery. During all six séances, meaningful responses were provided in adequate light on clean, marked slates (so as not to be switched). The arguments were unanswerable. Either these were facts, our humble ambassador asserted, or one is reduced to accusing nine worthy individuals of failing to perceive what was obvious to any person. To deny the truth was pitiable. "There is no loophole or crevice left for imagining fraud."

Alfred Russel Wallace was a strong supporter of independent slate writing. He acknowledged that a few slate writers might use trickery, but the practice had been confirmed in his mind by the "power of witnesses who cannot be deceived."[10] Wallace repeatedly made impassioned pleas for an acceptance of spiritualism based on the notion that trustworthy people had examined it and found it compelling. He argued that human testimony increases in value with the number of independent and honest witnesses. Regarding the miracles presented by spiritualists,

> they have been tested and examined by skeptics of every grade of incredulity, men in every way qualified to detect imposture or to discover natural causes—trained physicists, medical men, lawyers, and men of business—but in every case the investigators have either retired baffled, or become converts.

Wallace railed against those who disregarded eyewitness testimony. Should a person living at the equator reject the report of snow? Such examples had no end.

The pleas of Wallace were consistent with the development of modern science as articulated in the 1660s by Robert Boyle and the Royal Society.[11] Whereas secretive, sooty alchemists had once ruled science, Boyle insisted that science become an open process in which experiments be publicly conducted so that observers could confirm the results.[12] Gentlemen of all religious and political opinions could now agree about the facts of an experiment without becoming entangled in their allegorical meaning or attacking the experimenter. Boyle stressed that this new science would construct a body of knowledge based on direct observation and the convergent testimony of independent observers.

By 1886, Eglinton had received about a hundred endorsements for his spirit writing, and the numbers were growing every day.

DEMAND FOR ACTION

The public began to demand that the Society for Psychical Research investigate Eglinton, but the society was reluctant to become involved for a couple of reasons. It had previously decided not to test mediums selling their services to the public because any exchange of money created the temptation for a medium to cheat in order to retain the appearance of success. Additionally, they did not want to evaluate anyone previously caught cheating. Eglinton kicked over both hurdles. In 1876, Eglinton was giving séances in which persons from the other side materialized into visual presence. However, an archdeacon happened to peek into Eglinton's suitcase only to discover a beard and robe exactly like that of the materialized spirit in the previous séance. Eglinton had also conspired with Madame Blavatsky to obtain the "astral conveyance" of a letter from a ship at sea.[13] Finally, Professor Carvill Lewis noticed the tendons of Eglinton's wrist were moving while the spirits were presumably in the act of writing on slates.[14]

As the pressure was building to evaluate Eglinton, the society's directors were diverted by the surprising appearance of a slate writer who was even better. To understand this unfolding drama, I begin by introducing the relevant cast of characters.

ELEANOR SIDGWICK

Eleanor Balfour Sidgwick was born into an aristocratic English family where her mathematical precocity was evident from an early age. She conducted experiments in electromotive force with her brother-in-law, and she married the professor of moral philosophy at Cambridge, Henry Sidgwick. Eleanor and her husband were among the founders of the Society for Psychical Research. To protect herself against deception, she studied the tricks of conjuring and learned the methods of passing secret codes. When Mrs. Sidgwick was puzzled by the medium Anna Eva Fay, she personally took the opportunity to tie her up at her next séance.[15] Fay was unable to release herself and, unsurprisingly, was not able to perform her routine. Thereafter, Fay never again allowed any woman to bind her.

Mrs. Sidgwick was appalled by the fact that spiritualists could be caught in trickery yet be strongly defended by their supporters and continue practicing as if nothing had happened. Thus, she helped construct a policy for the Society of Psychical Research that rejected evaluation of anyone who had once been detected in fraud.

As the demand to investigate Eglinton escalated, Mrs. Sidgwick responded by charging Eglinton with fraud.[16] That false beard in his suitcase was sufficient reason for his evaluation to be forfeited. She offered her blunt opinion that Eglinton was nothing other than a conjuror, saying that his supporters were deceived by their own "mal-observation."

Eleanor Sidgwick: One tough woman when investigating psychic phenomena. *Proceeding of the Society for Psychical Research, 1939. Reprinted with permission from the Society for Psychical Research*

In a roaring response to Mrs. Sidgwick, attorney C. C. Massey said that observers of Eglinton's séances were not fools but pillars of the social, political, and scientific society.[17] No one would confuse a conjuror's trick with a genuine manifestation of an unknown force. In a detailed analysis of Eglinton's slate writing, Massey noted that conjurors rely on diverting the attention of spectators; multiple observers of Eglinton had watched him intently without distraction. In séances, there was no possibility of confederates, substitutions, or mechanical contrivance. Thus, the medium cannot depend on methods of trickery. Observers are right next to the medium; conjurors do not have questioning spectators on the stage with their eyes on his hands. A magician might be a successful deceiver using distraction and surprises, but sitters anticipate the writing on a slate, where and when it will occur.[18]

Mrs. Sidgwick had her back against the wall. The controversy was approaching showdown. Just when she needed rescuing, three people came to her support. The first she enlisted, the second individual was already a part of her social network, and the third person unexpectedly walked into her life.

ANGELO LEWIS

Barrister and amateur magician Angelo Lewis had written what would become a pivotal book in the history of conjuring, *Modern Magic*.[19] To get herself out of her mess, Mrs. Sidgwick asked Lewis to investigate Eglinton with specific attention to "points wherein the observation of the witnesses is likely to have been defective or misdirected." Lewis attended twelve sittings with Eglinton, but the meetings were worthless because no writing appeared. Instead Lewis composed a masterful essay about the problems of observation, an essay of stunning insight about deception that reflected his judicial standing and conjuring knowledge.[20]

Lewis noted that spectators describe conjuring effects, not according to what they see, but by what they believe they see, which is very different. Accurate descriptions of séances are particularly difficult if the method is not known. Thus, even the best magicians are often fooled when seeing a trick for the first time. During repeated viewings, however, the trained observer notices unnatural or unusual movements that cover the critical maneuver. For example, Lewis noted that Eglinton often dropped his slates, which was ignored. However, Lewis understood these "accidents" as a conjuror's artifice. It provided Eglinton an opportunity to switch slates or to turn the slate around, creating the impression that the writing occurred at the end of the slate opposite to the side he held.

Eglinton told Lewis that spirits could write on a slate if it was lying covered on top of the table. Why, then, Lewis asked, should a slate ever be held under the table? This common placement provided opportunity for deception. Mediums always set the rules for the times and conditions under which they produced their effects, and untrained people fail to appreciate this enormous handicap on meaningful observation.

Eglinton was wise not to produce writing on slates in the presence of Lewis.

Yet Lewis admitted that he had no plausible explanation for some of Eglinton's manifestations—as described by observers. Lewis noted, "The value of these cases will depend on the precise accuracy of the witnesses' testimony." His essay made clear that accurate descriptions would likely be difficult—even by the pillars of society.

RICHARD HODGSON

Richard Hodgson was a board member of the Society for Psychical Research who was convinced that the large mass of informed people had established slate writing as genuine.[21] Then he changed his mind. In June 1884, Hodgson and his friend R. W. Hogg had a sitting with Eglinton. Hodgson and Hogg were both convinced that Eglinton's results were not accomplished by trickery, but one small circumstance bothered Hodgson. After the séance, the two friends independently wrote a detailed report of their experience, and their perceptions were surprisingly different. One—or both—of them had malobserved.

Meanwhile, Hodgson gathered some knowledge about magic and concluded that Eglinton's slate-writing séances were staged in a manner consistent with conjuring. Further, after rereading all the reports on Eglinton, Hodgson realized that Eglinton failed to produce writing in situations that prevented trickery, like those monitored by Angelo Lewis. Thus, Hodgson began constructing theoretical categories of malobservation that Mrs. Sidgwick had suggested: omission, substitution, transposition, and interpolation.

It was still not clear where all this was headed until a young man with some surprising skills walked into the life of Hodgson and Sidgwick.

S. J. DAVEY

Early in his adult years, S. J. Davey had a serious lung ailment that sent him to an infirmary where he witnessed the deaths of fellow patients. This turned his mind, as one can imagine, to some serious introspection. Davey attended Eglinton's séances to learn about life on the other side, and Eglinton told Davey that he, too, might have mediumistic powers. Davey experimented with slates at home. One afternoon he was shocked when he suddenly found the word "beware" written on a slate that he had examined and found clean just moments before. He later discovered that some friends were hoaxing him; they wrote on his slates when he was not looking. He was forced to admit that the "clean slate" was a misperception—a malobservation.[22]

Davey began experimenting with slates to see what could be produced by conjuring, an exercise that convinced him that Eglinton was using trickery for his mysterious manifestations. Soon Davey was devising his own methods for writing on slates, including slates that had been locked together. One day he received a letter from an admirer who said: "I certainly think your slate-writing is quite equal to what we saw with Eglinton." Davey considered the implications: "As I went on, I was gradually forced to the conviction that my own reports about Eglinton were just as unreliable as those statements [made] about myself."

Davey decided to demonstrate his skills for the edification of the inner circle of the Society for Psychical Research. Mrs. Sidgwick and Richard Hodgson immediately saw the implications of the séance he conducted for them. Davey told them before he began that his entire performance was trickery, but his results were even better than those of Eglinton. At this point, Hodgson and Davey devised a plan to test Sidgwick's theory of malobservation. If true, they would have an explanation for why bright people were incapable of adequately reporting their observations at séances. Thus began what has become known as the first psychological experiment on the reliability of observation.

THOSE BAFFLING SÉANCES

Hodgson arranged to bring subjects to Davey for a séance. All participants agreed to write reports about what they observed as soon as possible after the session. In this way, they could match fresh reviews with what actually happened.

Hodgson did not make it easy for Davey. Only one person was inclined toward belief in spiritualism, and most knew that Davey would be using conjuring techniques. In fact, Davey told the sitters before each séance that nothing would be produced by spirits even though he was adopting the ordinary procedures of the spiritualistic medium. He urged his subjects to treat him as a conjurer, to use tests, and to take precautions against trickery. He recommended that they request changes in procedures if they observed any reason for concern. More-over, Davey decided to produce writing no matter if the imposed restrictions placed him at risk of exposure. Finally, Davey refused to create emotional ten-sion in his subjects by suggesting that any of his messages were from departed relatives because he was unwilling "to trade upon their emotions." On the other hand, he did not hesitate to use information about sitters that he had gleaned from outside sources.

Davey conducted sixteen sittings from which there were thirty-three written reports. He decided not to reveal his secrets, which any magician would have guarded. He believed that his sitters would not want to be annoyed by learning how they had been fooled. Moreover, he was concerned that people might gain a false sense of security if given knowledge of his methods, thinking they could now protect themselves against deception. He knew that mediums would certainly develop new strategies. Equally important, Davey had more ambitious plans; he had already designed a new study in which he would tell sitters precisely the things he intended to do. He hypothesized that specific knowledge about a forth-coming effect would still not enable sitters to provide reliable reports.

Then Davey died from typhoid fever at the age of twenty-seven after his first series of séances.[23] His untimely death continues to haunt me as I wonder what he might have accomplished as a stage magician.[24] How might psychology have evolved if Davey had moved to America and worked with William James at Harvard?

TRICKERY, TRICKERY, TRICKERY

The results of the Davey study were widely criticized, perhaps because no one anticipated that subjects could be so wrong on such a seemingly easy task as describing a few minutes of interaction. Of course, the whole idea of science is to criticize research so that new tests can be constructed to eliminate bias, self-deception, and other variables. However, the criticism was not fashioned to create a clearer understanding of malobservation. The critics complained that his conclusions were not relevant to their experience, just as Faraday's critics rejected his explicit demonstrations. Alfred Russel Wallace went so far as to suggest that Davey was really a medium.

[Davey's results] are claimed to be all trick, and unless all can be so explained many of us will be confirmed in our belief that Mr. Davey was really a medium as well as a conjurer, and that in imputing all his performances to "trick" he was deceiving the Society and the public.[25]

Wallace said that his experiences with slate writing were not like those Davey described because he had closely watched for trickery. Most spiritualists, he said, have encountered slate writing under quite different circumstances from those of Davey's performance. Hodgson responded to this assertion with something like an incredulous slap on his forehead. There was only one way to respond to this nonsense. Hodgson decided that because of Davey's death, he was free to reveal most of what actually happened during the séances. Thus, in his report, he juxtaposed the sitter's comments with Davey's procedures, illustrating how utterly overwhelmed these subjects were by Davey's performance despite the warnings he provided them.[26] Hodgson's revelations provided startling examples of malobservations and lapses of memory. For example, consider this simple statement by Mrs. Y: "I am perfectly confident that my hand was not removed from the slates for one single instant, and that I never lost sight of them for a moment."

The extensive literature of spiritualism is filled with literally thousands of similar statements of confidence. Davey's study showed why such compelling declarations have no evidential value. The difference between the events Mrs. Y reported and what actually happened is not trivial and deserves careful attention. To begin, the slate was under the table, so it *was* out of her sight! Mrs. Y mistook knowing *where the slate was* with her belief that it was *in her view*. Moreover, Davey had withdrawn the slate from under the table several times, during which she had completely relinquished her hold of it. And she also released the slate when it was retrieved to view the message. Her inattention provided Davey the opportunity to deceive her. This simple example elicits a powerful conclusion: her misstatement made it impossible to suggest a natural explanation.

Davey usually used simple sleight-of-hand slate switches and other standard techniques designed to create confusion and misdirection. However, he also used a thimble pencil[27] and some extremely clever moves that would have baffled magicians of the day—and even magicians of today.[28]

When a Miss Symons came to the séance, she brought her own slates. But Davey pilfered one of them while helping her hang up her coat. Then he excused himself under the pretext of gathering some slates of his own, using that ruse to write on the purloined slate before returning it to her parcel. Miss Symons then carried her own marked slate into the séance room with a message already in place. How could she explain the appearance of writing on her slate except by endorsing a miracle?

Substitution of slates was not possible in some cases. Davey's amazing ability can be appreciated only by providing some extensive quotations. For example, one subject reported:

> I now took the two new slates which I had purchased, and which had never for a moment passed out of my possession. I even [took] the precaution of sitting on them during the foregoing proceedings. I placed a piece of red crayon therein, and screwed them down top and bottom so tightly that by no possibility could even the thin edge of a penknife be introduced. I then corded the slates twice across and across, sealing them in two places with red and blue wax, stamping them with my own private signet. Mr. Davey placed the slates under the table, and requested me to name some word I would like written. I stipulated for "April." After a few minutes, during which I most carefully watched him, he returned them, and after 10 minutes work, so tightly were they closed, I found exactly what I had desired.[29]

Here's how the trick was done: Davey held this slate solidly under the table by trapping it with his knee against the leg of the table. From his pocket, he produced a brass wedge that was two inches long and half an inch wide. He forced this wedge between the slates at the point farthest from the screws, producing an opening of about a quarter of an inch. Into this opening, Davey inserted the rib of an umbrella that had a piece of chalk attached, which enabled him to inscribe the requested word inside the sealed slates. He used an elastic band to draw the gimmick out of sight when finished.

In another case,

> Mr. Davey asked my daughter to choose a book, which she did at random, he having his back to her and standing at some distance while she did it. This book was tied up and sealed by one of the party, Mr. Davey never touching it from first to last. Mr. Davey asked us all to choose in our minds two numbers under ten to represent a page and a line of the book. I locked two clean slates together and placed them on top of the table, which he and my daughter each put one hand on. I am confident that he could not possibly have manipulated that slate. We heard the pencil in that slate moving, and in a few minutes I unlocked the slate and found it covered on both the inner sides with writing. In the writing there were quotations from every page we thought of, but not always the line; in the case of my husband the line was correct but not the page. He had thought of page 8 but the line quoted was from page 3. Mr. Davey said this confusion frequently occurred because of the similarity of the numbers. This test seemed to me *perfect*.[30]

Although this test was described as "perfect," Davey failed on a previous try, which was forgotten or ignored. Thus, it took him two attempts to direct his subject to a preselected book. On the second try, Davey led his sitter to the bookcase and with a wide sweep of his hand, he said, "Take any book at random," adding, "don't take a book that anyone would take," pointing at a conspicuous book with a stance that psychologically blocked selections to the right while directing her eye to his target book. His involvement in drawing attention to this specific book was completely forgotten. His words suggested complete freedom of choice, which was remembered, but his behavior narrowed the options. Davey's target book contained poetry with repeated words and phrases, so a variety of pages presented opportunities for hits, and Davey's explanation of confusion about the similar appearance of the numerals 3 and 8 provided a satisfactory reason for the near miss. (The near miss was a miss, but Davey spontaneously described it as a hit after the fact.) Davey had previously written the target words on the slate, and he switched the slates while the sitters were writing down the page numbers and lines for the test. All of them failed to notice or forgot that the daughter, at that point, had taken her hand off the locked slates.

After another séance, a sitter wrote a report worthy of our attention despite the extended narration:

> He placed one of our slates on three little china salt-cellars that lifted it up about an inch from the table. Upon the middle of this he placed several pieces of different coloured chalks, and covered them with a tumbler. Then he told my husband to form a mental picture of some figure he wished to have drawn on the slate under the glass, and to name aloud the colour he would have it drawn in. He thought of a cross, and chose aloud the blue colour. I suggested that blue was too dark to be easily seen, and asked him to take white, which he agreed to. We sat holding hands and watching the pieces of chalk under the tumbler. No one was touching the slate this time, not even Mr. Davey. . . . With our own eyes, each one of us, [saw] the pieces of chalk under the glass begin to move slowly, and apparently to walk of their own accord across the space of the slate under the tumbler. My husband had said just before that if the piece of red chalk under that tumbler moved, he would give his head to anyone who wanted it, so sure was he that it could not possibly move. The first piece of chalk that began to walk about was that very red piece. Then the blue and the white moved simultaneously, as though uncertain which was the one desired. It was utterly astounding to all of us to see how those pieces of chalk thus walking about under the glass with no visible agency to move them! . . . When the chalks stopped moving, we lifted the tumbler, and there was a cross, partly blue and partly white, and a long red line marking the path taken

by the red chalk! We were impressed by this test beyond the power of words to declare. The test conditions were perfect, and the whole thing took place under our eyes on top of the table with no hands of anybody near the slate.

This report describes what the subject observed in accurate terms, and the expression of wonder makes the description compelling. Without any available explanation, Davey's sitter constructed an interpretation tinted by the lens of the predominant spiritualistic culture of that time. Davey accomplished this miracle through some clever (but standard) conjuring methods. He drew the cross on the slate before placement of the glass tumbler, which he did as soon as he heard the man's wife whisper her request. The image was not easily visible because the slate was about eye level with the observers and further hidden by the pieces of chalk.

Remember that only one of his sitters was a believer in spiritualism, so we cannot say that the participants were destined to a wrong conclusion by a faulty belief system. However, without an explanation for how a conjuror might accomplish these effects, only the context of the séance is available in the mind to formulate an opinion. These were worthy people trying to report their experiences. But they were not worthy eyewitnesses. When confronted with an anomalous event, people are uncomfortable until they have an explanation, and they are commonly left with no option other than something supernatural or occult.[31]

We have discussed people who experience an illusion because of suggestion, expectation, compliance, and manipulated neuropathways. Now we add malobservation. Davey's subjects considered themselves competent observers, so their testimony was expressed with sincerity and conviction. But Davey hid information from their awareness, and they made incorrect statements about what they observed.

One would presume that the publication of Davey and Hodgson had made it undeniably clear that honorable people—trained physicists, medical men, lawyers, and men of business—were not automatically qualified to detect imposture. This was a hard lesson that spiritualists did not want to hear, and many threatened to quit the society. Reviewing the implications of Davey's slate-writing research, Frank Podmore said that the task of observation seemed absurdly easy.[32] However, spiritualists had altogether underestimated its difficulty and overvalued their own competence.

It takes a special kind of philosopher to admit poor eyesight.

ADMITTING THE NEED OF AN OPTOMETRIST

Those who rejected the conclusions of Davey made excuses similar to those who rejected Faraday's table turning. In both situations, believers said their experiences were different. Table turners were unaware of their own contributions; those observing slate writing were oblivious to their own faulty observational skills. Both are examples of conclusions based on incomplete knowledge: not knowing what you don't know.[33] One could make the case, however, that some of these spiritualists were guilty of not *wanting* to know what they didn't know.

A couple of decades after the Davey slate experiments, his study was lost in the fog of history. One correspondent to the Society for Psychical Research complained that it appeared to him that its only reason for existence was to raise clouds of doubt and suspicion. All its interest in research resulted only in faulty observations by overeager critics.

> There has not been a proved case of fraud on the part of mediums in this country, but only the manifestation of incompetence on the part of observers who knew little about conditions under which the spirits operate—nothing more.[34]

That is a discouraging response, but it reminds us of the power of belief systems. Spirit mediums remained popular until better entertainment was available in motion picture theaters and on radio. Séances in the dark did not end because of exposures but because pocket flashlights became commonly available. Mediums could not risk being caught so easily.

It was time to move on to new delusions.

CONCLUSION

Does it not seem almost inconceivable, that credulity could manufacture such an immense amount of delusion out of so small an amount of actuality?

—Grimes, *Mysteries of Human Nature*, 1857

While reviewing sources on slate writing, I encountered "Dr." William Keeler. Like many other interesting characters, he unfortunately never made it into the pages of this book. He was investigated by American parapsychologist Walter Prince, who noticed that the different spirits who wrote on Keeler's slates all had the same handwriting. Moreover, all of the messages included a note about living past death, and these soapy sayings, Prince said, occurred ad nauseam.[1]

It occurred to me that maybe Keeler was the reincarnation of that priest of Klarnos who offered the same messages over and over. Oenomaus of Gadara was unsatisfied with these boilerplate responses. He visited his local optometrist to correct his visual defects and then set off to Delphi with his curiosity aroused. The oracles he received there were better focused with his improved vision. Now he saw them for what they were: nonsensical and utterly worthless.

I wonder what happened to Oenomaus. His book was lost and all but forgotten. I suspect it was harshly reviewed; maybe he was called an atheist and shunned by his friends. His criticism had no effect on the traffic up Mount Parnassus. Delphi remained popular until it was destroyed by wars, plague, looting, and, finally, the decree of a Roman emperor sympathetic to Christianity.

What did Oenomaus think about all the effort he expended in investigating oracles? Would he do the same thing again knowing the outcome?

Maybe a medium could grant me an hour to meet Oenomaus and some of the cameo players I most admire. I would like to tell them how much I appreciate their insights and ask—not about any historical enigma—but how they managed their disappointments in failing to stop the false beliefs they investigated. I would start with Faraday.

Faraday was a rock star in his day. He unraveled the mysteries of electricity, which changed our world more than we can imagine, and his adventures into the secrets of self-deception were a welcome revelation to skeptics. However, his findings were widely ignored if not scorned by those entranced with their presumed discovery of a new force. Those lynx eyes that conceptualized invisible electric currents immediately focused on the force causing tables to move: the fingertips. Typical of the careful scientist, Faraday was not satisfied with a single demonstration. Each test was conceived with dazzling insight that provided convergent validity for his theory. I particularly liked the upright straw on his table that moved whenever fingers produced lateral movement. As soon as participants noticed the straw wiggling, the table stopped moving. P. T. Barnum said it would be wonderful if some philosophic Yankee built a humbugometer,[2] and Faraday's devilishly clever straw is as close as you can get. I want to ask Faraday what he thought about all the resulting hate mail (or, more properly, stupid mail). Was he discouraged that no one identified the "important principle" missing in the educational system?

And speaking of clever, stage magician Maskelyne screwed two slates together, sealed the edges, and soldered them inside a canister.[3] No medium asked a spirit to convince skeptics by penetrating that container to write out an effective treatment for bacterial infection—or a deadly virus. Why not? Maybe Maskelyne is up in spiritland laughing with Cicero, who was similarly curious about why the gods failed to send straightforward messages instead of frightening tribulations and ambiguous dreams. And please contact Charles Page and Charles Dickens. Both wanted to know why spirits tease living humans with meaningless raps and useless words on a slate. Did they ever find out?

Please allow me some time with Eleanor Sidgwick. Did she laugh when Anna Eva Fay could not produce her manifestations after Eleanor tied her up? Please. Just one hour to hear some stories about her incredible long life of investigating psychics. Her rope-tying skills reminded me that the spirits refuse to bail out people entrapped in situations that prevent trickery. Why do they stop right when their intervention would be so convincing?

But I'll pass on meeting Reverend Coyne, the one-eyed evangelist who God refused to help when cheating was foiled. I already met him once at one of his religious charades and watched his unconvincing peek-down-the-nose routine. I can also forgo meeting the cute Kansas girl whose psychic ability was stymied when Randi cut the cards. And I would be too embarrassed to ask Professor Hare about his spiritoscope. But I would like to know if he is conducting chemistry experiments or still discussing moral platitudes on one of those seven spheres.

Bring me, instead, Stanley Krebs, that nemesis of Palladino. How did his eye catch that tiny shelf on the edge of Palladino's shoe? And did W. S. Davis, reformed medium and conjuring expert, take roses to Palladino when she returned to Italy? Was he discouraged that Palladino's failures did not slow the interest in spiritualism?

Put me down for a séance with C. F. Durant, that early American skeptic of clairvoyance. Mesmeric practitioners refused to believe the results of his experiments, and they made excuses for the failures of their own subjects when he tested them. As someone said, "There is a bad atmosphere about Durant."

His response echoes the conclusions of others we encountered in this book: "I was utterly amazed at the delusions by which so many around me were deceived and imposed upon."

Durant decided if mesmerists gave him bogus excuses, he would search the extent of their gullibility. When would they start questioning where subjects were getting their information? Even the title of his book is a joke, a hoax: *A New Theory of Animal Magnetism*. Charles Poyen proposed the theory that subjects received clairvoyant information from their magnetizers. Durant claimed he had discovered the magnetic cord that connected the brain of the mesmerist with his subject. In one lecture, he demonstrated how that attachment works. He sent the subject to the house at a specific address, "which she described correctly." Readers who consulted the associated footnote learn that Durant presumed that she described the residence correctly because he had never seen it. The address was imaginary, and no house existed there.

Durant concluded that he could not tell people what to believe. Instead, he wanted people to think for themselves and come to their own conclusions. He wrote his book with the same expectation.

I like that.

ACKNOWLEDGMENTS

I would like to express appreciation to my literary agent, Nancy Rosenfeld, for believing in this book and navigating it through the initial stages of publication. Jake Bonar at Prometheus Books gave me the courage to throw out several interesting but unimportant diversions that hindered flow. He was always right in his critical comments. Erin McGarvey was a master at the thankless task of editorial production and consistency of style.

Phillip Pirages and Scott Givens are both unique book dealers here in Oregon who have located amazing items for my library. Both are personal friends, so close, in fact, that each provided helpful editorial comments on this book. The literary skills of Pirages can best be appreciated by reading his rare book catalogs, widely considered the best in the business. Thomas Slate is also a personal friend, still reading books at home at 102 years of age. Through an unusual arrangement, he distributed his eclectic library to me for many years. Thanks, Tom, for all the historical information that influenced my thinking and supplied the necessary context for this book.

Jim Alcock, professor of psychology at York University, has been a faithful friend who introduced me to the importance of belief systems, which influenced my clinical practice and writing. He convinced me that the first lame drafts that I sent him would emerge as a book that others would want to read. He also restarted my thinking a few times when my mind was stuck, and he kept me writing when I was lost. Alcock introduced me to his colleague, mathematics professor Hal Proppe who made helpful suggestions about Cardano, particularly emphasizing the difficulty of solving statistical concepts at that time. Dan Wilson, Ron Friedland, Byron Walker, and Trevor Blake helped me with specific principles.

Michelle Ainsworth provided insightful comments to early drafts, especially for getting my timelines straight. My daughter, Beth Stockdell, helped with her good eye for typos. Nancy Koroloff, professor emerita of social work at Portland State University, kindly responded to my plea for a final review.

But it was Harriet Hall who caught blunders at the most crucial time during the editing process. Harriet was a flight surgeon in the air force, and she is still flying high against medical quackery and all things devoid of scientific evidence.

My wife, Ethelyn, the most important person in my life, helped me, as always, with clarity and charity.

NOTES

CHAPTER 1: HIDING A BLUNDER

1. Elliott (1852, 226).

2. Wormell (1963) says the Greek word "boil" is the same as "smelt," suggesting Croesus was forging currency. The turtle was an emblem on Greek currency, and the coinage of Croesus was marked with the image of a lion facing a bull. Also see Verneule (1950). The first known minting of coins occurred in Lydia, around 600 BCE, perhaps just a few years before the reign of Croesus (595 BCE–ca. 546 BCE), and these coins displayed a lion (Ferguson 2008, 24).

3. Croesus also gave two statues: a boy with water flowing out through his hands and a girl making bread (Poulsen 1920, 71). Croesus gathered enormous amounts of gold from alluvial river deposits in his own country, and he accumulated even more through taxes and plunder (Anonymous 1943). The saying "rich as Croesus" survives to the present day.

4. Parke (1985).

5. Leger (1846, 223) cites Diodorus of Sicily as source of the story about goats dancing and jumping in a most extraordinary manner at Delphi.

6. Pytho is where Apollo, according to legend, killed a dragon (or python) and thus the name. Poulsen (1920, 6) says "Pytho" is a Greek word meaning "to rot," and thus the stink of the dead dragon. Michelangelo painted five sibyls to accompany seven Old Testament prophets on the ceiling of the Sistine Chapel, and Mozart's *Requiem* says the world will dissolve in ashes, as foretold by King David and the sibyl.

7. For the history of various gods at Delphi, see Morgan (1990, 132). At least five different temples were built there (Beaumont 1724, 138). After one particular fire, an Athenian who was out of favor obtained the winning bid for temple reconstruction. To ingratiate himself, he promised to replace the limestone with marble. Only later was it

discovered that he counterfeited the marble, disguising the limestone to appear as the more desirable material (McConaughy 1931, 25).

8. Fontenrose (1978).

9. For attributions to the effects of a poisonous drug, see Burnett (1850, 282). For mesmerism, see Colquhoun (1851); Leger (1846, 222). More than one professor has chewed laurel leaves for the sake of science only to be left with a green taste in their mouths (Dodds 1951). Moreover, some varieties are poisonous (Flaceliere 1965). Jaynes (1976, 323) smoked laurel in his pipe and chewed it only to feel more sick than prophetic. For the theory on smoke, see Bevan (1929).

10. Flaceliere (1965, 50).

11. The complex work of this investigation has been delightfully documented by Broad (2006). Also see De Boer (2001).

12. Davy (1836, 1:92). Davy failed to appreciate the potential of this gas for surgical anesthesia.

13. Broad (2006, 208).

14. Lehoux (2007).

15. Bowden (2005).

16. Representatives of cities met to resolve disputes at Delphi (Poulsen 1920, 31).

17. Hutton (1928, 155) provides an excellent summary of Delphi's beauty as it existed over the centuries. Berve and Gruben ([1963]) list the many cities that built treasuries there.

18. A third maxim was also carved: give surety, get disaster. This has less meaning today because it was then associated with slavery, but a more modern version might suggest "debt in disaster." Flaceliere (1965, 58) translates this as "To commit oneself is to court misfortune." Also see www.swan.ac.uk/grst/What's%20what%20Things/delphic_maxims.htm (last modified January 30, 2006).

19. Robinson (1972). Also see Finley (1968, 91).

20. Seltman (1957, 102).

21. Flaceliere (1965, 76).

22. Eidinow (2007). Angus (1975, 11) suggests the moral life of early Greeks did not come from priests but from their philosophers.

23. Fiske (1921, 53) asserted that good myths are the result of imagination assisted by reason then followed by masterful construction. He proposed that the Greeks were the "inventors" of practical mythology because their stories are complete, admirably put together, and told in charming language. Even today, they retain their beauty, interest, and originality (Trivers 2011; Cheesman and Williams 2000, 150). Also see Rue (1994).

24. Compare the role of the Pythia as described by Bowden (2005, 25) and by Maurizio (1993, 53). Different oracle sites used various methods of divination. In 1621, about forty types of Greek divination were identified (Thorndike 1923, 8: 478).

25. Archaeologists have discovered more than sixteen hundred curses on amulets (Eidinow 2007).

26. For example, a robber who stole treasures from Delphi was killed by a wolf who then led priests back to his body to recover the booty (Poulsen 1920, 22). Josephus said that those who ridiculed oracles made themselves examples of the oracle's truth when they were killed in battle (Flaceliere 1965).

27. Books about punishment for doubt of the Christian message often promoted political agendas as well (Hunter 1990). For doubts about demons, see Davies (2017, 52). For revealing the secrets of alchemists, see Caron and Hutin (1961, 133). For scoffing at ghosts, see Bromhall (1658) and A Converted Infidel (1855). For the rejection of relics, see Sumpton (1975). For enquiry into forbidden things, see Coopland (1952, 34). For doubt about black magic, see Lapponi (1907, 35).

28. Brunvand (1981).

29. Flaceliere (1965, 53).

30. Coopland (1952, 95). "Loxias" has sometimes been translated as "devious."

31. Herodotus admitted Delphi tolerated influence peddling (Shimron 1989, 43). Bribery is documented by Poulsen (1920, 27).

32. Morgan (1990, 227); Berve and Gruben ([1963], 42).

33. Keegan (1994). H. G. Wells (1920/1971, 1: 257) informs us that Cyrus defeated Croesus because he learned that horses were afraid of camels. Therefore, Cyrus used his pack animals to lead the first charge against the cavalry of Croesus, which caused distress and disarray.

CHAPTER 2: DOUBTING THE ORACLES

1. Bowden (2005). Clement of Alexandria (1960, 91) reviewed Greek human sacrifice. Also see Howitt (1845, 39).

2. Plato (1909, 20).

3. James (1921).

4. Haidt (2012, 74). Haidt's book is required reading. Dolnick (2008) provides excellent examples of how confirmation bias works in specific real-life situations. Bright people maintained incorrect conclusions in their judgment of art forgery; they overestimated their knowledge, and were overconfident in that knowledge.

5. See, for example, the work of social psychologists Johnathan Haidt (2012), Keith Stanovich (1998), and James Alcock (2018); philosopher Lee McIntyre (2018); author Kevin Young (2017); magician and parapsychologist Richard Wiseman (2011); and science historian David Wootton (2015). The entanglement of emotions, an indispensable associate of reasoning, is a related topic but beyond the scope of our consideration (Damasio 2006).

6. Wolpert (1996) makes the point that science is an unnatural strategy of responding to the world, and people can live successful lives without knowing any science whatsoever.

7. McNeill (1976, 105).

8. Zinsser (1967); Dodds (1951, 193). The specific disease of this plague is unknown, probably because the first onset in a population is often different from later manifestations (Hecker ca. 1885). As for lawlessness, the same happened after the 1348 Black Death when people lost faith in the church. Barbara Tuchman (1978, 123) says that the Black Death created the stage for the development of modern man.

9. Hanson (2006, 271).

10. Kindt (2006); Maurizio (1997); Loyd-Jones (1976).

11. Africa (1968, 99) said that Herodotus never rejected a good story whether it happened or not.

12. Lucian (1961, 197) said that Herodotus was among those "who have somehow succumbed to this morbid passion for lying."

13. Cyrus must have been a genial guy. It was reported that he knew the name of every soldier in his army (Vergil 2002, 253).

14. Evans (1978) supports the opinion that Croesus committed suicide, perhaps by immolation, when his army fell.

15. Herodotus (1956, 442ff). The gold at Delphi was an attractive target for tyrants (Yonge 1909, 365).

16. Trivers (2011, 245). Institutions have similarly created myths; Oxford University was made older and more venerable by forgery of its charter (Tout 1934, 133).

17. Thucydides (1954, 21).

18. Middleton (1749).

19. Anonymous (1825, 3). For other fantastic stories, see Englebert's *Lives of the Saints* (1951).

20. Russell (1950, 99). Hanson (1970, 16) says that historians have commonly pointed out the mathematical slips and blunders of scientists, like Newton, but ignored their misuse of words. And a recent article reviewed the surprising number of Nobel laureates in the sciences who appear to have embraced weird ideas, seldom acknowledged (Basterfield et al., 2020).

21. Lewes (1864, ix).

22. At the beginning of the war, Athens had a large treasury in contrast to that of the Spartans who desperately needed assistance from Persia and Delphi (Hanson 2006, 271).

23. Dodds (1951). Obviously, astrology was not the only cause of their moral failures.

24. Cicero (1997).

25. Oenomaus is believed to have been a philosopher who lived early in the second century of the Christian era. His rant against oracles, sometimes translated as *Detection of Deceivers*, is known from the writings of Eusebius. A modern translation has been provided by Parke (1985).

26. Fontenrose (1978, 237).

27. Berve and Gruben (1963, 36).

28. Plutarch (1993, 369).

29. Hutton (1928, 157); also see Petrakos (1977, 9).

30. De Camp, Sprague, and de Camp (1966, 37) say that the tripod of Delphi remains in Constantinople to this day.

31. Greary (1990).

32. On the other hand, the sanctuaries for Asclepius were especially designed for healing, with some spectacular results attributed to the site (Kerenyi 1959; Marsh 1854, 91).

33. Thompson (1946).

34. Leff (1976).

35. Festinger (1956).

CHAPTER 3: CHAMBER OF COMMERCE PROMOTIONAL SECRETS

1. Glucklich (1997).

2. Van Dale's book on oracles also inspired a monumental four-volume treatise *The World Bewitched* by Dutch Reform theologian Balthasar Bekker. Like Van Dale, Bekker argued against the earthly activity of evil spirits, rejected the reality of witchcraft, and debunked spirit stories by suggesting natural causes (Fix 2010). This work had an extensive influence, especially in Germany and France, but only the first volume was published in English in a timely manner.

3. The first English edition of Fontenelle's book was translated by Aphra Behn in 1688, and the second edition was translated by Stephen Whatley in 1750. My comments are all based on the 1753 English printing of the second edition (Fontenelle 1753).

4. Leger (1846, 225).

5. Flint (1999, 282).

6. White (1896).

7. Thomas (1971). Some philosophers argued that catastrophes were chastisement and testing by God, thereby preserving his absolute authority in the universe (Webster 1982, 80).

8. Davies (2017, 50).

9. Benedict (2001). Falconer (1685, 109) indicates that Prometheus being gnawed by a vulture is understood as representing the reward (consequence) of too much curiosity.

10. Peters (1978).

11. Fanger (1998, 254); Davies (2017, 56). In 1319, John XXII had a rival bishop tortured, convicted, and burned (Bailey 2007, 118).

12. Yates (1979b, 375). Also see Arikha (2007, 176). Urban VIII later turned against Campanella.

13. Shea and Artigas (2003, 149). The edict extended to threats against anyone in the pope's family.

14. Fontenelle was familiar with the work of French clergyman David Blondel (1591–1655), who unraveled the pseudo-Isidorian Decretals, showing them to be forgeries designed to enhance the power of bishops (Williams 1954). Blondel also identified

an early forger of the Sibylline Oracles, an author named Hermas who was "pestered with fantastic imaginations," which he expressed in "wretched Greek" (Blondel 1661).

15. Miles (2011, 291).

16. Bowden (2005, 143). Curnow (2004) mentions the importance of oracle reputation.

17. My friend, retired Multnomah County judge Ron Cinniger, flew a practice mission in an A-4 Skyhawk off USS *Saratoga* aircraft carrier over Delphi in the 1960s. He still has vivid memories of the beautiful area despite his blazing speed. And, incidentally, he would like to provide this late apology to those he frightened because of his unacceptably low-flying altitude.

18. Finley (1981, 131).

19. Berve and Gruben ([1963], 38).

20. On the statues and cave, see Fontenelle (1753, 65 and 107). On natural beauty and theater, see Flaceliere (1965, 33 and plate 11). On pictures of Hades in the temple, see Luck (1985, 188).

21. French archaeologists found extensive records kept by priests on events and politics, even for far-off lands (Poulsen 1920, 26).

22. Fontenelle (1753, 26).

23. A Priest of the Church of England (1709).

24. Beaumont (1724). Other sources identify Baltus as a Dominican.

25. A Priest of the Church of England (1709, 49).

26. Marsak (1959).

27. Alexander encouraged Severianus to invade Armenia, and his army was cut to pieces. Alexander rewrote his advice, just like the priests at Delphi rewrote the story of Croesus (Lucian 1961, 234).

28. Another translator of Lucian noted this as "shearing the fat-heads" (Lucian 1949, 184).

29. Also see Connor (1999).

30. Robison (1822, 4: 454). Timbs (1856, 42) indicates that Hippocrates invented an ear trumpet that was lost for usage until the improvements by Moreland. Also see Moffat (1842, 312). Aristotle purportedly invented a megaphone for Alexander the Great that was capable of conveying orders to his generals at the distance of 100 stadia (about 12 miles), but there is no evidence that the instrument was ever used. Timbs (1860, 102) described one of Morland's many instruments as capable of carrying the voice about a mile and a half. The first scientific explanation of the speaking trumpet I have found is in Ozanam, Montucla, and Hutton (1803, 4: 250).

31. Thanks to Dan Wilson of Bishop's Stortford, Hertfordshire, for helping me better understand the transmission of sound and the science of mysterious noises. The resonant and reflective properties of a speaking tube may serve to amplify to a negligible extent, but principally the shape of the tube creates directionality that maintains the sound.

32. Poulsen (1945).

33. Abbott (1908, 291).

34. A rubber speaking tube, handy for transporting from venue to venue, was used to connect an observer with the performing clairvoyant (Pinetti 1905 and Hunt ca. 1890, 20; Robinson 1898, 67; Sylvestre 1901).

35. Alexander, "the man who knows," received information from his informant through a speaker hidden inside his turban (Alexander 1921). A wide variety of battery-powered tricks were exploited by spiritualists and performing clairvoyants, everything from rapping devices to talking teakettles (Chislett 1949). For additional electrical apparatus, see Dr. X (1922).

36. Timbs (1856, 112) mentions a statue of Aesculapius and another at Alexandria that delivered messages through the windpipes of cranes.

37. Melton (1620). Not all reports of talking statues can be taken at face value. An articulated statue created by a carpenter was described as a device of deception by Protestants eager to show that Catholics were trying to fool the simpleminded. But those who displayed the statue always described it as a prop used for Easter festivals (Bridgett 1890).

38. Sounds can also be mysteriously transmitted by hidden parabolic sound reflectors. One strategically placed near the ear of a statue transmitted questions (through a thin cloth disguised as a painting) to the ear of a hidden confederate. After hearing the question, he responded into a second reflector that projected his voice back to the statue, making it appear that the statue was speaking (Anonymous 1826, 47). Also see Anonymous (1821, 123); Cremer (1892, 76); Enfield (1821, 149).

39. Hopkins (1898, 102). Also see Anonymous (ca. 1827) and Gale (1838). Eventually I discovered that Sir David Brewster said the trick had been described in so many old magic books that he declined to reveal the secret again, so I stopped looking for additional references. Jay (2001) provides several illustrations of this illusion and mentions various mannequins that appeared to speak based on the same principle.

40. Granger (1803, 1: 38–41).

41. Matlock (1996). Whaley (1989) identifies Robertson as a Belgian optician-turned-showman who exhibited his "Fantasmagorie" in his native Liège in 1784.

42. Heard (2006).

43. Schmidt (2000, 155).

44. Anonymous (1871, 160) indicates that a series of mirrors at right angles revealed the spectators below. A description of the method and a "curious dialogue" between a traveler and the Invisible Girl is described in Anonymous (1855, 199). A speaking tube was also used to give the impression that a suspended wax head was talking (Lindhorst, ca. 1930s).

45. Hopkins (1898, 103) says E. J. Ingennato, a physicist (pseudonym of Robinson?), revealed the mystery in an 1800 pamphlet, which he said was now rare. I find no reference to this work in WorldCat or any of the magic bibliographies. Partington (1825) says the illusion was first explained in 1806 by Professor Millington in his lecture, but this was certainly not the first correct explanation. Joyce (1815, 117) provided an early explanation. For the Paris display, see Matlock (1996).

46. Many variations of this illusion appeared, including one in Boston in 1804. Emmons called it "The Extraordinary Aerial Phenomenon," but William Pinchbeck (1805) identified it as "The Acoustic Temple." By the end of the nineteenth century, many of the principal museums in the United States were exhibiting this illusion, and the secret was widely exposed in books of that era (Emmons 1853, 280).

47. A Priest of the Church of England (1709, 135).

48. A Priest of the Church of England (1709, 144–145).

49. Vandenberg (1982, 156) suggests that Delphi sent spies to Croesus to see what he was doing on that critical day. The delegates of Croesus were told the Pythia had spoken, but they delayed revealing the answer until the spies had returned.

50. Beaumont (1724).

51. Parke (1967, 126).

52. Lucian (1961, 231) described three methods for opening sealed messages, but he said there were even more.

53. Beckmann (1817, 1: 209). Also see Calmet (1850, 91).

54. Hippolytus (1868).

55. Porta (1658, 351).

56. Saberi (2016).

57. See Pieper (2013, 31–234) for a translation of this interesting manuscript.

58. Spiritualists used the same method (a warm wire) to surreptitiously open slates sealed together (Mann 1919, 28).

59. Groebner (2007).

60. Pybus (1810, 214). A similar method of compounding different colors of sealing wax was recommended in Anonymous (1803).

61. The methods described by Hull (1920) are amazingly ingenious.

62. Hull (n.d.). Many books on mediums and for mentalists describe methods for reading sealed messages. See, for example, Jewett (1873); Carrington (1907); Jones (1911); Dexter (1956).

CHAPTER 4: EXPLOTING THE MYSTERIES OF NATURE

1. Marsh (1854).

2. Calmet (1850, 84).

3. Wilkins (1691, 173). Also see Howe (1856, 450).

4. Sharpe (1992) describes several ingenious delayed-response methods. The same effect could be accomplished by having a ball bearing roll through thick oil, tipping the center of gravity. For additional information on Hero's adaptations by modern magicians, see Sharpe (1991) and several ingenious devices by Harbin (1979).

5. The Greek spelling of Hero was Heron, but the Latin form, Hero, is more frequently used. The exact dates of his life are uncertain. See Landels (1978) for the complete list of Hero's likely publications.

6. Hero's steam-driven engine is known as an aeolipile. By attaching whistles used by hunters, the device created frightening, horrible noises (Anonymous 1847a, 60–61). Steam was not used for industrial purposes until the manufacture of iron pipes.

7. Pollard and Reid (2006). As used by Hero, the term "pneumatic" referred to the use of any type of pressure.

8. Cooke (1903, 35). For a good illustrated explanation of Robert-Houdin's automaton, see Dunninger (1967, 66) and Dunninger (1926, 63).

9. Ewbank (1850, 387).

10. Hopkins (1898, 219).

11. Woodcroft (1851, 1).

12. Drachmann (1948). Hero acknowledged that some of his presentations were previously known and not all were fully functioning. His creations served various purposes: ornamental, entertainment, utilitarian, and mystifying (Landels 1978).

13. Boas (1966).

14. de Caus (1982).

15. Bate (1654). The first edition appeared in 1634 and contained information on hydraulics, fireworks, drawing, engraving, etching, and "sundry experiments."

16. McKenzie, Gibson, and Reyes (2004).

17. Christopher Wordsworth, bishop of Lincoln and amateur archaeologist, discovered this in 1832 (Eidinow 2007, 57).

18. Hopkins (1898, 208).

19. Later magic books describe this same method as a way of producing the sound of rain or hail, depending on the size of the shots of lead (Raymond 1875; Signor Blitz 1889).

20. Smith (1987, 3). I currently have two of these "lota bowls" on my library shelf in memory of a magician who amazed me more than sixty years ago.

21. Moehring (2013).

22. McCormick (1976, 3) relates this story to Egyptian priests who challenged Chaldean priests, but he gives no reference.

23. Plutarch (1993, 353).

24. Ozanam, Montucla, and Hutton (1803, 4: 496ff).

25. Salverte (1846).

26. Platts (1875).

27. Kirby (1815, 5: 150).

28. Phin (1906, 103).

29. Mayor (2009). In 332 BCE, Alexander the Great and his Macedonian army suffered the effects of a chemical incendiary that caused terrible casualties during their seven-month siege of Tyre. Thucydides (1954, 143) described the devastation created by sulfur and pitch.

30. The formula for phosphorus was probably known and lost during several periods of time. When the Byzantines used a version of this "Greek fire," they said that angels had conveyed the formula to Emperor Constantine (Volkman 2002, 54). The

ingredients and method for manufacturing phosphorus entered general knowledge only at the end of the seventeenth century after publication by Robert Boyle.

31. Godwin (1834, 107).

32. Paul (1845, 154).

33. Badcock (1828, 17).

34. Anonymous (1803, 24); Signor Blitz (1889, 17); Badcock (1828, 136).

35. Badcock (1828, 17).

36. Paul (1845, 154). Also see Hippolytus (1868, 32). A 1905 catalog issued to secret societies and fraternal organizations offered a 24×32-inch sheet of metal to shake to represent thunder intended to frighten the initiate (Lilley & Co. ca. 1905).

37. Badcock (1828, 75, 74).

38. Anonymous (1803, 14). Also see Anonymous (1835, 10); Bacon (1659/2011, 24); Dean (ca. 1739, 131); Signor Blitz (1889, 36).

39. For creation of calluses, see Beckmann (1817, 3: 276–77) and Brewster (1839, 351), who recommended rubbing the skin with sulfuric acid. For chemical preparations, see Scot (1584/1930, 279), Dean (ca. 1739, 67), and many other early magic books. Hippolytus (1868, 100) said the skin should be smeared with a salamander, which was then believed to be able to live in fire. For asbestos, see Brewster (1839, 352), Timbs (1856, 111), and Adams (1864, 193). For soot, see An Old Boy (1868, 26); Paul (1845, 47); and Anonymous (1889, 86). For chemical unguents, see Adams (1864, 1: 193).

40. Harry Price described this event in his popular books (1936a, 1939, and 1942).

41. Price (1936b). Also see Brown (1938).

42. Pankratz (1988).

43. Galileo was one of the first to explore the conductivity of heat; he invented a primitive type of the thermometer about 1597 (Burr 1933).

44. Cause No. 02-CV-1688; Towne v. Robbins. (My report on Robbins and his deceptions is in the public domain of the Federal Court.) Physicist Bernard Leikind walked on the embers at Robbins's seminar without attending the expensive lecture and without following the bogus guidelines Robbins told his believers they had to follow. Robbins encouraged people to look up and think "cool moss," but Leikind said he followed the advice of his mother and watched where he was going (Leikind and McCarthy 1985; Garrison 1985).

45. Salerno (2005) describes the enormous sums of money exchanged in the self-help, motivational speaking business. His chapter on Robbins is informative.

46. Eric Kurhi and Mark Gomez, "San Jose: 21 People Treated for Burns after Firewalking at Tony Robbins Appearance," *Mercury News*, July 20, 2012, www.mercurynews.com/crime-courts/ci_21125630/san-jose-21-people-treated-burns-after-firewalk.

47. "Tony Robbins Believers Get Burned during Fire Walk in Dallas," *Star-Telegram*, June 24, 2016, www.star-telegram.com/news/local/community/dallas/article85740932.html. Robbins downplayed this disaster by saying that it was a small

number of individuals among the seven thousand who attended his seminar. (It is of small concern if your feet are burned; it's a bloody disaster if my feet are burned.)

48. Pendergrast (2003); James and Thorpe (1994, 248–54).

49. Beckmann (1817, 3: 162).

50. We know of reading spectacles from a fresco of 1352 (Hirshfeld 2001). However, glass was not sufficiently clear for a good mirror or telescope until the time of Galileo. Then, larger sheets of glass were possible and thus the construction of larger mirrors, although they were expensive.

51. Cohen and Drabkin (1958, 261–68). One of Hero's illusions was used for an entertaining exhibit at the World's Fair in Paris in 1900 (Guillemin 2002). Magicians have devised endless deceptions using reflection (Sharpe 1985).

52. Cohen and Drabkin (1958, 365).

53. Brewster (1839, 146).

54. Anonymous (1803, 18ff).

55. In 1969, Bimbo's 365 Nightclub displayed a nude "mermaid." The woman was on a couch, and her image was projected into the tank. The illusion was previously featured at the 1933 Chicago Century of Progress Exhibition, then moved to a night club in New York. The display was greatly improved by Guy Jarrett for the Steel Pier in Atlantic City (Steinmeyer 1981, 101). Also see Garenne (1886, 264). A building plan for this illusion was available for $5 from a company that provided products for carnivals, fairs, and amusement parks (Brill 1975). Also see U.S. Patent 4,094,50; Romano (2006). Many patents using mirrors have been granted for optical illusions, toys, stage magic, display devices, advertising cabinets, and theatrical apparatus (Rees and Wilmore 1996).

56. Despiau (1801, 266) described a method for displaying the image of a specter on a column of smoke, which he said exhibited very distinctly. Also see Anonymous (1803, 21).

57. Similar methods of illumination were used in religious plays in the fifteenth century (Butterworth 2005, 80).

58. Pearl (1999). The last execution for witchcraft in France was 1745.

59. Davies (2017, 95).

60. The French Academy of Sciences has had three lives, becoming the National Institute of Sciences and Arts after the French Revolution, then being reconstituted in 1816 as the Royal Academy of Sciences (Snyder 2011, 131).

61. Brewster (1882, 14).

CHAPTER 5: UNCOVERING A FORGERY

1. For example, Eugenius tried to dissolve the Basel Council, but the delegates confuted his authority to dismiss them (Norwich 2011, 238).

2. Renaissance humanists threw out medieval Latin as if all of it was immature, and they favored earlier Greek and classical Latin as more sophisticated. They often made eloquence the sole test of learning (Lewis 1954).

3. Christie-Murray (1989). Valla probably did not know when the Apostles' Creed first appeared, but he knew it was after the life of the apostles.

4. Jardine (1996).

5. Wilberforce (1898).

6. Many authors have cited the extensive technological accomplishments of the medieval period (Africa 1968; Gimpel 1976; Hart 1963). Singer (1958, 59) suggests that we should not focus on how much we assent to the opinions of the prescientific era but how much influence certain individuals exerted in their own period.

7. Malone (2012). G. A. Williamson notes with irony that the last emperor, who abdicated, was named Romulus (Procopius 1981, 11).

8. Jarrett (1942, 39). Some believed Pope Sylvester II (999–1003) had sold his soul to the devil because of his extensive worldly knowledge, but he had lived among Spanish Arabs who saved the intellectual treasures that had vanished from Rome (Kelley and Sacks 1997). Clergy and cardinals were so illiterate that a mathematics book in the possession of Sylvester was mistaken for a treatise on necromancy (Headley 1990).

9. Donagan and Donagan (1965, 2).

10. Daston and Park (1999).

11. Popkin and Vanderjagt (1993).

12. Clark (1997, 153).

13. Lewis (1954, 9) points out that philosophers believed in the existence of many invisible beings that were theologically neutral.

14. Weill-Parot (2012, 219–93).

15. Malone (2012).

16. Daston and Park (1999).

17. Dante lamented the current irreverence for the ruins of Rome, and Poggio Bracciolini compared it to a gigantic cadaver killed by misfortune. The remedy was a return to ancient wisdom (Lefaivre 2005, 37).

18. Kahill (1995) describes St. Patrick as a far more liberal, forgiving Christian. Lefaivre (2005, 76) says Valla scandalously defended the principle of free love. At this time, it became popular to support an idea by writing a treatise against it with weak arguments. Bernardino Ochino (1487–1564), a Capuchin friar turned Protestant (or maybe atheist), wrote a dialogue in which he forcefully expounded on a topic he claimed to oppose, while countering it with unconvincing rebuttal (Minois 2012, 53).

19. Lefaivre (2005).

20. Rummel (1998).

21. Menchi (1994, 271–82).

22. Cameron (2011, 119).

23. Greenblatt (2011, 138). Also see the comments of Horace Bridges, who described Rome at that time in similar terms of disgrace (Erasmus 1925, xvi). Moreover,

clergy were immune from secular justice and responsible only to corrupt ecclesiastical courts.

24. Burckhardt (1929, 168) translated this as "the chamber of lies."

25. Valla wrote a textbook on Latin grammar that remained the authority for the next three hundred years. Fifty manuscript copies and 150 printed editions still exist in libraries throughout the world (Johnson 2000, 58).

26. Symonds (1909, 175). Bracciolini was particularly noted for his ability to write invectives, a style perfected during the Renaissance, in which a victim would be accused of any crime, no matter how implausible. And the demand to write flawless classical Latin was such that the slightest slip might ruin a man for life (Goldschmidt 1951). Bracciolini was employed by the Curia, which he excoriated in his writings (Menchi 1994, 273).

27. Kells (2017, 59). Minois (2012, 41) says he fathered only fourteen children, but maybe it was hard to keep track.

28. Inquisitions in the Middle Ages were a process of legal inquiry, nothing more. But as heresy became of more concern, Pope Lucius III decreed (1184) that all bishops conduct an inquiry into heresy at least once or twice a year. In 1252, Pope Innocent IV permitted inquisitional torture and trial by ordeal to extract information (Bailey 2007).

29. At the beginning of the Christian movement, Jesus and his disciples were considered lower class individuals not worthy of intellectual attention (Celsus 1978). Jones (1979) argues that early Christians did not use disputation to gain Greek converts, but Greek Christians incorporated logic and sophistry into Christianity. Either way, the introduction of intellectual conflicts quickly set one sect against another (Jones 1979, 149; Celsus 1978, 18).

30. Rue (1994, 23). For a poignant review of what happened when the church focused on belief while ignoring behavioral improprieties, see Lea (1922, 3: 616–50).

31. Simonetta (2008). Churches were popular places for assassinations during this period (Burckhardt 1929, 76; McConaughy 1931, 155; Ralph 1974, 89).

32. Sentences of severe physical punishment, like being burned alive or quartered, were sometimes reduced to pillory confinement on the grounds of mercy and charity (Groebner 2007).

33. Barnett (1999). Wolf (2015) provides a good understanding of the workings of more recent inquisitional trials.

34. Coleman (1922, 2). The Donation was part of a larger body of forged works known as the pseudo-Isidorian Decretals (Kerchever 1891/1970).

35. Norwich (2011). Cardinal Nicolas de Cuse had declared the document a forgery in 1431, a few years before Valla (Fontenelle 1990, 77n10).

36. French canonist Johannes Monachus (d. 1313) was the first to conclude that no judge, even the pope, could come to a just decision unless the defendant was present in court. Through some unusual reasoning, he concluded that even God, who is presumably omniscient, had presumed that Adam was innocent until proven guilty before evicting him from the Garden of Eden (Pennington 2003).

37. Stillman and Gordon (1972).

38. De Rosa (1988, 43).

39. Valla had a thorough understanding of the New Testament and an admiration for the teachings of Jesus (Celenza 2004).

40. Barraclough (1979). Other scholars give the production of the Donation to other popes of the eighth century, such as Pope Zachary (Gardner 2005). There is still agreement, however, that it was constructed around the eighth century.

41. Modern historian Moss (1964, 219) was overly generous when he suggested this action may have been more of a rationalization than a forgery.

42. Gardner (2005, 76).

43. Siricius (384–399) was the first Roman bishop to use the title of pope (Barraclough 1979).

44. Gardner (2005, 76).

45. These claims of Farrer (1907, 137) are startling yet worthy of serious discussion.

46. Jerome acknowledged difficulty in interpreting the multiple fragments of scripture that did not always agree, some with significant consequences. But over time his translation became revered as inspired and was declared so at the Council of Trent (Ehrman 2005, 2010).

47. Ginzburg (1992a).

48. Hall (1975). Roger Bacon was condemned for "suspected novelties" (Goldstone and Goldstone, 2005) and Copernicus was reluctant to publish because of the "hard-to-understand novelty of my theory" (Adamczewski ca. 1973, 144).

49. Eamon (1994).

50. *Foucault's Pendulum* (Eco 1988) is a wonderful spoof of "scholars" chasing metaphysical nonsense, attempting to create meaning where none exists.

51. Valla (2008, x).

52. Grimm (1949). Bracciolini was accused of writing or at least spreading the ideas in *The Three Impostors*, a book denouncing Moses, Jesus, and Muhammad. This phantom book was first mentioned in 1239, but the "reference edition" was compiled from various texts in the early 1700s (Minois 2012). See Nasier (1904) for the English translation of the French edition. It was dangerous even to express shock and horror about the evils of the book because it spread the arguments.

53. Dickens (1967). For Luther's praise of Valla, see Cassirer, Kristeller, and Randall (1967, 153). Even Cardinal Bellarmine, at the time of Galileo, was still angry at Valla, saying he was the one who opened the way for Martin Luther (Bronowski 1975, 172). Luther's revolt was also laid at the door of Erasmus, a devout Catholic associated with the counterreformation (Erasmus 1925, xxi).

54. Cassirer, Kristeller, and Randall (1967, 4).

55. Many Renaissance scholars were forced to dissimulate their true beliefs because of persecution (Zagorin 1990). Although Valla admired the teachings of Jesus, his faith was obviously complex.

56. www.newadvent.org/cathen/15257a.htm (accessed December 13, 2019). It is particularly ironic that the morality of Valla is questioned given the state of the Vatican at that time.

57. Erickson (1976).

58. Toland (1747, 350–403). See his essay titled "A Catalogue of Books Mention'd by the Fathers and Other Ancient Writers, As Truly or Falsely Ascrib'd to Jesus Christ, His Apostles, and Other Eminent Persons."

59. Ehrman (2010, 2011).

60. Wilcox (1987, 240). Galileo reflects a similar sentiment about describing nature.

61. Milman (1862, 123).

62. Bracciolini (1928). Inquisitors and censors did not laugh at his jokes about lecherous priests. And Bracciolini called Valla an immoral writer!

63. The days of safety for humanists were ended with the installation of Pope Paul II. After his coronation, academic humanists, including students of Valla, were imprisoned and tortured. At least one died from the mistreatment (Dunston 1973; Norwich 2011, 256).

CHAPTER 6: ASTROLOGY AND PROBABILITY

1. Mathematical problems were sometimes posted on handbills as intellectuals challenged one another to solve problems and debate in public competition (Aczel 2005, 44, 118).

2. Cardano (2002, 5).

3. Siraisi (1997, 216); Ernst (2010).

4. Burckhardt (1929, 490). Cecco also has the dubious distinction of being the first university scholar to be burned by the Inquisition.

5. Naudé (1657, 164).

6. His book was titled *On the Bad Practice of Medicine in Common Use*. At a later date, Rheticus, the assistant of Copernicus, was sent by a Nuremberg publisher to obtain books from Cardano. This was a successful relationship most notably because they published his pioneering book on algebra in 1545, only two years after the publication of *De revolutionibus* by Copernicus (Cardano 2002, xi).

7. Wootton (2007) provides a depressing review of the dangerous treatments that physicians have inflicted on humanity.

8. Siraisi (1997, 123).

9. For his admission to the College of Physicians, see Cardano (2002, 154). French physician and astrologer Nostradamus published his predictions for the same reasons that Cardano published his book on physician errors, namely for a better reputation and more clients (Bailey 2007, 188).

10. Before 1600, an empiric was someone who tried things by guessing because he was not trained in reasoning. Thereafter, the empiric gains more respect, denoting

someone who honors systematic observation above truth from reason alone (Wootton 2015). In a statement uncharacteristic of physicians at that time, Cardano once said that he was mistaken in his assertion that he had cured consumption. "I was deceived by hope" (Morley 1854, 1:168).

11. Wootton (2015) makes a strong case for this viewpoint.

12. The content of these early magical texts, as well as descriptions of their use, are well illustrated in Fanger (1998).

13. Hermetic texts received considerable interest after being translated from the Greek to Latin when brought to Florence in the 1460s by a Byzantine monk (Faivre 1995; Bailey 2007). Philosophers sometimes distanced themselves from new discoveries or beliefs by ascribing their ideas as originating with Solomon or Hermes Trismegistus, which diverted the attention of inquisitors. In 1614, Isaac Casaubon conceded that these documents were from the second and third centuries, not earlier, as commonly believed. Writings associated with Trismegistus were discovered in the Nag Hammadi Library (Robinson 1981, 9.)

14. Page (2012, 79–112). A belief that Adam had special knowledge existed as late as the last part of the seventeenth century. Joseph Glanvil, a member of the Royal Society, claimed that Adam was the first natural philosopher who understood the Earth's motion, circulation of the blood, and the effect of the moon on tides. Also see Schaffer (1998, 83–120).

15. Scholars of that time believed history descends a path downward from Eden to Armageddon (Milis 1999). Some Muslims also believed that all the knowledge that God wants people to know is contained in the Koran, a perspective to which some current Muslim sects subscribe (Al-Khalili 2011, 125).

16. Debus (1992, 4).

17. Fleming (2013) suggests that "kabbala" describes the Jewish tradition whereas "cabbala" reflects the Latin form as adopted by Christians. Kabbalists abandoned the ordinary meaning of words and assigned them values and metaphysical numbers that could be turning into new words and meaning. Some Kabbalists used this strategy to produce anti-Catholic propaganda and to give non-Christian literature a Christian meaning (Wrightsman 1975, 225).

18. Floyer (1713); McKnight (1992, viii).

19. Eamon (2010).

20. Palter (1961, 82). Vesalius noted the irony that he was not allowed to cut a dead man because people were created in the image of God, yet men were butchered like animals in wars endorsed by the church.

21. In the chapter on Galileo, I provide additional discussion on this position promoted by Wootton. For here, I suggest that the advances of civilization during the medieval period were by artisans, not natural philosophers, reflecting advances in technology and trades rather than science.

22. Van der Waerden (1963).

23. Singer (1958, 64).

24. Greenblatt (1991).

25. Whewell (1837, 1: 253) and others reviewed the problem of persons inhabiting the opposite side of the Earth who presumably would be living upside down.

26. Oresme said that no one could live in the antipodes because they would not be subject to the Roman Catholic Church (Coopland 1952, 33).

27. Wootton (2015, 128). Although physicians maintained their belief in Galen, sailors demanded maps based on direct observation. If Galen is wrong, the patient dies but the physician still lives. If a map is wrong, the sailor has more to lose by ignoring reality.

28. Ferguson (1953, 58).

29. Daston and Park (1999, 166). For his praise of new discoveries, see Cardano (2002, 166).

30. Cardano's life work was encyclopedic, totaling more than two hundred works on such topics as astronomy, physics, music, chess, death and dying, and the obligatory topic of the day, secrets.

31. Augustine (1958, 100). "But, what kind of rule over men's actions is left to God if men are necessarily determined by the stars?"

32. Thorndike (1923).

33. Smoller (1994).

34. Morley (1854, 5).

35. Burckhardt (1929).

36. Oresme's *Book of Divinations* presented evidence for and against divination, showing that the Bible and church fathers promoted a full range of opinions. He laid out the opinions on both sides of the issue before presenting his own conclusion.

37. Popes and inquisitors often held contradictory positions about whether to censor authors' beliefs, individual works, or passages of text (Baldini 2001).

38. Centuries later, experts in military deception used the same concept, namely that a purposely leaked piece of information should contain a discussion of a planned attack so that after the fact the source is not rejected as unreliable (Holt 2004).

39. The whole of the insurance industry emerged from maritime expeditions. Park (1799) said the first court settlement of a policy dispute occurred in the thirtieth year of Queen Elizabeth's reign (i.e., 1588). Pirenne (1937) says Venetians were investing money in maritime ventures, and by the eleventh century, members of the nobility of Genoa and Pisa cooperated to distribute risks. By the twelfth century, commercial credit and marine insurance were in operation.

40. Thomas (1971).

41. Grafton (1999, 44).

42. Melton (1620, 22–23), spelling modernized.

43. Anonymous (1680, 13).

44. Which reminds me of Smullyan (1983, 23), who said he did not believe in astrology because he was a Gemini.

45. Partridge (1703, 3). Astrologer William Lilly described the comet of 1680 as a portent of his own death, and he obligingly died the following year at the age of eighty (Webster 1982, 41).

46. Curry (1989).

47. Morley (1854, 138).

48. Cardano (2002, 66) said he was particularly under the grip of gambling during bad times or when impoverished.

49. An extensive history on early gamblers, con men, and lowlifes can be found in what are known as the "coney-catching" pamphlets, a series of about twenty-five books from 1552 until the early part of the 1600s. These are described and reprinted in a variety of sources including Chandler (1907), Fuller (1936), Kinney (1990), and Judges (1930). Based partly on these documents, Pankratz (2014) proposed that the development of con games emerged from cooperation among beggars and thieves.

50. Diaconis and Graham (2012).

51. Morley (1854, 1: 275). *The Spiritual Magazine* (vol. 4 [1863]: 4) says that Hutton's *Mathematical and Philosophical Dictionary* (1796) gives credit to Cardano for sixteen of the chief improvements in algebra.

52. For a more complete history of bookkeeping, see Beckmann (1817, 1: 1–9). Also see Isaacson (2017, 25), who confirms Cardano as the first to describe double-entry bookkeeping.

53. Ore (1953).

54. Turnbull (1993, 82).

55. Hacking (1984, 92).

56. Professor Hal Proppe pointed me to the post-Cardano work on probability by the Chevalier de Méré (aka Antoine Gombaud [1607–1684]). See https://introductory stats.wordpress.com/2010/11/12/one-gambling-problem-that-launched-modern-prob ability-theory (November 12, 2010). This history emphasizes the complexity involved in developing the mathematical concepts that are the foundation of modern probability and statistics.

57. Lovell (2003) identified three types of bar bets: (1) *proposition bets* in which the hustler proposes to do something that seems impossible, (2) *odds bets* in which the probability, unknown to the sucker, is heavily in favor of the swindler, and (3) *not-a-prayer bets* in which the mark has no chance of winning. This book was reprinted for general readership (Lovell 2007). "Bar bet" books have been popular from the sixteenth century to the present day (Wiseman 2016). Canadian attorney and magician Bob Farmer (2004) created *Rules for Fools*, which provides an excellent summary of cognitive and emotional pitfalls that exist because of our difficulty estimating probability. In a later variation, Farmer showed that the combination of statistical ignorance and cheating (marked cards, crooked dice, etc.) leaves victims helpless (Knuckles 2018).

58. Stafford and Webb (2005).

59. Plotkin (1997, 184).

60. Snyder (2011, 141) presents a fascinating history of how statistics entered into science in early nineteenth-century England, especially transforming economics.

61. Kucharski (2016). Also see King and Read (1963). Bruce (2001) wrote a wonderful book that teaches how to solve mathematical conundrums.

CHAPTER 7: NATURAL MAGICK

1. Porta (1957, vi). All my discussion about Porta's book is taken from the first English edition (Porta 1658).

2. Changeux (2004, 234) says science is dependent on a social milieu of free discussion, uncensored publications, tolerance, and intellectual competition.

3. Wootton (1985).

4. Lucas-Dubreton (1966, 288).

5. Cornelius Agrippa said that magic was the absolute perfection of natural philosophy; Roger Bacon said it was the most elevated and rewarding focus of knowledge; Pico della Mirandola said magic was the sum of all-natural wisdom (Clark 1997). All of them attempted to change the concept of magic into natural philosophy, which was the initial term for science.

6. Porta (1658, 3).

7. *Meteorologica* covered all atmospheric phenomena including everything from the Earth's surface to that which intruded into the sky, like comets. Meteorology was still without the thermometer as used by Galileo in 1607 or the barometer as used by Torricelli in 1643 (Heninger 1968). A translation of Aristotle's book can be found here: https://web.archive.org/web/20040125012342/http://etext.library.adelaide.edu .au:80/a/a8met/index.html (January 8, 2000).

8. Coopland (1952, 101).

9. Willey (1949).

10. Gentilcore (2000). Alessio Piemontese's *Secreti* of 1555 went through seventeen Italian editions by the end of the century.

11. Eamon (1994).

12. Felix was an "antipope" (Crisciani 1999).

13. Wounds were presumably cured when "weapon salves" were applied to the weapon that caused the damage. Knives were annealed in radish water because the radish is sharp. Similarly, "sympathy" proposes a complex bond or interconnection of nature. In 1660, members of the English Royal Society were treated to a lecture on magnetic cures then asked to bring their "powder of sympathy" to the next meeting (Mottelay 1922, 126).

14. Read (1937, 97).

15. Youngson (1998) argues that alchemy is not the forerunner of chemistry because its methods and findings were incidental to the search for personal accomplishment. Chemistry, in contrast, is the quest for developing scientific knowledge. Dobbs (1992),

on the other hand, argues that alchemists, or at least Newton's alchemy, was a search to understand the vital reactions of all substances. Although they came to a dead end, their goal was understanding nature and not limited to the quest for gold.

16. For example, compare Agrippa (1898), whose *Natural Magic* was published only a couple of decades earlier. Also see Agrippa (1575).

17. Secretly pour water in the vessel. Then pour in wine while your victim watches. The wine stays on top of the water, which you drink. Then hand the rest to your unsuspecting guest; only water remains.

18. Old wives' tales are difficult to extinguish. John White (1677) repeated this nonsense about garlic and magnets in his book, which promised to reveal secrets and promote modern ideas.

19. Some, including Gilbert, concluded that Porta may not have performed all of the experiments he claimed (Wootton 2015, 274, 328).

20. Mottelay (1922).

21. See the illustrations in Godwin (2009).

22. The presentation of these tricks reflects Porta's interest in theater. He authored at least seventeen successful stage productions (Kodera 2014).

23. Magicians quickly devised other variations of this trick by inserting iron filings into beans, chickpeas, almonds, and baked dough. A recently discovered seventeenth-century manuscript by an unknown author described placing iron filings in paper figurines of humans or insects. The insects moved about on a table, and the human figures stood on command because of filings embedded in their feet (Pieper 2013, 136).

24. I confess that I inserted a high-powered magnet behind the drywall in my living room during the construction of my current residence. I have never actually used this to amaze anyone, perhaps because I am concerned about being accused of invoking the devil.

25. Huber (2007).

26. For the learned swan, see Anonymous (ca. 1827), Anonymous (1835), Anonymous (1824), An Old Boy (1868), and Elliott (1873). For magnetic swan, see Pepper (1861), Wood et al. (ca. 1866). For educated swan, see Pinchbeck (1805). For wonderful swan, see Anonymous (1826). For sagacious swan, see Hooper (1794, 3: 200) and Devant (1931, 154).

27. Bailly (1987).

28. Hooper (1794, 3: 128). Toys with magnets have been popular from the earliest times (Anonymous 1824). Magnets are still used in magic, most effectively in situations in which no one would think of magnetism as part of the method. The current high-powered magnets provide amazing possibilities.

29. Lamont and Steinmeyer (2018, 22, 24).

30. We experienced a vestige of this history in 2013 when our family-chartered van stopped at a checkpoint entering Florence. The driver asked for our tax money to enter the city, which we did not understand because he spoke no English. Eventually, the checkpoint attendant, a young man who said he had lived in Chicago, explained, in utter

disdain of his own role as the fee collector, that the Italian government taxes everything, even breathing. He let us through without cost. I suspect that every visitor to Italy has a story to tell.

31. Thomas (1971).

32. Webster (1982, 89).

33. Gentilcore (2000, 294).

34. Gentilcore (2000, 313).

35. Their repertoire included such common effects as sleight of hand with a ball, a penetration trick, some fire-eating stunts, and a fake knife trick (Gentilcore 2000, 313). There were no unusual tricks here that would ordinarily suggest demon assistance.

36. Kieckhefer (1989, 91).

37. Maxwell-Stuart (2003, 14). Maxwell-Stuart notes that a seventeenth-century French attorney proclaimed that no magic is simply an illusion. His book aroused the ire of Gabriel Naudé, librarian of Cardinal Barberini, who responded with a fierce defense of many notable figures who have been accused of consorting with demons (Naudé 1657). Naudé's book illustrates the pervasiveness and diversity of demon accusations.

38. Massironi (2016). Horatio's book contained the first description of a stacked or prearranged deck of cards.

39. This deception was confirmed in 2016 using evidence of inconspicuous defects in Stagliola's typeset that are present in Galasso's book. Stagliola later became a member of the *Academy of Lynxes*, whose membership included Galileo.

40. James (1924, 22); James (1616, 107). These two versions, presumably of the same text, are somewhat different, but the strong language communicates the same conclusion.

41. James participated in torture of accused witches (Lacroix 1878). All sorts of civil and religious activities created risk for citizens. For example, one of the translators of the King James Bible, Richard Bancroft, archbishop of Canterbury, was responsible for the imprisonment of fifty-two Puritan separatists, some kept in filthy solitary confinement for three years before their eventual execution (Nicolson 2005, 90).

42. James was proud of his cleverness, and later in his life he saved many from false accusations. As an example, one woman failed to exhibit demon possession when the scriptures were read to her in Greek, thus suggesting no demon was present (Notestein 1911, 140; Webster 1677, 275).

43. Wootton (2015, 314). Also see Black (2009, 182). Taut (2020) cites the torture and death of a juggler (magician) accused of witchcraft as late as 1737.

44. Weir (1583/1964); Weyer (1998). Reginal Scot, no friend of Bodin, said that Bodin got all of "his Woonderfull tales of witchcraft" from his host at an alehouse (Scot 1584/1930, 279). In other words, it was all gossip gleaned during bouts of drinking.

45. Yates (1979a, 79ff).

46. Francis Bacon also considered salve a more likely explanation: "for it is certain that ointments do all (if they be laid on anything thick), by stopping the pores, shut in the vapours, and send them to the head extremely" (quoted in Smith 1831, 278). The

whole idea of witches' visions and ecstasies caused by salves or other natural means has a good foundation in the ergot poisoning and related toxins present in common grains (Ginzburg 1992b; Fuller 1968; Matossian, 1989).

47. LaVater (1804).

48. Black (2009, 182).

49. For insights into the troubles of Porta, I am particularly indebted to modern historian William Eamon (1994).

50. This is an example of magical thinking, namely that certain words or rituals by themselves can bring about a change.

51. What one priest identified as superstition might be acceptable to another. For the extremely confusing task of separating licit from illicit magic, see Bailey (2013).

52. A brief Google search suggests that the tarantula bite is not known to be fatal. Not considering this full range of natural responses is still a problem in modern medicine; incomplete information often leads to wrong conclusions (Dear 1990; Herbert et al. 2002).

53. Menghi (2002). The church maintained enormous power by managing a health care system and an army that fought invisible forces.

54. Bungener (1852).

55. Gilbert (1600/1958) said the anthropomorphic language of Porta was "the height of absurdity . . . the maunderings of a babbling hag." Gilbert also pointed out that Porta got his ideas about magnets from Sarpi (Mottelay 1922, 110). Tragically, Gilbert died in the London plague of 1603.

56. Harre (1970, 26). Robert Norman was the first to describe magnetic declination in 1581.

57. Gilbert's analysis made it possible for Kepler to use magnetic force as a guiding principle for his analysis of planetary orbits, namely to think about the physics of force rather than simple geometry. Similarly, Newton's theory of gravity was based on a force that could be measured. Although Gilbert criticized Porta, Galileo criticized Gilbert for not being well grounded in mathematics (Burtt 1955, 76).

58. Browne (1646) himself lived in both worlds of science and metaphysics: half-scientific, half-magical and half-skeptical, half-credulous (Willey 1949, 41). The dual habitat of Browne is evident even with the most cursory examination of his book.

59. Barfield (1985, 151).

60. Hall (1975, 312).

61. Butterfield (1962, 39).

62. Wootton (2015, 341) suggests the correct translation should be the "Academy of the Sharp-Eyed."

CHAPTER 8: EXTRAORDINARY VISION

1. Clavelin (1974, 119). Some philosophers argued that mathematics had no relevance to the physical world and natural science and, therefore, was not important to the

educational curriculum. Christoph Clavius, a critic of Galileo, countered this contention by citing the examples of timekeeping, surveying, and accounting (Lattis 1994, 36).

2. Some historians, like Ferris (1998, 87), have accused Galileo of falsely claiming creation of the telescope because he used the term "inventor" in the title of his book, *Sidereus nuncius*. But Hall (1975, 330) says an inventor then referred to anyone who made something for himself. The "primus inventor" was one who said nobody made this before. Also see (Rosen 1965, xviii).

3. Fermi and Bernardini (1965).

4. This brief quote is from a Stéphane Clavreuil Rare Books catalog. I was surprised to discover this book by Galileo's father for sale at the Boston International Antiquarian Book Fair in 2018. I was unaware that Galileo's father, Vincenzo, had actually published his research (in 1581). Stillman Drake cites Vincenzo's investigations as the inspiration for his son to experiment in a similar way and report in a similar style.

5. Drake (2001, 19–20).

6. This complex history is documented in the valuable reviews of Dijksterhuis (1964) and Lattis (1994).

7. Eisenstein (1980, 636). Jean Bodin, the witchfinder whom we met as an accuser of Porta, flew off the handle again when told of the heliocentric theory (Hirshfeld 2001, 42).

8. In the same year, Vesalius published his empirical study of the human body.

9. Boas (1966, 81).

10. Temkin (1975).

11. Whiston (1737).

12. Golinski (1999, 23).

13. Thorndike (1923).

14. Drake (1978, 19). In contrast to Drake, some believe the ball-dropping event is a legend. Typically, doubters concur with Ferris (1988, 84), who says, "it is almost certainly apocryphal."

15. Some Jesuits believed that Newton's conceptualization of gravity eliminated the need for God, thus construing Newton as an atheist (Fara 2002, 104).

16. Dijksterhuis (1964, 484). Isaac Newton credited Galileo for his ability to develop the laws of gravity.

17. Wootton (2015). Alchemists may have conducted experiments, but their investigations were usually uncoordinated and problematic in other ways. Many impressive technological accomplishments occurred during medieval times, especially in agriculture, horology, architecture, metallurgy, warfare, and the building of dams and windmills (see James and Thorpe 1994). Gimpel (1976) also makes a strong case for the advancement of technical skills during the medieval period, but science as conceptualized since the time of Galileo was absent with the possible exception of work by Grosseteste. Teresi (2002) and Al-Khalili (2011) have championed non-Western cultures for the beginnings of several scientific enterprises. However, it is unclear how many of these early discoveries were the product of science versus trial and error. I em-

phasize here that Galileo provided a new conceptualization of science to a broad spectrum of the population. He changed the discourse about the strategy for finding truth.

18. Occultist Robert Fludd (1574–1637) hated the idea of measurement; he believed secrets only resulted from inspiration (Vickers 1986, 11).

19. Grosseteste (1175–1253) and his Oxford colleagues designed some primitive experiments intended to falsify theories (Wallace 1981, 30). Additionally, Galileo conducted thought experiments similar to Einstein's (Ferris 1988, 90).

20. Wrightsman (1975, 235) points out that a hypothesis was not judged on its ability to match physical reality but on its internal consistency and on its agreement with accepted authorities. A theory's philosophical virtues were sometimes more important than how well it matched reality (Lattis 1994, 113).

21. At that time, the universe was referred to as "the world." The nova of 1572 and the comet of 1577 were important challenges to the Ptolemaic theory. Hawkins (1983, 177) says the 1604 supernova observed by Galileo can still be photographed as an expanding cloud of gas. A series of comets in the 1680s also terrified the public (Fara 2002, 7).

22. My favorite prediction of doom is the one that claimed London would be flooded in 1524, destroying ten thousand homes. That year brought a terrible drought (Research Officer 1930, 40). Thomas Pullein, an English Puritan, said that anyone who failed to see the working of God's divine will in the plague of 1603 was an atheist or an ignorant person (Nicolson 2005, 25). For modern nonsense, the predictions of Criswell (1968) should receive the award for most outlandish forecasts made by anyone not confined in a mental institution.

23. Adamczewski (ca. 1973, 85).

24. Hirshfeld (2001). Parallax measurement involves spotting a target from two different locations to determine the angles and calculating the distance using standard geometry.

25. Author's poetic license.

26. Freedberg (2002, 89).

27. On the power of an emotional story, see Konnikova (2016); Jackson and Jamieson (2007); Ariely (2008). Plato, of course, also used dialogue for teaching. But so did Bruno (1964), which might have been in the minds of inquisitors when they read Galileo's writings.

28. Langford (1966, ix).

29. Duncan (1998). Catholics were reluctant to accept heliocentricity because Protestants were accusing the Catholics of not following scripture, and the Bible says in several places that the sun moves, not the Earth.

30. Porta's important contribution in optics was the idea of placing a lens in a camera obscura. Van Helden (1977) has an excellent review of the early history of the telescope.

31. Thanks to Ron Friedland for pointing out that an observer would need to stand nearly forty-five miles high on land to observe a ship six hundred miles out at sea.

32. There were massive attacks on Italian cities in the last part of the sixteenth century by the Turks seeking captives to maintain their slave culture (Beeching 1983, 181; Egginton 2016, 76). The telescope was a welcome early warning system.

33. Freedberg (2002, 102).

34. Galileo was one of the few philosophers able (or willing) to work with his hands. He later gave up grinding lenses because others quickly surpassed his ability. Lens making soon became more dependent on machines than on individual grinding skills (Singer et al. 1957, 233).

35. Galileo said the six stars in Pleiades were actually forty, Orion had five hundred, and the Milky Way was "a congress of innumerable stars grouped together in clusters" (Panek 1998, 37). Theologians believed no unseen stars existed because they would have no use or purpose (Hofstadter 2009, 129).

36. Crombie (1959).

37. The moon was associated with symbols of Mary, the unblemished mother of Jesus (Shea and Artigas 2003, 44).

38. Galileo originally called these moons "stars" then "planets."

39. The movement of the Earth is even more complicated. Newton made the startling deduction that the Earth swiveled on its axis every twenty-six thousand years, like an unstable top. This was "nature's forces deeply hidden—and by Newton revealed" (Alder 2002, 92).

40. Randall (1963).

41. This argument was countered by noting what happens when a ball is dropped from the top mast of a ship moving in the sea.

42. Galileo's telescope was sufficiently powerful to determine that Mercury and Mars show obvious phases as they circle the sun. The stars are so far away that their distance was first measured in the nineteenth century. The angle is about the same as that created by the two sides of a small coin when viewed from three miles away (Gingerich 1992, 213).

43. The correspondence of Kepler back to Galileo is effusively congratulatory (Rosen 1965).

44. Kepler had polyopia, seeing multiples of a single object.

45. Hankins (1995).

46. Freedberg (2002, 108).

47. Shea and Artigas (2003, 41).

48. Galileo wrote a letter to Sarpi saying he wished they were together so that they could laugh at Cremonini's stupidity (Burtt 1955, 77). Also see Sharratt (1999, 87). Wootton (2015, 74) identifies several philosophers who refused to believe that Aristotle was wrong even when evidence was handed to them. In a letter, Vatican diplomat Piero Dini said: "Every day Galileo converts some of the heretics who did not believe him, although there are still a few who, in order to escape knowing the truth about the stars around Jupiter, do not even want to look at them" (Shea and Artigas 2003, 41).

49. Cremonini was himself pursued by the Inquisition for teaching that Aristotle denied the immortality of the soul (Wootton 2010, 106).

50. Freedberg (2002, 314). Emsley (2000, 4) says that the Bologna stone was not reported publicly until about 1640 by Vincenzio Casciorolo, an alchemist in Bologna, a couple of decades after Galileo had obtained some for his own investigations.

51. Phin (1906); Beckmann (1817, 4: 422).

52. As early as 1917, a sunspot was measured as rising 140,000 miles above the sun's disk (Gingerich 1989, 77). An early-twentieth-century book supporting spiritualism described sunspots as mountains on the sun (Duguid 1914, 537).

53. Anagrams were then commonly used as a means of establishing priority of discovery without precisely revealing what had been found (Drake 1978, 163).

54. Ptolemy explained the retrograde movements of planets as a complicated system of epicycles and eccentrics.

55. I accept Freedberg's version of this incident as more credible than the description in De Santillana (1961, 34).

56. Freedberg (2002, 121).

57. Bruce (1978).

58. Gilson (1938).

59. Grafton (2009).

60. In his book, Galileo acknowledged his debt to the Academy of Lynxes, which arranged publication. From that time on, Galileo proudly identified himself as a Lynx (Middleton 1971).

61. Hawkins (1983, 183); Wootton (2010, 121).

62. Freedberg (2002, 129). Neher (1986, 203) argues that Jewish exegesis of the scriptures was sufficiently flexible to accommodate Copernicus. Thus, no Jew was ever excommunicated or had his writing placed on the Index of Prohibited Books. Galileo pointed out in his letter to Duchess Christina that the Bible never mentions the planets except Venus, so opinions on them could not be heretical.

63. Freedberg (2002, 130).

64. Although Sarpi was a priest, Wootton (1983) makes a strong case that he became an atheist. Because of his important political and religious responsibilities, he was just one more Renaissance scholar who developed skills in dissembling.

65. Pope Paul V was not one to show mercy. He had a minor author beheaded on a bridge for treason just to demonstrate that he would show mercy to no one (Ranke 1847–1848, 2: 109). The word "congregation" refers to any committee of cardinals charged with an administrative task. Known by various names, the Congregation of the Holy Office has sometimes been called the Congress for the Propagation of the Faith, from which we get the word "propaganda."

66. Freedberg (2002, 131). The Council of Trent specifically condemned idiosyncratic interpretations of the Bible, a response intended to combat Protestant expansion.

67. The doctrine of "blind obedience" proposes that followers must believe what the church says to believe. For example, because everything in the Bible is true, then it

is necessary to believe all of it, even if not necessary for salvation. Bellarmine supported this theory when he interacted with Galileo.

68. Controversy still remains about exactly what Galileo was told and thus how truthful he was in his denial at the trial of 1633. Bellarmine was unavailable for clarification because he died in 1621 (Langford 1966). Weinberg (2015, 184) says that Galileo was given two confidential orders: a signed document that ordered him not to hold or defend Copernicanism and an unsigned one that went further by ordering him not to teach it. The second trial in 1633 was focused on whether Galileo had been loyal or disobedient to the demands of Bellarmine (Blackwell 1991, 131).

69. Sobel (2000, 256). Bronowsky (1975, 180) says that Galileo's book sold out before it could be seized by the inquisitors.

70. Langford (1966, 132). Galileo may have had the pope in mind as well, but his primary model for Simplicio was Cremonini. For a good source of information about other limitations faced by Galileo, see Shea and Artigas (2003). Blackwell (1991, 55) makes the case for Ludovico Colombe, who disputed Galileo's mountains on the moon, as the prototype for Simplicio.

71. Hofstadter (2009). Urban was concerned about being assassinated by pro-Spanish factions (Ferris 1998, 99).

72. Many of the archives of Galileo's trial were not made available until 1998, and Hofstadter provides insights not available elsewhere. Regarding torture, the transcript reads: "And he was told to tell the truth, otherwise recourse would be had to torture" (Hofstadter 2009, 186). The use of torture was so thoroughly accepted that Sebastian Castellio wrote under a pseudonym when criticizing the practice as "un-Christian." Even Castellio's publisher was sufficiently intimidated and the book was printed with a false imprint (Grafton 2009, 25).

73. Kenyon (1972) noted that seventeenth-century trials were not an attempt to ascertain the truth or to administer justice. They were morality plays, staged as a demonstration of institutional power, an affirmation of authority, and a warning to the unwary.

74. Galilei (1914).

75. Finocchiaro (1991, 14). Finocchiaro provides an excellent review of the many controversies raised by Galileo and the historians who interpreted these controversies.

76. Drake (1978, 436). Sarpi said "To give us the science of motion, God and Nature have joined hands and created the intellect of Galileo" (Burtt 1955, 74).

77. The Romans said, "What the Barbarians did not do, the Barberini did." Urban was of the Barberini family (Shea and Artigas 2003, 137; De Rosa 1988, 227). Ranke (1847–1848, 2: 264) says that Urban VIII spent his papacy strengthening military fortifications, creating a vast arsenal, and asserting a formidable military presence everywhere. He also commissioned multiple statues and portraits of himself (Baldinucci 1682/1966, 15).

78. Norwich (2011).

79. DeRosa (1988, 230) says Galileo's body was in the cellar of the bell tower, and I failed to verify his previous location when I visited this magnificent church.

CHAPTER 9: FIGHTING FOR MEANING

1. Hacking (1984, 47). Singer (1958, 164) describes and illustrates how early botanical texts emphasized the presumed utilitarian purpose of each plant. The results are often humorous and bizarre to the modern eye.

2. Shapin (1996, 18).

3. Hankins (1991, 145).

4. Findlen (1990, 43, 292–331).

5. Boas (1966, 51).

6. Freedberg (2002, 113).

7. Bishop James Ussher calculated that God created the world on October 23, in the year of the Julian Period 710 (i.e., 4004 BC). Wilcox (1987, 187) suggests this was not as much from religious zeal as an attempt to create a more absolute dating system so that Renaissance scholars could organize ancient historical sources.

8. Freedberg (2002, 151). The device was called an "occhialino," a name given to any instrument with a lens. A fellow Lynx called this inverted telescope a "microscope," and the name has endured. Gingerich (1999, 109) says Galileo's telescope was called a perspicillum and was first designated a telescope at a banquet in his honor in Rome in March 1611 (Sharratt 1999, 92).

9. Cornell ecologist David Wolfe (2001) has described the enormous variety of microscopic life that exists beneath our feet that we never see but rely on for our existence on Earth. His book changes our understanding of life.

10. Hutchison (1982).

11. Wootton (2007, 138).

12. Rowland (2004).

13. Printers facing censorship commonly backdated items, used pseudonyms, and omitted or altered the place and name of the printer. For Germany, see Crane, Raiswell, and Reeves (2004). For England, see Johns (1998). For France, see Darnton and Roche (1989).

14. Rowland (2004, 49).

15. Smith (1953).

16. Annius created eleven "wildly ambitious" chronicles attributed to eleven separate authors for what has been described as the most complex of pseudo-identities the world has seen. Olds (2015, 11) suggests that perhaps two-thirds of all ecclesiastical documents issued before 1100 were wholly or partially forged. The Annius forgeries promoted the idea that older societies were more advanced (Grafton 1992, 32).

17. For forgery and the Metropolitan Museum, see von Bothmer and Noble (1961). For the British Museum, see Jones (1990). For the Getty Museum, see Hoving (1996).

18. Jeppson (1971, 165).

19. Hoving (1996, 107).

CHAPTER 10: HIDDEN AGENDA

1. Ferrone (1995, 4). To his discredit, Galileo failed to cooperate with Kepler.

2. Kuhn (1970); Kuhn (1975). For a compelling criticism of Kuhn, see Sanbonmatsu and Sanbonmatsu (2017). Horgan (1986) offers the opinion that Kuhn and Popper were so highly admired that their ideas were taken to extremes that can no longer be supported.

3. Webster (1982, 59) identifies Porta's Academy of the Secrets of Nature as the first scientific society. Porta's academy had the spirit and goal of science but did not work under the constrictions of the scientific method or lay out the strategies subsequently accepted as those of science.

4. Beretta (2009, 1–16). Marchetti's treatises are now lost.

5. Campanella was imprisoned for about thirty-two years but also conducted magical rites of protection for Pope Urban VIII, who was terrified of eclipses (Wootton 2010, 194). When Urban turned against him, Campanella fled Rome, convinced that the Devil was more powerful than Christ's ability to protect him (Zagorin 1990, p. 257).

6. Beretta (2009, 152).

7. Weinberg (2015, 69).

8. Many of these ingenious devices are now on display in the Museo Galileo where you can still appreciate their intricacy and unexpected beauty. Pictures of their instruments are available online. Modern historian Liane Lefaivre (2005, 186) suggests that an appreciation of beauty was necessary for the development of science because it took attention away from the ugliness of personal sin and refocused minds on the mysteries of nature.

9. Middleton (1971, 69).

10. Clericuzio (2009, 17–44).

11. More recent studies show a small decrease in volume that would not have been evident with their instruments. What happens when you mix water with alcohol or acid? "If the sum of the volumes of the two separate liquids is 100, the volume of the mixture will be only 94" (Hopkins 1890, 3).

12. Middleton (1971, 158).

13. Some have blamed this misperception on Galileo's bad eyesight (Mueller 2013).

14. Fermi and Bernardini (1965, 52).

15. Turnbull (1993, 105). Huygens improved the telescope by discovering how to correct chromatic aberration, distortions caused by the variable diffraction of different colors of light through a lens. Huygens had a complex relationship with the Academy of Experiment because he was accused of plagiarizing Galileo's development of the pendulum clock (Landes 1983, 116).

16. Lanners (1977, 145).

17. In a recent study of optical illusions, Richard Gregory (2009, 162) proposed an additional reason for the confusion of Galileo and other early observers: they did not see rings because there was no concept of rings existing in space.

18. Haidt (2012).

19. For an utterly fascinating discussion of lone experimenters, see Honig (1984).

20. Vickers (1984).

21. Middleton (1971, 92).

22. Putnam (1962).

23. Bernardino Telesio (1509–1588) founded an academy whose purpose appears to have been the detection of Aristotelian errors in science. I find little information on this academy, but Telesio was interested in direct observation and empirical investigation as a reliable method of discovering truth. Similarly, the academies of Porta and Cesi deserve credit for their historical importance.

24. Shapin (1996, 134).

25. During the Renaissance, cardinals became rich from taxation, whereas parish priests made about the same as unskilled laborers (Dickens 1967). Being a Medici, Leopold did not need money, but the pope wanted Medici support.

26. The torture of Borelli was not related to his scientific ideas but his sexual misadventures.

27. McCartney (1920). For example, Aristotle presumed that lice could create nits.

28. Lloyd (1973).

29. Corrington (1961).

30. Beretta, Clericuzion, and Principe (2009, 209). Despite some of Malpighi's amazing discoveries, none of this knowledge led to better medical treatment (Porter 2000, 52).

31. Wilson (1995).

32. Phillips (1950, 18). Also see Hope (1961) for problems in Aristotle's physics. Fleming (2013) reminds us that Aristotle had some political ideas about slaves, women, and so forth that are shocking to our modern sensibilities.

33. Beckman (1971).

34. Sambursky (1963, 83).

35. I am omitting here the contributions of Descartes (Shorto 2008) and the French Academy of Sciences founded in 1666.

36. Zagorin (1990, 257). Dissembling was a common practice during the Renaissance. For discussions of concealment, see Long (2001); for lying and storytelling, see Black (2009). Tacitus, an early Roman historian, said "those who know not how to dissimulate, know not how to rule." Gabriel Naudé (1600–1653) stated something nearly identical (Minios 2012, 92).

37. Ornstein (1938, 41). Also see Fleming (2013, 46).

38. Kessler (1990).

39. Dickens (1969) concludes that scholasticism eventually died from its own sterile, abstract style because it was unable to incorporate the new findings of Galileo and the Experimenters.

40. Beretta, Clericuzio, and Principe (2009, 157).

41. De Santillana (1961, 259). Merton (1970) emphasizes the important influence of Puritans in the Royal Society who championed the idea that the study of nature was equivalent to glorifying God.

42. Bronowski and Mazlich (1960, 190).

43. Brockman (2017, 315).

CHAPTER 11: SOME ENCHANTED EVENING

1. I was distressed to learn that one survey suggested 55 percent of American adults do not know the Earth revolves around the sun once a year. Only 5 percent of eight year olds understand the Earth is a globe (Garwood 2007, 361). Pythagoreans understood the Earth was a globe, and belief in a flat Earth was rejected until the time of "parallax" (1873) in the middle of the nineteenth century. It has been commonly thought that Columbus believed the Earth was flat, but he did not (Russell 1991).

2. There were one or two "woefully flat and pedestrian tracts" intended for popular reading that appeared previously (Fontenelle 1990, xlv). The importance of Fontenelle was also endorsed by early American astronomer and physicist Samuel Langley (1877).

3. Fontenelle was not the only Frenchman to express anxiety about the new universe of Galileo. Descartes reluctantly acknowledged that he was not certain that the universe was made for man. No moral imperative appeared to be embedded in its structure, and it did not appear to provide a meaning for life (Shorto 2008). Pascal said that "the silence of these infinite spaces terrifies me" (Neher 1986, 187).

4. Panek (1998).

5. Toland (1747, 347). Bruno's mention of a dungeon was perhaps ironic because his heretical ideas earned him years of confinement in a dark cell before his blazing death.

6. Wootton (2010) says the exact charges of Bruno are unknown. Bruno's life was filled with dissimulation, even during his periods outside of Italy (Bossy 1991).

7. Fontenelle (1990).

8. Galileo's discovery of mountains on the moon created extensive discussion about the possibility of habitation but he avoided this controversy.

9. Ekstein (1996).

10. Ernst Mach, a pioneer in the philosophy of science, discussed the ways of instructing people about the forces in science. "Someone who knew the world only through the theater, and who came across the mechanical contraptions behind the scenes, would likewise come to think that the real world needs a backstage" (Sigmund 2017, 19). Fontenelle gave the marquise a backstage tour.

11. Fontenelle closely follows the arguments that Galileo provided.

12. Fontenelle (1753, 63).

13. Monet (1930, 211). Casanova told Voltaire that everyone was talking about Fontenelle's book on plurality. Timbs (1856, 8) cites *Edinburgh Review*, no. 208, as saying

that Copernicus acknowledged that other worlds are capable of affording conditions for the highest kinds of life.

14. Ferrone (1995, 2).

15. Ferrone (1995, 7).

16. Porter (2000, 72).

17. Darnton and Roche (1989). Protestant England was not always a model of tolerance. The Church of England and dissenters expressed violent hostilities toward one another. Eight thousand dissenters died in prison during the reign of Charles II. In his discussion of book burning, Farrer (1892, 132) says that Protestant England was much the same as Catholic France. Also see Nicolson (2005).

18. Eisenstein (1980).

19. Fontenelle (1990, xviii).

20. Adkins (2000).

21. Wootton (2010, 249); Wootton (1983, 10).

22. Others had difficulty keeping religious issues out of the plurality debate. As late as the 1850s, William Whewell, an Anglican who denied plurality, and David Brewster, Scottish Presbyterian, created a storm of controversy about life on other planets (Whewell 2001).

CHAPTER 12: MEN ON THE MOON

1. De Bergerac (1962). In 1661, after the time of Bergerac, at least one author reported that the telescope of Galileo showed the moon to be "another America, full of pleasant rivers, hills and dales, and also well inhabited with people (such as they are) namely, Lunatick people" (Anonymous 1661).

2. The works of Cardano inspired Bergerac to write his fantasy travel (Howgego 2013, 108).

3. I took these quotes from scholarly journals, and I'll spare the authors some embarrassment by not providing their names. Take my word for it: don't read this stuff unless your dissertation is on Bergerac. Perhaps these authors would have lived much happier lives if they had joined the generations of academics who analyzed Winnie the Pooh (Sibley 2001, 123).

4. Swedenborg (1840, 4).

5. Swedenborg (1875).

6. Maudsley (1897).

7. Johnson (1994).

8. A biography published by a denomination honoring Swedenborg omits any mention of his visions of other planets. They hail some examples of his clairvoyance but ignore his embarrassing psychological difficulties (Trobridge 1928).

9. Similarly, Kant (1900) considered alternative ways our world might be perceived on Jupiter or Saturn.

10. Quoted in Owen (1860, 21).

11. Raspe's first edition did not have the extraterrestrial travels, but these adventures appeared in Raspe (1811). The baron was based on a real-life individual that Raspe knew from his college drinking days in Germany (Carswell 1950a, b).

12. The Munchausen tales were not immediately successful but soon became widely popular in England and Germany. The appellation Munchausen was subsequently attached to any extraordinary liar and then associated with outrageous medical deceptions (Pankratz and McCarthy 1986). Munchausen by proxy became the label for mothers who harm their children or lie about their medical condition (Pankratz 2006, 2010).

13. Trobridge (1928).

14. Fleming (2013, 381). For a brief but informative biography of Jung Stilling, see *The Spiritual Magazine* 3 (1862): 289–309. Johann was sometimes known as Heinrich Jung and sometimes by his assumed name, Heinrich Stilling. He became well known for his cataract operations.

15. Jung-Stilling (1851, 227). First published in Germany in 1808. Although not published in America until 1851, it conveniently fit into the cultural landscape in a way that helped the Fox sisters' stories of spiritual communication appear natural.

16. Jung-Stilling (1851, 46).

CHAPTER 13: SPACE RACE

1. Watts (1883).

2. Kerner (1845).

3. To learn how to mesmerize water, see Davey (1889, 47, plate 7). To magnetize and identify, see Deleuze (1886, 292).

4. Anonymous (1837).

5. Werner (1847, iv). The English translation contains only two cases of his visionary girls.

6. Cahagnet (1851).

7. Podmore (1902, 82).

8. Cahagnet (1851, 207).

9. Sandby (1850).

10. It has long been known that several brain abnormalities are prone to experience visions and paranormal experiences (see chapter 18 on mesmerism). However, even certain normal brains are more likely to make paranormal interpretations of ambiguous events. See the fascinating research of Swiss neuropsychologist Peter Brugger (Brugger and Taylor 2003). Society is better with different types of brains so that we don't all think alike.

11. A repeated theme of my book on patient deception is that the flaw of the impostor is overacting (Pankratz 1998).

12. Hockley (1850).

13. Davis (1855). The formal education of Davis was scarcely more than reading, writing, and the basics of arithmetic, which his supporters insisted verified his claim of receiving revealed wisdom.

14. Crabtree (1993, 210).

15. I believe I have read that Jung-Stilling was a direct ancestor of Jung, but I have no reference for it.

16. Jung's dissertation (*Zur Psychologie und Pathologie sogenannter occulter Phänomene: eine psychiatrische Studie*) was published as a book in Leipzig but never translated as far as I know.

17. Noll (1997, 41).

18. Druckman and Bjork (1991, 89). Also see Pittinger (1993).

19. Corson (1919). Indian guides often spoke in laughable accents (Jewett 1873, 51); one gave war whoops that startled everyone (White 1854). One little Indian girl spoke fluent English and provided no indication of any Indian language or culture (A Plain Citizen [1918], 146). Crowell (1879) gives Indians their own place in heaven "where no white man robs [him]." To read the teachings of Silver Birch, see Austen (1938). Sadly, the portrait of Silver Birch on the frontispiece is not a photograph but a painting. Harding (ca. 1930, 9) says that the red Indian makes a perfect spirit guide because "before the coming of the whites, the minds of the native Americans were attuned in a simple and genuine harmony with the Infinite Spirit in all things." She must have been reading Rousseau.

20. Anonymous (1858a); Hensley (1921); Petersilea (1889).

21. Flournoy (1900, 194).

22. Sarah Thomason, a professor of linguistics at the University of Pittsburgh, reviewed similar claims of ability to speak an unlearned language (see her summary under "xenoglossy" in Stein 1996).

23. Corson (1919).

24. Lender (1903).

25. Smith (1860).

26. Davy (1836).

27. Spiritualists made many scientific pronouncements that were not useful or meaningful. For example, no benefit came from being told to study the vibrations in ether (C. E. D. 1922), which was not surprising because ether had been dismissed by the Michelson-Morley experiments of 1887. Ben Franklin was a popular person to contact in séances, but his messages were mostly about how to create a new social order (Sconce 2000).

28. Herschel had a twenty-foot mirror that he transported to South Africa (Snyder 2011, 159). Today, astronomers are looking for objects that are some ten billion times fainter than what can be seen with the unaided eye. The Hubble telescope would be able to see the flicker of fireflies in the dark recesses of the moon (Kanipe 2006, 5, 26).

29. Locke (1859, vi). Kevin Young (2017) provides a compelling case for racial subtexts of this and other American hoaxes.

30. Griggs (1852, 30).

31. The writings of Herschel (1849) show how far astronomy had advanced by this time.

32. No one noticed that Locke inadvertently used an older name of the journal. False stories consistently attract individuals willing to confirm the event, no matter how implausible.

33. Snyder (2011, 305). We should be equally tired of the hoax perpetuated by conspiracy theorists that the Apollo moon landing never happened.

34. Locke (1835). WorldCat shows only one London and one New York reprint. People threw them away and libraries considered them unworthy.

35. Sharps (2018).

36. Flammarion's novel *Urania* was first published in 1889 with the first English edition in 1891. He provided several reasons why civilization on Mars was superior to ours (Flammarion 1923, 118).

37. Crossley (2008).

38. As late as 1926, the Second International Psychical Congress in Paris debated the possibility of contacting martians. In England, solicitor Henry Mansfield Robinson invented a psychomotormeter to contact the martians. English mediums, scientists, and crackpots (one wrapped copper wire around his body) all attempted contact with the red planet (Morris 2006, 83).

39. Lowell's books all promote his findings of canals, and they were all beautifully bound (Lowell 1895, 1906, 1908).

40. Blackwelder (1909). Also see Crossley (2000).

41. Lane (2006, 207). Some astronomers continued to believe in the existence of canals as late as the time of Webb (1957). Lowell made significant contributions to astronomy, but his legacy is more literary than scientific.

42. Taylor (1922).

43. David Langford (1990) confessed that he forged a book that detailed the appearance of extraterrestrials in 1871. This fiction then became incorporated into the works of other authors, like Whitley Strieber (1988), and promoted as authentic.

44. Vallee (1979).

45. Telano (1960, 41).

46. The fascinating book of Donna Kossy (1994) defines a kook as "a person stigmatized by virtue of outlandish, extreme or socially unacceptable beliefs that underpin their entire existence."

CHAPTER 14: POSTER BOY FOR CLAIRVOYANTS

1. Gauld (1995, 235).

2. Many have written about the abysmal hit rate of psychics. See for example Nickell (1994).

3. Winter (1998).

4. Méheust (2007). http://bertrand.meheust.free.fr/documents/conf-esalem.pdf (May 14–18, 2007). Later I discovered a poor-quality video of Méheust giving a speech on Didier: http://vimeo.com/37790155_(February 17–18, 2012).

5. Devant (1931, 174) quoting the *Illustrated London News*.

6. Wallace (1898). This quote was also reproduced in a pitchbook of Will Goldston (1905, 22), a contemporary of Wallace and the most popular conjuror in England. The endorsement remains in the awareness of the magic community to this day (Hansen 1990).

7. Gauld (1968).

8. Dawson (2012). Townshend also presented mesmerism to the faculty at Harvard (Putnam, [1874]).

9. Townshend (1852, 407).

10. *The Zoist* was published in London in thirteen volumes from 1843 to 1856. Didier arrived in England in 1844. Marcillet and Didier made an additional trip to England (mostly Brighton) in 1849, and reports of those séances can be found in Lee (1866, 255–80).

11. Elliotson (1845).

12. Prevost (1851–1852).

13. Elliotson (1852).

14. Sims (1845).

15. Hart (1894). Subjects were at risk from personal strains and from unstable conditions that could cause a fall. Knowles (1914) admitted that accidents resulted from a failure to select sufficiently strong furniture, and, indeed, many pictures in these old pamphlets show unstable situations. Herbert Flint accidentally killed an assistant who was helping him with the rock-breaking stunt (Price 1985, 477). Some pamphlets on "how to hypnotize" were probably sold to gullible spectators, and the authors would not want to tip the reality of the planking stunt. Bonomo (1952) provides specific techniques and tricks for less painful and dangerous demonstrations.

16. Baldwin (1895) describes acrobats performing this feat in England not many years after Didier, and he understood the critical physics. Also see Trudel (1919); Anonymous (1847b). Ricky Jay (1986, vii) says that an East Indian magician managed a seven-hundred-pound stone on his chest, which was smashed with a sledgehammer. Loyd (1927, 100) indicates that magicians performed this feat with a large rock on their *head* with only a small blanket between for support.

17. Burlingame (1895, 104). The Nelson Enterprise catalog no. 12 (believed to be his first) offers to sell the "complete a to z dope on this pseudo-hypnotic test." For $5, the instructions include methods that allow the performer to post a reward of $100 to anyone who can duplicate the feat. He may have included a method for constructing the corset, which spectators would not know (Nelson 1929).

18. Beckmann (1817, 3: 284). Also see Arnold (1862, 334) and Tegg (ca. 1860, 176).

19. Fisher (1980). For strongman feats, see Trudel (1919), Coulter (1952), and LaVelle (1981); for sideshow tricks, see Carrington (1913), Dean (1937), and Shaw ([1896]); for fire eating, see Barnello (ca. 1880); Barnum (1919); for heat resistance, see Anonymous (1859) and Houdini (1920); for endurance and torture feats, see Scot (1651), Carrington (1909c), and Hunt (1934). The thirty-six issues of *Swami: A Monthly Magazine of Exotic Mysteries* (1972–1974) and the thirty-eight issues of *Mantra* all contain exquisite details of traditional Indian feats of torture and endurance. These volumes were reprinted by Richard Kaufman in 1997.

20. Forbes (1845b, 64). I am surprised that Forbes didn't find some fit person and test his belief.

21. For a review of modern misuses of applied kinesiology, see various examples at www.quackwatch.org.

22. Hurst (1897). Her real name was Lula, and she died in 1950 at the age of eighty-one in Madison, Georgia. Lulu said she started performing at the age of fourteen, but Wiley (2004) concluded she was fifteen.

23. Lulu offered minimal explanations for her powers, so people provided their own theories including electricity, human magnetism, Reichenbach's odic force, or assistance from spiritual powers. Some were certain they saw sparks emanating from her feet. Most resorted to supernatural or occult explanations, which is common when confronted with an anomalous event. These false attributions disturbed Lulu, which contributed to her decision to stop performing.

24. Dexter (1956); Joseph (ca. 1940); Hull (1946). Robinson (1898, 121) even has a blindfold method for jugglers.

25. Anonymous (1810).

26. Kirby (1815, 6: 353).

27. Sandars (ca. 1917, 22)

28. Rauscher (2011). Harry Price (1936a) tested Bux only to conclude "how extremely difficult it is to blindfold a person, using ordinary methods." Blindfold performers easily fool individuals who examine them. Jules Romains endorsed Bux, in part, based on his testing before physicians in London and Montreal. (First published in Paris in 1921, then in London in 1924, and then reprinted in Romaines 1978.) Tarbell (1954, 6:251) exposed multiple methods of eyeless sight that fooled Romains.

29. A portion of his act can be seen here: www.youtube.com/watch?v=q6x_zO0IIsE (March 8, 2007). Another version is here: www.houdinifile.com/2012/06/ghosts-of-magic.html (June 28, 2012).

30. Sperber (1982, 22).

31. Gardner (1981).

32. Coyne (1954). Coyne enjoyed shocking his audiences by removing his glass eye, holding his eyelids apart, and walking into their midst so everyone got a closer look at the grim details.

33. www.youtube.com/watch?v=_zv7ZdlO0pk&feature=relmfu (May 7, 2012). There are similar clips of Coyne's evangelistic meetings online.

34. www.jehovahs-witness.net/watchtower/scandals/146208/1/Eyeless-vision -Ronald-Coynes-eyeless-vision-act (accessed December 13, 2019). This same site reproduces an essay by Martin Gardner that ridicules Coyne's act.

35. Méheust obtained a PhD from Sobornne University in 1997 but lived most of his life outside the academic setting as a popular writer. His paranormal views are well reviewed in Kripal (2010).

36. "There are no stones in the sky; therefore, none can fall upon the earth" (Hubbell 1901, 112). Then in 1803, one fell in France, witnessed by several thousand people.

37. Lapponi (1907, 15).

38. Goldsmith (1934, 144).

39. Gauld (1995, 137).

40. Berna declared that "conditions agreed upon between us as necessary were constantly violated, and they slyly avoided under one pretext or another, in spite of my earnest reclamations" (Leger 1846, 177).

41. For the French prize money, see de Courmelles (1895, 17). For the English offer, see Carpenter (1877).

42. Forrest (1999) provides an extensive review of testing eyeless vision.

43. Kripal (2010, 219).

44. Prater (1846).

45. Wootton (2007) makes the case that physicians have mostly been dangerous to the health of patients and that modern medicine did not begin until Lister's germ theory in 1865.

46. Prater (1851).

47. Prater (1851, 27).

48. Dr. John Forbes noted the same facial movements when he tested Didier's brother, Adolphe (Lamont 2012).

49. Prater (1851, 25).

50. Despite Prater's warning, observers continued to insist their subjects were properly blindfolded. For example, "Her eyes were, without doubt, IN TOTAL DARKNESS —in regard to that, there is no possible mistake; but notwithstanding all our precautions in bandaging, SHE DID SEE" (Newman 1850, 164, emphasis in the original). In a similar test, a Mrs. Blair was able to paint beautiful flowers while blindfolded. Unfortunately, her success was thwarted when a card was placed between her eyes and the canvas (Grimes 1885, 328).

51. In 1858, French clairvoyant Prudence Bernard, who was reported to be even better than Didier, confessed that she colluded with her magnetizer through the use of silent communication (Dingwall 1967, 157).

52. Townshend (1944, 170).

53. Dingwall (1967, 104, 130, 143, 165). Also see de la Motte (1843).

54. See "Blindfold" in Whaley (1989).

CHAPTER 15: MIRACLES OR ENTERTAINING TRICKS?

1. Kalush (2002).
2. Giobbi (2010).
3. Kaufman (2011).
4. Farmer (2014).
5. Forte (2020).
6. Those who doubt that cards can be manipulated such that someone could know the opponent's hand should watch the TED presentation by Lennart Green. Green shuffles and bungles a deck of cards so thoroughly it appears that no single card could be at any known position. Yet Green deals five poker hands, each with a higher value, ending with a royal flush for himself. His style is so engaging and impossible that his audience is reduced to laughing in disbelief. See www.ted.com/talks/lennart_green _does_close_up_card_magic.html (February 2005). Also see Stone (2012).
7. Randi (1982, 254).
8. Randi's reward for any paranormal, supernatural, or occult power was subsequently raised to $1 million. Because of Randi's advanced age, his offer was withdrawn; however, he cited nine other current offers of considerable money for any paranormal demonstration (see https://cfiig.org/ for the Center for Inquiry Investigations Group). Randi has been criticized for his sometimes abrasive and confrontational style, and Storr (2014, 279) describes his manipulation of an agreement that violated the spirit if not the content of one challenge. Randi acknowledged that he lied and became angry when confronted.
9. Ray Hyman (personal communication, February 5, 2013) told me that this is the only time, as far as he knows, that Martin Gardner ever directly participated as an investigator in testing a claimant of psychic ability.
10. Walker wrote what is considered the first of the coney-catching pamphlets: *A Manifest Detection of the Most Vile and Detestable Use of Diceplay, and Other Practices Like the Same* (1552). At that time, gambling was so common and pernicious that Walker said, "the contagion of cheating is now so universal that they swarm in every quarter, and therefore ye cannot be in safety from deceit, unless ye shun the company of hazarders, as a man would flee a scorpion."
11. Burlingame (1891, 146).
12. Cumberland (1918, 108). For other methods of separating the color cards, see Karr (2019).
13. Wilson (1877/2015) reprinted this rare book, which is in the collection of the Stanford library.
14. Little (1899). I'm certain there are many similar effects in nineteenth-century books on card magic.
15. Musson (1937).
16. https://m.youtube.com/watch?v=bRgCvCTG_XQ (October 21, 2016).
17. Dingwall (1967, 169).

18. Forbes (1845a, 33). If Didier could read the back of some cards, why would he not be able to read all cards? Mesmerist John James (1879, 56) said that "his power seemed to come in flashes and then suddenly depart." Invoking "power" explains nothing. A better way of explaining this is to say that he was successful when he could cheat and then his "power" failed when he could not.

19. Forte (2004); Forte (2006).

20. Dingwall (1967, 104, 130, 143, 165). Also see de la Motte (1843), who noted that when a piece of paper was placed between the eyes and the designated object, the subject could not identify it.

21. I said that Townshend indicated Didier read eight pages ahead. Actually, he said eight *leaves* ahead, so my example does not specifically explain Didier's method. Townshend provided other examples that are not easily explainable given his description, and they become doubtful when you consider the test of Forbes described below. Lee (1866, 163) says that Didier was frequently requested to read ten, fifteen, or any number of pages further on. He was "sometimes wrong" about the pages, but he was given credit for reading the words.

22. Agnew (2002).

23. Bleeding was still practiced for gas attacks in World War I. Wootton (2007) says the idea of counting success and failure of bleeding as a treatment is so simple, yet doctors were slow in adapting because it required a cultural shift.

24. April 1845 and October 1846.

25. Carpenter (1877, 85).

26. Carpenter (1877, 77).

27. Forbes (1845b, 6).

28. Bernstein (2012).

29. In a private magic publication, a mentalist described a situation in which he obtained knowledge of a young lady from her father, who wanted to play a practical joke on her. The revelations seriously frightened her. They were not entertaining.

30. Placing text on the stomach provides the perfect position for the magician's peek-down-the-nose; see Durant (1837/1982, 59).

31. Lee (1866, 108).

32. Jung-Stilling (1851, 32). Barrett (1911) described a clairvoyant who apparently had unclosed skull sutures, through which he supposedly read messages.

33. Forbes (1845a, 79) declared this was the method used by George Goble, a pretending clairvoyant whom he tested.

34. The *Oxford English Dictionary* suggests that this is one of the earliest uses of the word fishing in reference to searching for answers, but the term was presumably first used with reference to lawyers.

35. Elliotson (1845).

36. Dingwall (1967, 192).

37. Craft (1881). Burlingame ([1905]) describes how the mentalist can determine answers by the sound of the pencil.

38. J. D. (1886).

39. Forbes (1845b) ensured that boxes would not be opened by screwing them closed and covering the screws with wax. Even more cleverly, he filled one box with pieces of cork that flew out when opened. Using this strategy, he obtained a confession of cheating from pretending clairvoyant George Goble.

40. Burlingame (1891).

41. Wiley (2009).

42. Wiley (2005).

43. Burlingame (1905).

44. Beard (1882, 14). For more on muscle reading, see Burlingame (1905), Fitzkee (1935), Hume (1886), MacAire (1889), Masters (1882), Morrow ([1914]), Schatz (1974), and Skinner ([1895]).

45. Bishop (1880) sometimes exposed spiritualism and mental tricks in his lectures, often making more money through pretended skepticism. Frederick Wicks (1907), his ghostwriter, later made some minor revisions and published this book under his own name. A pitchbook is any pamphlet or printed material sold by performers as a way of making extra money. They were usually thrown away, making them highly desirable for modern collectors.

46. Cumberland (1918); Cumberland (1919).

47. Cumberland (1888).

48. The next issue of the *Proceedings* declared the Creery sisters as the real thing. Their father later admitted that the family communicated through simple codes. To close the story on Bishop, he fell into a deep coma after a strenuous performance in 1889. He carried the equivalent of a medical alert card that directed his care in such situations, but the two doctors attending him pronounced him dead and performed an autopsy while his body was still warm. He had bragged that his brain was like no other, but they found nothing unusual. Magic historians have speculated about the gruesome details because Bishop had stated he was aware of everything around him but unable to move or speak while in his coma (Jay 1986, 159; Price 1985, 455; Christopher 1970, 58).

49. Hall (1964); Gauld (1968, 179). Blackburn said that the Society for Psychic Research was as easy to deceive as children (Hall 1980, 92).

50. Lee (1843). Going back over the records, I was surprised how often I missed the fact that Didier was noted to be holding someone's hand. James (1879) repeatedly mentions Didier's hand holding. Later spiritualists also were known to hold the hands of sitters to obtain information (Brandon 1983, 211).

51. American humorist Artemus Ward (1865, 142): "Calls himself Perkins, or Simpkins, or something like that," replied Mr. Wilder. "Some name that don't amount to shucks. We must make a 'Professor' of him. I don't go on any conjuror if he's not a professor."

52. Telepathist (1926) reported a Baldwin séance from 1899. Baggally (1917, 64) provides a good review of the Baldwin show.

53. Baldwin (1879); Baldwin (1895). Also see Cadwell (1883), whom Baldwin endorsed.

54. Anonymous (1893).

55. I obtained this information from a single-sheet description with no identification as to author or publisher. Byron Walker searched several of his older magic catalogs and suggests it was probably written by Ralph Sylvestre (pseudonym for H. J. Burlingame). Sylvestre's catalog no. 16 lists "Professor Samri S. Baldwin's Complete Manifestations" for $25, a considerable amount. Walker is the owner of the original Burlingame library, which is the world's oldest collection of magic (Whaley 1990, 58).

56. Anonymous ([1865], 287).

CHAPTER 16: REMOTE TRAVEL

1. Du Prel (1889, 155) said that "somnambulism is not the cause of clairvoyance but the condition without which it cannot arise."

2. Ogden (2018) provides an excellent example of how William Stone unintentionally provided information to travelling clairvoyant Loraina Brackett, who then convincingly elaborated those details back to him. For example, Stone asked, "do [the statues] resemble lions?" Brackett agreed and then expanded the narrative by saying the statues were *bronzed* lions. Stone became convinced of her clairvoyance after a series of similar interactions, completely unaware that he had been leading her, providing the information she needed. The same process still occurs with psychic readers today.

3. The list of references on the history of visions could be endless. For a sense of the range, see Anonymous (1707), Anonymous (ca. 1866), Cheiro (1928), Davis (1866), Mother Shipton (1641), Nixon (ca. 1818), and Oldmixon (1746). During the late nineteenth and early twentieth centuries, a flurry of books of prophecy were published based on measurements of Egyptian pyramids. The twenty I have must be only a partial list; see, for example, Casey (1876). Prophesies based on the Bible are continuous throughout the Christian era, none of which has ever been fulfilled so far as I know.

4. Second sight was described as a gift or a curse, not a skill. For a small sample of nineteenth-century books explaining how second sight can be so convincing, see Anonymous (1874), Anonymous (1884), Anonymous (1899), and Hunt (ca. 1890).

5. Bond (1728, 286). Campbell may have obtained information by lip reading, which his subjects would not likely have encountered.

6. Harry Price (1934) attributed this book to Defoe. Also see Watt (1965, 102) and Lee (1866, 293).

7. Defoe (1835).

8. Insulanus (1763). Reprinted edition, Insulanus (1819). Even earlier, the credulous John Aubrey (1857) provided examples of specific visions and predictions of highlanders.

9. When Dr. Johnson and Boswell toured the Western Isles of Scotland in 1773, Boswell remarked that their adventure was much the same as being with a tribe of Indians in America because of the villagers' wild and savage appearance (Burke 1981, 8).

10. Lee (1866, 34).

11. Gregory (1877, 111).

12. Colquhoun (1851, 16ff).

13. Goldsmith (1998, 22). Horace Greeley offered $2,500 if someone would provide the daily London news headlines every day for a year (Mattison 1853). For excellent descriptions of the telegraph as a metaphor in spiritualism, see Luckhurst (2002) and Sconce (2000). Some spiritualists were convinced that clairvoyance was as certain and reliable as the electric telegraph (Barkas 1862, 21).

14. Townshend (1854, 215).

15. Townshend (1844, 361).

16. Lee (1885, 115).

17. *The Mesmerist* 1 (1843): 150.

18. *The Zoist* 2 (1845): 481.

19. Lee (1866).

20. Carpenter wrote some inaccurate and unfair accusations against Sir William Crookes in 1871 (Wyndham 1937, 245; Lamont 2006, 218). His criticism of Didier, however, was fair and based on observations.

21. *The Zoist* 2 (1845): 491.

22. *The Mesmerist* 1 (1843): 153.

23. E. W. C. N. (1844).

24. Ashburner (1867, 171). "The party assembled had scarcely a good head among them." Ashburner had confidence in phrenology because he confirmed that each "organ" of the head had its unique swing or rotation when tested with a hand-held pendulum (67).

25. I'll be watching when the BBC produces a Masterpiece Theatre series entitled *Reverend Sandby, Vicar of Flixton.*

26. Also see Newnham (1830), who said clairvoyance was a dysregulated state of the brain, making it unreliable. He cited numerous mistaken predictions arising from dreams and clairvoyant declarations.

27. Quoted in Wootton (2007, 145).

28. I found one report on Didier that explicitly described multiple failures in his remote travel attempts (Lee 1843).

29. Beard (1879b)

30. Prater (1851, 58).

31. Carpenter (1877, 94).

32. For testing psychic claimants, see Wiseman and Morris (1995).

33. Durant (1837/1982, 116) noted that whenever the descriptions were wrong, the clairvoyant was told he was in the wrong house. Thus, he went outside and entered a

different house to find the right place. We do not know how many of these false starts were just ignored.

34. Rice (1849, 252).

35. Rice (1849, 265). Soal (1937) said his telepathist's "statements are small masterpieces of evasion." James Randi recorded the psychic readings of famed clairvoyant Peter Hurkos, who claims his powers are 87.5 percent accurate (Randi 1980, 7). When asked, his subjects gave glowing accounts of his success. But when the tape was played for them in a neutral setting, the hit rate was one in fourteen. In an early American setting (Durant 1837/1982, 115), a man said the clairvoyant was correct but later admitted she was so completely wrong that he was embarrassed to say so. It might feel impertinent to tell someone they are wrong in a group setting, thereby leaving observers with a false sense of the clairvoyant's ability.

36. Perhaps this is similar to how people today ignore the long list of potential side effects of advertised medications. The screen shows pictures of individuals in glowing health engaged in pleasurable activities while the voiceover describes unending potential risks.

37. Dingwall (1967, 155). Also see *The Mesmerist* 1 (1843): 88.

38. Hardinge (1870, 532) describes a few "useful inventions" that emerged from communications with spirits. These are nothing more than the next evolution of mechanical devices improved through Yankee ingenuity. Amanda Jones (1910) shared credit with spirits for her patents on an oil burner and food processing method, but the spirits failed to assist her in her business enterprises, which mostly failed. I have seen other discoveries attributed to spirits (like oil wells in Pennsylvania), but these are not much different from attributing any successful outcome to God. Wolfe (1875, 257) said that inventions on Earth (like the telegraph) are available in heaven, but "we have them more perfect than you." Wolfe claimed that spirits gave Morse information so that he could construct his successful machine. But Morse was dead by then, so no one could check that small detail.

39. Hypnotists used the term "horse" as a reference to their shills.

40. I was not surprised to discover that Didier became a stage comedian after his life as a clairvoyant was over (Lee 1866, 164). The point here is that he was always a good entertainer. He died in 1886 at the age of sixty, probably from liver cancer.

CHAPTER 17: WHO IS BEING TRICKED?

1. Robert-Houdin's sobriquet, father of modern conjuring, is not without critics (Lamont and Steinmeyer 2018). He may not have been first in originality, mode of dress, or presentation style; however, he designed a performance for the audiences of his day with unprecedented success. Houdini (1908) bestowed the "father of modern conjuring" moniker on Wiljalb Frikell (1818–1903), a Prussian conjuror who began touring the world at the age of sixteen.

2. Ether was introduced as an anesthesia in the 1840s, so the implications would be fresh in the minds of his audience.

3. In his autobiography, Robert-Houdin says he introduced this routine in February 1846 (Robert-Houdin, ca. 1903). However, Fechner reprinted a broadside that dates it to November 1845.

4. Fechner (2002, 261).

5. Scot (1584/1930, 191).

6. Wiley (2012, 174).

7. C. S. Lewis, usually after considerable consumption of alcohol, was known to regale his friends with the idea that it was terrible to remember nothing but even worse to forget nothing. He then solicited the most skeptical individual to select any book from any of his bookcases, which included works in several languages. The skeptic opened to the page of his choice, read aloud, and stopped at any place. Lewis then quoted the rest of that page from memory (Como 1994). Several mediums and entertainers were known to have prodigious memories and unusual skills that baffled their audiences.

8. Stagnaro (2004).

9. This simple example was taken from Morrow ([1914]). When categories are numbered, the psychic can similarly identify months, days of the week, colors, the value of money, articles of clothing, and so forth.

10. Despiau (1801, 278),

11. Fulves (1979) describes an easily learned nonverbal method of communication using simple finger signals. Verbal and nonverbal coding systems were described by Wilkins (1641) and Falconer (1685), but these were primitive and cumbersome. Giovanni Della Porta wrote an important work on codes in 1563. Garenne ([1886]) devised a silent code using hand motions that is surprisingly efficient.

12. With practice and a system for association, an amazing amount of information can be quickly memorized. Memorization was more common in the days before the printing press (Wright 2007, 126).

13. Fechner (2002, 286–89). Also see Karr (2006).

14. Dingwall (1967, 156).

15. We should not be too hard on believers who failed to understand the meaning of Robert-Houdin's endorsement. Some notable skeptics also believed that Robert-Houdin had fallen for Didier's tricks (Mann 1919). Robert-Houdin was apparently once stumped by a particular psychic trickster (Dingwall 1967, 157). The Penn and Teller show, *Fool Us*, illustrates the ability for cheats to confound the most informed.

16. Robert-Houdin (ca. 1878).

17. Robert-Houdin (n.d.).

18. Why would Robert-Houdin mention several ways that cards could be faked for reading by touch alone? Perhaps he was trying to convey to Mirville a way the feat could be accomplished. Didier could have been using—or substituted—a prepared deck, even with the help of his manager, Marcillet.

19. Robert-Houdin (ca. 1903).

20. Robert-Houdin's endorsement of Didier was presented to the French as well as English (Flammarion 1900a, 248; Flammarion 1900b).

21. Mann (1919, 6–7). Similarly, David Hume said that if we witness a miracle, then we should be prepared to doubt the testimony of our own senses (Mangan 2007).

22. Reverend A. M. Creery (1887) confessed that his daughters cheated, but he concluded his letter by mocking the members of the Society for Psychical Research for its inability to detect a simple code used by his children. For an extensive review of the Creery sisters, see Coover (1917, 463–77).

23. Luckhurst (2002, 127).

24. Committee on Thought-Transference (1883). Several authors have reviewed the sorry history of the society's investigation of Blackburn and Smith (Wiley 2012, 113; Gauld 1929, 126; Langham 1951, 38).

25. Colonel Austin Kibler of the Advanced Research Projects Agency (ARPA, a division of the Defense Department) sent Hymn to the Stanford Research Institute in 1972 to evaluate Uri Geller. Hyman reported that Geller was a charming con man with no psychic power but with a lot of chutzpah.

26. Hyman (1957). Reprinted in Hyman (1989).

27. Hyman (1985b).

28. Honorton, Charles, et al. (1990).

29. Hyman (1994). The psi ganzfeld controversy is still alive (Blackmore 2018).

30. These studies were carried out over many years at SRI International (formerly the Stanford Research Institute, not associated with Stanford University) and continued by Science Applications International Corporation (Hyman 1996).

31. Hyman (2007).

32. Although popular writers have appealed to quantum physics as an explanation, *Scientific American* columnist Lawrence Krauss says that no area of physics stimulates more nonsense in the public arena. He said that Deepak Chopra, alternative medicine advocate, could not pass an undergraduate course on the topic. See www.scientific american.com/article.cfm?id=a-year-of-living-dangerously (September 1, 2010). For an excellent review of physics and the paranormal, see Rothman (1988).

33. I reviewed the first volume of ten journals to get a sense of the quality of information published on mesmerism, spiritualism, and associated topics. All of them published the most improbable nonsense. Ann Braude (1990) identified an astounding 214 American periodicals on spiritualism published from 1847 to 1900. I conclude, therefore, that if my ten journals were in any way representative, an amazing amount of false information was flooding the general public.

34. Wolffram (2012) makes a case that false diagnoses account for cases in which patients recovered from incurable diseases and disorders.

35. Elliotson was following in the footsteps of others who were unable to discourage bad theory and dangerous practices in medicine, as vividly documented by Wootton (2007). Mesmerism was perhaps less harmful but the ability to connect cause and effect

was still missing. In 1826, a patient of Puységur published twelve hundred pages of presumed cures attributed to magnetism.

36. John James (1879) provides an obvious example of bad research, which modern readers can easily understand. His chapter on the miracles of Didier is followed by his "proof" of phrenology. A phreno-mesmerist merely points at a specific place on the head of a blindfolded subject, which causes a behavioral response congruent with the phrenological map. This nonsensical overacting by the subject is clearly bogus. Thirty years previously, the editor of the *Phreno-Magnet* (Hall 1843, 128) had to admit that phrenologists were providing clues to their subjects, and he recommended such demonstrations stop. *The Zoist* published similar procedures by phrenologists that are laughable at best. The point here is that if you accept the wild reports on Didier, then you incur an obligation to explain why similar descriptions by the phrenologists are not equally valid.

37. Lee (1866).

38. Stagnaro (2004).

39. See Project Alpha for the story of how two teenage magicians manipulated the research of psychic investigators (Richards 1982; Randi 1983a; Randi 1983b).

40. Elliotson (1856, 444). Both quotes from the same page.

41. Durant (1837/1982) provided a crisp history of mesmerism that ridicules Poyen and his colleagues for their credulity. See Stone (1837) for more of the clairvoyance controversies among these men.

42. Poyen (1837). William Stone (1837) flew into a rage against Durant's criticism.

43. Carlson (1960a).

CHAPTER 18: MESMERISM

1. Full disclosure: I am a past president of the Portland Academy of Hypnosis. In my clinical work, I used hypnosis mostly as an efficient way to desensitize fears and conflict through imaginal rehearsal, like a psychodrama of the mind. I helped patients experience themselves in problematic roles and practice better social responses.

2. Campbell (1853, 102).

3. Although Mesmer's book was not translated until 1948, his propositions appeared earlier in some books (Leger 1846, 291). Watts (1883) called these "Mesmer's aphorisms."

4. Mead's work was originally published in Latin in 1707 with the first English translation in 1708 (Pattie 1956).

5. Thorndike (1923, 7: 228). Van Helmont was the first to attach magnetism to the trance state. He also described transposition of the senses, which in his case was vision transferred to the epigastrium (Leger 1846, 251).

6. Thorndike (1923, 8: 571).

7. Back in Vienna, Mozart had played the glass harmonica at Mesmer's estate. Ben Franklin once went to hear Mesmer play this instrument, which Franklin had redesigned, but Mesmer only wanted to discuss animal magnetism.

8. The first commission was composed of members of the Paris Faculty of Medicine and the Royal Academy of Sciences. Benjamin Franklin notably served on this committee representing the Royal Academy.

9. Donaldson (2005).

10. Gauld (1995).

11. Podmore (1909).

12. In England, Professor Zerffi (1871, 67) said that "hundreds of trustworthy witnesses have asserted facts that somnambulists exhibit clairvoyant powers."

13. Mathiesen (1998, 156) provides insight into how books of that era likely produced the guidelines for hypnotic visions, provoking powerful experiences.

14. Thurston (1951).

15. Summers (1950). Even Summers's biographer is uncertain about his personality and spiritual beliefs (Jerome 1965). "Summers exhibited a marked tendency to untruthfulness . . . also given to alternating manifestations of near-blasphemy and piety" (76). "[H]e seemed unable to express with moderation his disagreement with the views of others . . . frequently passing the bounds of civility" (84). It is likely that many of his controversial religious expressions were sarcasm and hidden expressions of his opinion that Christianity is ridiculous and false.

16. Summers (1950, 39).

17. The migraine connection was first suggested by Singer (1928/1958, 232) and has been endorsed by Sacks (1986). Migraine aura has also been associated with a specific style of art that features stars and dancing lines (Fuller and Gale 1988). Fox (1987, ix) notes that Bingen has been called an "ideal model of the liberated woman."

18. Begbie (1851); Brittan (1868); Clarke (1880); Craft (1881); De Boismont (1853); Ireland (1885); Madden (1857); Maudsley (1887/2011); Parish (1897); Raue (1889); Sully (1881). These early books show a startling spectrum of ways that our brains and bodies are externally influenced, often beyond our understanding and awareness. If Freud had taken the time to read one of these books, he may have been more cautious in assigning psychological etiology to some of his featured patients who were suffering from organic disorders. See Webster (1995) for examples.

19. Colquhoun (1839, 41).

20. Supporters of recovered memory therapy used the same argument, rejecting the research of Elizabeth Loftus because she was an experimental psychologist, not a clinician. Mesmerists and recovered memory therapists believed that their experience verified the theory.

21. Poyen (1837, 32).

22. Mesmer visited England in 1785, but he received a cold reception (Leger 1946, 350).

23. See Wootton (2007) for the sad state of medicine during this time.

24. Williams (1947, 33).

25. Winter (1998) presents the full complexity of this controversy based on original documents.

26. Rosen (1946).

27. Elliotson expected his colleagues to force the hospital to reinstate him. It never happened.

28. *The Spiritual Messenger* 1 (1858): 45.

29. Kaplan (1974) says that Wakley alleged that Elliotson was guilty of sexual improprieties associated with the O'Key scandal, but I have not verified this. Wakley (and others) have complained that "quacks and impostors, called mesmerists," have a notorious history of sexual passions (Elliotson 1849).

30. *Spiritual Messenger* 1 (1858): 14.

31. *The Phreno-magnet* 1 (1843): 27.

32. *Buchanan's Journal of Man* 1 (1887): 2.

33. Didier (1856, 21).

34. Reichenbach (1851, 349–50, xi). Reichenbach compared od to electricity, and his long list of od's characteristics are an amazing collection of impossible conditions.

35. Braid (1899).

36. Crabtree (1993, 158).

37. In a similar vein, George Beard (1877a) reminded people that no human being has any qualities different in kind from those that belong to the species in general.

38. In modern terms, Forbes was describing somatoform disorders in which symptoms, for example, those symptoms called conversion disorders, express themselves as physical dysfunctions without sufficient underlying organic explanation.

39. For the natural endurance of surgical procedures and other examples of pain insensitivity, see Gould and Pyle (1897). For rituals of torture, see Hunt (1934). For abdominal self-surgery and other sensitive procedures, see Pankratz (1986).

40. The modern concepts of hysteria are now classified under somatoform disorders (Creed and Guthrie 1993).

CHAPTER 19: MODERN HYPNOSIS

1. Elliotson changed his mind about spiritualism near the end of his life after attending a couple of séances by D. D. Home and lived his final days as a Christian (Home 1872, 61). Elliotson's biographer never mentions his conversion, but Home makes a convincing case with the reprint of obituary notices.

2. I have suggested that hypnosis is so poorly understood that therapists are easily drawn into accepting any new theory to guide their therapeutic strategy (Pankratz 2003a).

3. Charcot's biographer, Georges Guillain (1959), declared that "to take from neurology all the discoveries made by Charcot would be to render it unrecognizable."

4. See Carey and Perry (1948) for a wacky version of this entangled with astrology, organ systems, and color. Also see Carey (1918) and Carey (1932).

5. Harrington (1988).

6. Bernheim (1884/1965, 89). Bernheim's book further insults Charcot by mostly ignoring his work.

7. Charcot (1987).

8. Munthe (1930).

9. Borch-Jacobsen (2000, 24).

10. Webster (1995). Trivers (2011, 317) says that psychoanalysis is not subjected to research because it is "a full-time hoax" supported by psychiatrists sharing clinical lore with one another over drinks after a day's work. Moreover, Freud's theories tend to blame the victim and violate expectations of evolutionary biology.

11. Borch-Jacobsen (2000, 22–23).

12. Bernheim (1884/1965, 165).

13. www.psych.upenn.edu/history/orne (updated May 1, 2010).

14. Orne (1951).

15. The National Vietnam Veteran Readjustment Study (Kulka et al. 1988) was a four-year research project that cost $9 million, yet no one checked the military records of any subject. Dallas stockbroker B. G. Burkett (Burkett and Whitley 1998) identified multiple methodologic errors in the study. My favorite was the six women who claimed posttraumatic stress disorder secondary to being a prisoner of war. Apparently, no one noticed that no American woman ever became a prisoner of war in Vietnam.

16. In the middle part of the twentieth century, police were trained in hypnosis to help victims recall details of a crime. Orne's research helped stop that dangerous practice.

17. Orne (1979). Orne's work on recovered memories was important in ending the disastrous era of recovered memory therapy, especially as it pertained to courtroom procedures (Orne et al, 1996). Full disclosure: I was on the advisory board of the organization until it closed in 2019.

18. Orne (1962).

19. See my discussion of this research problem in Pankratz (2003a, 2003b).

20. For example, Orne noted the difficulty in studying the historical controversy about whether someone could be mesmerized into performing a dangerous stunt or a harmful or immoral act. Subjects know the experimenter is testing that possibility and implicitly trust the experimenter to protect them from untoward consequences. Professionals endlessly debated the issue as courts considered defendants claiming they had been mesmerized by criminals. See Andriopoulos (2008) for a historical review. Many early books on hypnosis devote full chapters to the dangers of hypnosis; see, for example, de Laurence (1902), de Courmelles (1895), Young (1928), Sandby (1848), Bramwell (1903), Deleuze (1886), and Hart (1896). For legal perspectives, see Hammerschlag (1957) and Scheflin and Shapiro (1989).

21. Pankratz (2002).

22. Ginzburg (1983).

23. In the early 1990s, a whole industry emerged to help patients recover memories of forgotten childhood sexual abuse, often exhumed with the assistance of hypnosis. For recovered memory and suggestion, see Ceci and Bruck (1995) and Loftus and Ketcham (1991). On trauma, see McNally (2005) and Leys (2000); for multiple personality disorder, see Piper (1997) and Spanos (1996).

24. Bowman (1997). For additional reviews on the inappropriate use of posttraumatic stress disorder, see Burkett and Whitley (1998) and Young (1995). Trauma does not automatically create posttraumatic stress disorder; that psychiatric diagnosis is appropriate only with an *abnormal* reaction to stress, a *pathological* response to trauma. The inappropriate use of the label may explain why treatment of trauma is so unsuccessful despite enormous amounts of money spent on therapy. For comprehensive reviews, see Litz et al. (2002) and Arendt and Elklit (2001).

25. Orne (1952).

26. Carlson (1960b); also see Bernheim (1884/1965, 116). Wootten (2007) provides the disconcerting opinion that analgesics were slow in being adopted because the self-image of surgeons was that of being experts in rapid surgery and impervious to pain in their patients.

27. The mesmerist for this procedure was Phineas Quimby, known for his treatment of Mary Baker Eddy, the founder of Christian Science, who later became exceedingly fearful of "malignant animal magnetism" (Dresser 1921). Milmine (1971) has an excellent review of plagiarism in the writings of Mary Baker Eddy. Interestingly, her *Church Manual* says that members "shall not plagiarize her writings" (Eddy 1920, 43). Eddie was likely addicted to morphine, which may have triggered or enhanced her paranoia (de Camp and de Camp 1966, 217).

28. Elliotson (1851).

29. In 1842, the report of a leg amputation using mesmeric anesthesia was read before the Royal Medical and Chirugical Society of London. The paper was met with critical comments resulting in the passage of a resolution that "no record should be kept of any painless operation having been performed." A Dr. Copeland contended that "pain is a wise provision of nature, and patients ought to suffer pain while their surgeon is operating, as they are all the better for it" (*The Spiritual Messenger* 1 [1858]: 3–4).

30. The Bengal Commission said that only seven out of ten Esdaile patients could be mesmerized, and one patient endured a procedure without hypnosis. Three patients said they had no pain but appeared to be in agony, and two more had an erratic pulse. The commission said Esdaile's hypnosis was not as powerful or effective as he had reported, but this conclusion cannot be considered unbiased (Wenegrat 2001). Elliotson's patient received twenty days of mesmeric preparation before his leg amputation (Leger 1846, 362).

31. Rosen (1946).

32. Humphry Davy experimented extensively with nitrous oxide but failed to anticipate its subsequent use as an anesthetic, which began in 1844 (Holmes 2010). Ether and chloroform were introduced in 1846, but surgery with chloroform resulted in many

deaths. Volume 7 of *The Zoist* has five articles that review mortality associated with chloroform.

33. Bargh and Chartrand (1999).

34. Wegner and Wheatley (1999).

35. Macknik and Martinez-Conde (2010) present some amazing studies demonstrating that the brain decides things before the conscious mind is aware.

36. McGlashan, Evans, and Orne (1969).

37. Spanos (1996, 35).

38. Montgomery, DuHamel, and Redd (2000).

39. Kinnunen, Zamansky, and Block (1994).

40. Martinez-Conde and Macknik (2017).

41. Underhill (1868, 35).

42. Burlingame (1891, 99). Lustig (1930) said the successful vaudeville hypnotist needed publicity and shills, five to ten of them. Eventually, hypnotists realized that every audience had individuals willing to participate in anything suggested, and good entertainers learned to identify these unpaid assistants quickly. Some hypnotists demonstrated pain tolerance on themselves after they became aware of the possibility of unintended infections and other liabilities with volunteer participants.

43. Parkyn (1900, 353) described the methods spectators never see that enable a confederate to endure a seven-day window sleep (i.e., to sleep for seven days visible to the public through a window).

44. As early as the sixteenth century, Reginal Scot (1930) illustrated modified knives and awls that feigned mutilation. Walford Bodie (1905) exploited his knowledge of voltage and current to endure what appeared to be terrible electric shocks. For explanations of his methods, see Karlyn (ca. 1912), Cannon (1953), Jay (1986), and Woods and Lead (2005). My book on medical deception has a full chapter on illness illusion (Pankratz 1998). For tolerance of pain by individuals, see Jay (2001), Hunt (1934), Carrington (1909c), and Gould and Pyle (1897).

45. Burlingame (1891, 107).

46. Ovette (1947).

47. Gresham (1948) was a biographer of Houdini and the husband of Joy Davidson, who later married C. S. Lewis. Gresham's *Nightmare Alley* (1946) describes a carnival geek driven into despicable activities because of drug and alcohol addiction.

48. Burlingame (1891, 103); Carrington (1913, 15); Carrington ([1928], 11).

49. Allison Stewart, "David Blaine Is Going to Go Himself One Better, Which Is Saying Something," *Chicago Tribune*, June 13, 2018, www.chicagotribune.com/enter tainment/ct-ott-david-blaine-chicago-0615-story.html.

50. Congenital pain insensitivity has been studied for some time by physicians (Crichley 1956) and is now understood as a genetic disorder (Wang et al. 2018). I followed a couple of children who were thought to be victims of abuse before their pain insensitivity was identified. Their lives over the past thirty years have not been easy because of accidents occurring secondary to their inability to appreciate the dangers of injury.

51. For information on Minnock and his congenital pain insensitivity, see Rinn (1950). Jay (2001) provides a list of hypnotists Minnock worked with and some of his daring feats. Minnock claimed that he did a burial stunt with Charcot, who left him money in his will that allowed him to return to America. I obtained Charcot's will, and it contained no mention of Minnock.

52. Santanelli (1902) is the stage name of J. H. Loryea. Minnock says in the *St. Louis Star* (September 4, 1904) that he found Santanelli in Denver doing a fire-eating act under the name of Stewart. Santanelli was once arrested and fined in Fort Wayne, Indiana, on charges of assault because one of his subjects had been sleeping five days, presumably without food or water (*New York Times*, April 4, 1896). It is amazing that doctors would participate in such inhumane bets as those with Minnock, which only shows how primitive and unregulated the profession was at that time.

53. Parkyn (1900).

54. Price (1985, 447).

55. Pankratz, Hickam, and Toth (1989) were the first to define drug-seeking behavior, anticipating the serious addiction crisis secondary to excessive prescription of narcotics. We identified patients at risk for addiction and described appropriate clinical intervention.

56. Letter dated November 20, 1928. His mimeographed pamphlet *Lessons in Hypnotism* was undoubtedly from about the same time. Lonk (1947), among others, also promoted an instantaneous method of trance induction that involves occluding the carotid arteries of resistant subjects. Also see Anonymous (1901), Calvert (n.d.), and Dr. Q (n.d.).

57. The distinction between prescription and over-the-counter drugs was murky until the federal Durham-Humphrey Amendment of 1951. See Young (1967).

58. I have in my private collection the business cards of hypnotists that advertised their availability to give lectures, provide hypnotic entertainment, and render psychological treatment. Some carried two separate cards, each offering a different service.

59. Flint (1912b).

60. Dumont (1917). Theron Q. Dumont was the alias and pen name of American writer William Walker Atkinson (1862–1932); he was also editor of a couple of "new thought" magazines. Lonk (1947, 66) has a similar picture of two entwined adolescents; try not to laugh.

61. Arons (1948, 1953); Powers (1961, 1964). Also see seductive covers on McEwen (1912), Lustig (1930), and Tracy (1952). Lurid pictures of hypnotic seduction on dime novels might be unending. Joan Brandon (1956) has an illustration in her book of herself and another woman standing on a man stretched between two folding chairs. This and other pictures of her suggests that she is completely comfortable touching men she has hypnotized.

62. These sparks from the eyes are undoubtedly related to the theory of ocular or visual emissions, the belief that the eyes send out emanations (rays) that capture images brought back to the eye. A surprising number of adults endorse this false belief (Winer

et al. 2002). For books on hypnosis showing pictures of visual emission on the cover, see Flothmann (ca. 1916) and Holt (1947).

63. Wolffram (2012).

64. My experience with the Ethics Committee of the Oregon Psychological Association gave me insight into the prevalence of sexual misconduct by unlicensed hypnotherapists.

CHAPTER 20: RAP IF YOU BELIEVE IN SPIRITS

1. Dewey (1850, 16). This pamphlet with Houdini's note went from his possession to the library of Milbourne Christopher. I purchased it from David Haversat, who brokered the Christopher collection.

2. Shepard (1984).

3. I seriously doubt that the birth of my mother was officially recorded in 1906 when she was born at home on a rural Oregon homestead to immigrant parents with limited English language ability.

4. Cadwallader (1922).

5. "The other girl, who is in her fifteenth year" (Dewey 1850, 16).

6. These quotes from the statements taken by Lewis are from Dewey (1850). No library holding the Lewis pamphlet would loan its copy, and I have not discovered a reproduction. However, other early authors quoted Lewis, and their congruence gave me confidence that I could rely on Dewey's version as authentic with the later ones representing modification of the Lewis original.

7. Dewey (1850); Anonymous (1850); Anonymous (1851); Elliott (1852); Vizetelly [1853a]; Spicer (1853); Capron (1855); Gordon (1856); Daniels (1856).

8. Owen (1860) said he possessed the Dewey pamphlet, so he had a couple of sources for the correct ages of the girls.

9. Clanny (1841, 10).

10. Houran and Lange (2001). Poltergeist manifestations are almost invariably associated with the presence of female adolescents. Perhaps spiritualism's beginning could be viewed as a result of poltergeist pranks.

11. Lewis (1796/2002, 138). Even before this, Pierre LeLoyer (1605) said that the souls of those who are happy do not return to Earth, only those whose souls are languishing in purgatory. Hall (1980, 74) said it best: "No haunted house is complete without a legend of a crime, or a tragedy, or a badly-spent life, to explain why the ghost walks."

12. Capron (1855).

13. Weisberg (2004).

14. See, for example, National Spiritualist Association (1948, 8).

15. Mattison (1855, 57). In the early days of email transmission, one letter at a time appeared on the cathode ray screen; it took longer to download a message than to write it.

16. Dewey (1850, 38).

17. Anonymous (1850, 79).

18. MacWalter (1854, 60).

19. Wegner, Fuller, and Sparrow (2003). Also see Novella (2012); Lilienfeld et al. (2015). The International Society for Augmentative and Alternative Communication issued a position statement noting at least fifteen different professional organizations rejecting the procedure: www.tandfonline.com/doi/pdf/10.3109/07434618.2014 .971492 (November 7, 2014). This includes the American Speech-Language-Hearing Association: www.asha.org/News/2018/ASHA-Discourages-Use-of-Facilitated-Com munication-Rapid-Prompting-Method (August 8, 2018).

20. Podmore (1902, 4).

21. Post (1852).

22. Mattison (1853, 115).

23. Hammond (1852). Dr. Bell described the essays in Hammond as absurd, puerile, disgusting, and infinitely below the intelligence of the alleged authors (Hare 1855, 64).

24. Vizetelly ([1853a], 82). A twenty-eight-year-old Philadelphia store clerk wrote a massive 537-page book through automatic writing (Linton 1855). Books were also dictated by spirits through the Ouija board, the most famous being those of Pearl Lenore Curran, who was supposedly contacted by a spirit named Patience Worth (Prince 1964; Yost 1916). See Worth (1917, 1918, 1923, 1928). The most controversial was that by Emily Hutchings (1917). She said her book, a novel from a Ouija board, came from the spirit of Mark Twain, which outraged his widow but won the support of James Hyslop (Litvag 1972, 167).

25. Anonymous (1850, 51).

26. Anonymous (1851).

27. This controversy was laid out by Capron, and sections not otherwise referenced were taken from him.

28. Spicer (1853, esp. 390).

29. Vizetelly ([1853a], 252).

30. Mattison (1855).

31. Devant (1931, 187) reported that fellow magician Maskelyne became a confirmed skeptic after repairing a rapping table for a medium. Also see an electromagnetic rapper inside a table (Collins 1939, 108) and another with a telephone mechanism that can be used for emitting a sepulchural voice (Hopkins 1890, 588). For an improved version, see Hopkins (1911, 101). Goldston (1921, 105) illustrates an ingenious mechanical table and a rapping device attached to a belt.

32. Wyndham (1937, 215). Also see Committee of the London Dialectical Society (1873, 225) and Committee of the London Dialectical Society (1871, 225). The earliest electromagnetic rapping device I have discovered is extensively described in Wylde (1861, 332).

33. Hoffmann ([1887], 156) said the toy was known in France as a "cri-cri" and in England as a "distinette." He said the device was originally brought out as a substitute for the castanet but was speedily abandoned.

34. Robert-Houdin also produced music from this empty box by attaching the hook to a mechanism like those inside music boxes (Robert-Houdin 1900, 133). Vere (1879, 105) shows how to start and stop a music box suspended in air, and Cremer ([ca. 1874], 159) describes a method for making a drum beat while hanging from the ceiling.

35. Variations of the Fox sisters' routine became known as "The Living and Dead Test" among magicians and mentalists who devised infinite variations. See, for example, Abbott (1908), Dr. X ([1922]), and Booth (1931).

36. Page (1853).

37. The classic demonstration of the difficulty of locating sound is to tap two coins together over the head of a blindfolded person. The person will have no idea where the sound is coming from but will usually endorse a particular location that has been suggested.

38. Jackson (1972, 104).

39. Anonymous (1851, 5).

40. Anonymous (ca. 1870, 16). Mrs. Henry Sidgwick (1886b) described the same sound when Kate was in London.

41. The affidavit of Mrs. Culver's involvement can be found in numerous places (Elliott 1852, 160; Grimes 1885, 321; Capron 1855, 421).

42. Weisberg (2004, 129).

43. Elliott (1852, 150). This "recognition reflex" is commonly spontaneous in children.

44. Kane and Fox (1866).

45. The tragic final years of Maggie Fox are well documented in Chapin (2004).

46. Kane and Fox (1866, 26). This book had little impact due to the long gap between its publication and Kane's death. That delay also disguised the false age given to Maggie.

47. Readers in 1866 may have felt sympathy for Maggie, but the book is disturbing today for anyone who thinks about Kane's interest in a thirteen-year-old girl. However, in fact, she was nineteen.

48. Underhill (1885, 7). Sociologist and religious historian Todd Leonard (2005, 226) quotes this same passage by Leah, whether by ignorance or preference for promoting the spirit world.

49. Goldsmith (1998); Owen (1990). For an association of spiritualism with healing and miracles, see Britten (1884). Spiritualism was also associated with women's right to divorce, which sometimes became indistinguishable (in the minds of some) from free love and prostitution (Farmer 1886, 86; Doesticks 1859; Moore 1977). Also see Lehman (2009) and Kerr (1973).

50. By 1860, *The Spiritualist Magazine* listed sixteen broad categories of manifestations that spiritualists would encounter in séances (Editor 1860), but similar lists occurred even earlier. See Ballou (1852), Marsh (1854), and Mattison (1855).

51. Moore (1972).

52. Abbott (1908, 218).

53. Seybert Commission (1887, 48).

54. Weisberg (2004). Tromp (2006, 165) says it is difficult to exaggerate the extent of alcohol in the séances and lives of the Fox sisters.

55. Rinn (1950) says Kate was often clumsy in her séances, and her fishing for information was obvious. Sir William Crookes, among others, was disappointed in her séances.

56. Weisberg (2004, 222).

57. Kate encouraged spiritualist D. D. Home when she learned that he was ready to expose the tricks of spiritualists (Home 1877, 182). This adds one more piece of evidence that she recanted her life of fraud, a fact often disputed by spiritualists.

58. Davenport (1888, 17).

59. Hardinge (1870, 72). Hardinge had a background in acting, which she hid after becoming associated with spiritualism. Natale (2016, 38) makes a case that spiritualism was religious entertainment that thrived on controversy for publicity. The object was to excite fascination with the lure of occult possibilities; to be entertained was perhaps equivalent to belief.

CHAPTER 21: THE TURING TEST

1. The failure of mediums was usually blatantly obvious because they did not know that Houdini's mother spoke five languages but never learned to read, write, or speak English (Kalush and Sloman 2007, 444; Wolf 2017, 79). The most laughable was a materialization of Houdini's mother at a séance of Nino Pecararo. She introduced herself in a heavy Italian accent (similar to Pecararo's) by saying "Hey, Houdeen. Dis is you mama." Incidentally, Pecararo was completely helpless in displaying his spiritualistic phenomena after Houdini tied him with short pieces of fishing line (Gresham 1959, 236).

2. Descartes discussed a similar issue regarding automata (Dear 1998, 51–82). Turing's accomplishments and influence are of legendary proportions (Dyson 1997).

3. Trevor Blake helped me articulate this "Turing test" issue. Also see Hoffman (1988, 143).

4. Page (1853).

5. Ray Hyman has written extensively on the techniques of cold reading (Hyman 1977; Hyman 1981, 169–81). Many cold readers have revealed their strategies (Earle 1989; Hester and Hudson 1977; Nelson 1968; Riding 1989; Saville 1984; and Webster 1986). There are also ways of answering questions that appear direct but avoid commitment that might later reflect poorly on the reader (De Lawrence 1927).

6. Schlesinger (1886).

7. Hare (1855)

8. Hare (1855, 51).

9. Hare and Hare (1901). For other spiritualist bibles, see *New Testament* (1861) and Northrup (1885).

10. Podmore (1897, 49) kindly says that Doctor Hare's book "is not such as to inspire confidence in his judgment." The editor of *Two Worlds* and *Spiritual Quarterly Magazine*, Will Phillips, wrote vicious responses to Podmore suggesting that the messages to Hare could not be the result of fraud (Phillips ca. 1902).

11. Mystic Helper (1924) transcribed messages of Michael Faraday from beyond. Unfortunately, what Faraday told us is mostly nonsense. The book was probably written in 1909 but not published until 1924, by which time most of the scientific errors would have been evident. Nevertheless, David Starr Jordan (1851–1931), spiritualist and founding president of Stanford University, gave this book a rave review. He noted that all of Faraday's discoveries in heaven were without the need of mathematics, "which indicates that formulae and equations are emanations of the material mind" (*Science*, September 19, 1924, 269). My conclusion is that Mystic Helper and Jordan had no mathematical ability.

12. Hare (1855, 64).

13. Dawson (2012, 124).

14. We know from Jones (1861, 323) that Mrs. Hayden had her English subjects hold an alphabet card as she questioned them. This gave her the advantage she needed to directly observe her sitters' responses.

15. Wyndham (1937, 27).

16. Anonymous (1853, 183).

17. Spicer (1853, 374, 390).

18. De Morgan's picture in a publisher's catalog shows him looking as wacky as the stereotype of an absent-minded mathematics professor (Anonymous 1909, 89). Despite his goofy appearance, the reviewer described him as a genius, and, indeed, his *Budget of Paradoxes* shows the common reader his expertise with numbers (de Morgan 1872/1954).

19. C. D. (1863). The 1871 yearbook of spiritualism noted that Mrs. de Morgan was now in ill health, but she provided a page-long message she had received from a spirit. The content is mostly about the beauty of the spirit world, but where the words are not confused, they are shallow and meaningless (Tuttle and Peebles 1871, 177).

20. Fournier d' Albe (1923, 197).

21. The discovery of a new island was probably a poor choice as an example of what everyone should believe. Brooke-Hitching (2018) shows a plethora of islands falsely reported. The same could be said of reports on exotic animals; for example, nearly everyone who went to Africa was expected to describe the sighting of a unicorn (Shepard 1982). Descriptions of werewolves, bigfoot, mermaids, sea serpents, and other mythi-

cal animals are too numerous to name. For a few examples, see Dance (1976), Gould (1886), Krutch (1969), and Lewinsohn (1954).

22. Fara (2002, 223).

23. Owen (1836, 1860).

24. Quoted in Podmore (1910, 18–19).

25. Ashburner (1867, 315, 81). Ashburner predicted that spiritualism was about to astound the best philosophers into silence, and he could hardly wait.

CHAPTER 22: TABLE TURNING

1. Godfrey (1853).

2. Vizetelly ([1853b, 8]).

3. Anonymous (1853, 49).

4. Levi (1914, 480).

5. Marsh (1854, 76).

6. John White (1677, 7). A French magic book from the mid-seventeenth century also described this pendulum effect (Pritchard 2018). A later magic book described the same effect but admitted not having an explanation (Anonymous ca. 1840, 2).

7. Spitz and Marcuard (2001). Chevreul was persuaded to publish his work with the encouragement of Andre Ampere. He introduced his findings by saying this information "might be of some interest for psychology and even for the history of science."

8. Mayo (1851, 200).

9. A Brighton homeopathic physician was another victim of self-deception. He evaluated the pharmacopoeia with his swinging magnetometer only to later discover his extensive work was the result of his own anticipation (Carpenter 1886, 286).

10. Johnson Smith (1941, 382).

11. Charvet and Pomeroy (2004). Alexander also separated people from their money by selling throwaway pamphlets promising occult secrets (Alexander, n.d.; Alexander 1924a; Alexander, 1922). For big spenders he sold a small five-volume treatise on the "inner secrets of psychology" (Alexander 1924b) and a matching companion on oriental wisdom (Alexander 1924c). Few knew that the real secrets of Alexander included bootlegging, extortion, bigamy, womanizing, and even murder.

12. Kreskin (1991) and Jermay (2007) came up with the brilliant idea of using a Slinky in place of the pendulum for stage presentations. The volunteer holds one end of the Slinky with the other end on the floor. The results are surprising to the participant and impressive for the audience.

13. Abbott (1908, 29).

14. For a good description of multiple pendulums, see Minch (1975). Magician Will Dexter (1979) exposes some deceptive methods for creating the swing of a pendulum in a bottle, but he has no explanation for why it sometimes works without trickery.

15. Popular Mechanics Press (1915, 280). Although early twentieth-century boys were informed about the physics of the pendulum, mystical attributions persist. Archdale (1961) recommended the pendulum for discovering your car's mechanical malfunction. Good luck with that. For similar modern nonsense, see Blackburn (1983).

16. Goldston (1921, 87); Behnke (1990, 473). This same method, which I am not exposing here, was revealed in a popular magazine-style publication by Dunninger (1928, 78). Bordelon (1711) describes a gypsy who had a statue that held a silken string with a shiny iron fly on the end that hung down inside a tumbler. She used the swing of the fly striking the tumbler as an oracle. Her secret was a magnet attached to a ring on her finger that she held next to the glass. Porta would have been proud. (My edition of Bordelon has pagination problems, but the critical text follows page 96.)

17. Carpenter (1852).

18. Carpenter's ideas were also based on previous English and German research (Stock and Stock, 2004).

19. Because the response is outside of awareness, unscrupulous magicians and Milton Bradley were able to make a few bucks selling useless pendulums. However, this same ideo-motor response allowed con artists to sell bogus bomb detectors to the international arms trading market for an estimated $100 million in profit (https://sites .tufts.edu/corruptarmsdeals/the-worldwide-fake-bomb-detector-scam/).

20. Page (1853).

21. Lang (1894, 312) recommended sprinkling flour on the table to see if any fingers touch it.

22. Edmonds and Dexter (1853). Edmonds was a prolific writer of spiritualistic tracts (Edmonds, 1858).

23. Some enlightened nineteenth-century Englishmen were undoubtedly offended when Campbell (1853, 155) said that those who saw a table moving without hands were no different than those who had observed a witch fly on a broom.

24. Someone applied oil to the hands of Mrs. Hayden, who then could not move her table (Linton 1855, 494).

25. There are several ways of tipping tables using trickery including the use of black thread, suction cups, and slotted finger rings that attach to the table (Dr. Q, 1941), which are also used for levitations. Table lifting is only accomplished by intentional deception. For methods in magic books, see Anonymous (1858b), Collins (1939), Lunt (1903), Karlyn (ca. 1912), Ovette ([1921]), Pinetti (1905), Reilly (1957), and Shaw (1893). Table-lifting devices were available in catalogs for spiritualists (Sylvestre 1901) and magic catalogs, the earliest probably being Rebmuh (ca. 1889). For methods specifically used by mediums, see Anonymous (1860) and Cumberland (1918).

26. His laboratory has been reconstructed to its original 1845 condition and is adjacent to a museum dedicated to his accomplishments.

27. Winter (1998, 290). McKenzie (ca. 1910, 13) suggests that table movements are most successful when two sitters make a good positive and negative battery: a vigorous and healthy man with a healthy lady with a passive disposition.

28. Hankins (1991). The Latin root of experiment means "experience."

29. Faraday (1853).

30. Other strategies to test the ideo-motor effect included placing talcum powder on the table, which allowed fingers of sitters to move but not the table itself. Similarly, oil on the hands stops table tipping. People engaged in conversation fail to move tables, presumably because multitasking is forbidden by the spirits (Podmore 1902). Cumberland (1918, 69) placed smoked glass on the table, which left telltale signs of finger movement.

31. Faraday was invited to investigate D. D. Home, who produced some manifestations that were never explained—if they really happened as stated. However, Faraday first wanted a "program" of the miracles that would be presented. This was refused, so he rejected the offer, which was appropriate.

32. Anonymous (1853).

33. Podmore (1902, 10). Another example of the principle "too stupid to know you are stupid" (McIntyre 2018).

34. Townshend (1854, 157).

35. Samson (1869).

36. Sandby (1853).

37. Wallace (1875).

38. Editor (1864).

39. Davies (1867, 248).

40. Borel (1879).

41. De Gasparin (1857).

42. Borel (1879, 91).

43. I like the definition of metaphysics provided by Davies (1867, 273): "If you are talking to me of what you know nothing about, and I don't understand a word of what you are saying—that's metaphysics."

44. Godfrey (1854).

45. See Oppenheim (1985) for an excellent review of this history.

46. Gillson ([1853a]).

47. Morgan ([1854]) quoted Nahum 2:3, which says in the last days there will be chariots with flaming torches, and Morgan blasted out other scriptures on fire and brimstone.

48. Gillson ([1853b]).

49. Dibdin (1853).

50. Magee (1854, 19). Similarly, Laicus (1854) declared that Godfrey and Gillson were deluded. Also see Clay ([1853], 16) who tried to answer critics.

51. Kerner (1845).

52. Cahagnet (1851).

53. Godfrey (1854, 84).

54. Vincent (1853).

55. Close (1853).

56. Gillson ([1853c]).

57. Anonymous (1853).

58. "Table Turning" (1854).

59. Glazebrook (1854).

60. An Anxious Enquirer (1854).

61. A. B. (1853).

62. Interest in satanic intrusions emerged again at the end of the twentieth century when police agencies and psychotherapists began promoting the ideas that satanic rites were being promoted by secret organizations. Ken Lanning (1992), FBI crime expert, documented that these presumptive organizations did not exist but were urban legends (Victor 1994).

63. Mattison (1855, 232) says that table tipping was dying away because of Faraday's explanation, taking its place in history with the Salem witchcraft craze and similar mental epidemics. Surprisingly, the Toronto Society for Psychical Research conducted a study in the 1970s in which table turning emerged as a prominent feature. Despite its claim of being a "serious scientific investigation organization," it failed to take the most basic precautions. Its "Philip experiment" is like a parody, concluding that belief in something imaginary creates its reality (Owen and Sparrow 1976). Martin Gardner (1989) has described the inadequate evaluations that occur when those promoting psychic phenomena conduct the tests.

CHAPTER 23: PALLADINO: YOU JUST HAD TO BE THERE

1. One source says: "Her husband is a carpenter, and she is an ironer" ("Science and the Spirits" 1893).

2. These features of Palladino's manifestations were taken from Wiseman (1992).

3. Lombroso (1909). The preface to Lombroso's book is dated October 29, 1908, and he died about one year later.

4. Robinson (1935).

5. Lombroso (1909, figure 24, but also see figure 23 and 39). The three small photographs of table lifting reproduced in figure 39 are completely unconvincing if not evidence of her cheating.

6. Munsterberg (1910). Also see Price (1939, 190).

7. Lombroso (1909, 306).

8. Dingwall (1967, 157).

9. Flammarion (1907, 8).

10. An apport is any object materialized in the presence of a medium.

11. Podmore (1893).

12. Eleanor Sidgwick unraveled the mystery of how medium Henry Slade unsealed and resealed tied knots (Brown 2017). For more on Mrs. Sidgwick, see chapter 25 in this book and Johnson (1936), which is an offprint from *Proceedings of the Society for Psychical Research* 44 (June 1936).

13. Eric Dingwall (1962) said Palladino was characterized by vulgarity, lust, and cheating. Among her many improprieties, she sometimes put her bare foot in the crotch of men during a séance, and she allowed personal inspection of her body for the purported reason of showing that she was not hiding ectoplasm (Brandon 1983, 149). Gauld (1968, 224) mentions that the purses of sitters and other valuables sometimes disappeared during séances with her, and she was practiced in telling "naughty Neapolitan stories."

14. Maxwell (1905, 408). I presume these comments in appendix A were written by Charles Richet.

15. Maskelyne wrote a couple of books exposing the tricks of spiritualists, but these were published before his contacts with Palladino (Maskelyne 1876; Weatherly and Maskelyne 1891).

16. Editor (1895). Dr. Hart said they should stop wasting their time when so many serious problems await solutions. This editorial is reprinted in Hart (1896, 202).

17. Hart (1896).

18. Dingwall (1967, 259). I have been unable to determine whether Hart or Flammarion was the first to interact with the young actress. Either way, Luys's commitment to truth was obviously tenuous. Among his many questionable findings, Luys concluded that a drug sealed in a tube placed around the patient's neck produced the same favorable effect as if taken internally (Alpheus 1902).

19. Sidgwick (1909a).

20. Lachapelle (2011).

21. The annual pay of an experienced French judge or professor would have been about 10,000 francs per year. Carrington raised the money for Palladino's visit to America through donations by members of the committee he selected, and some came from McClure.

22. Sidgwick (1909b).

23. Complete descriptions of the Naples séances are found in Feilding, Baggally, and Carrington (1909) and Feilding (1963).

24. Feilding, Baggally, and Carrington (1909).

25. Gauld (1968, 244).

26. Aronson (1990, 172, my emphasis).

27. Tamariz (2014).

28. Magicians sometimes discuss this as the "too perfect" theory, when the audience has no conceivable explanation (Stone n.d.).

29. Gibson (1928).

30. To review the genius of Annemann, see Abrams (1992) and Karr (2019).

31. The trick actually fooled the show's magic consultants, as well, who adjudicate the conclusion if conflict arises. They, like Penn and Teller, presumed to know the method so never bothered to ask and were then fooled during the live performance. The solution is revealed in *Genii* magazine (Kopf 2018). You can see the act on YouTube: www.youtube.com/watch?v=xfQEkPM3nq8.

32. Andrus (1978). Andrus denied that his sleight of hand was difficult, and young magicians that he taught were able to master the moves.

33. Hyman (1981).

34. Baggally (1917, 77) noted that the Zancigs were so efficient and quick in responding that their communication could only be explained by "a supernormal faculty alone." Baggally also endorsed telepathy as the explanation of the Baldwins' act. Also see Fixen (1912) and Zancigs (1907).

35. Gregory (1985); Price (1930); Price (1933).

36. Wiseman (1992).

37. Polidoro (2003). Surreptitious observations under the table revealed that Palladino was using her feet in unusual ways consistent with Polidoro's conclusions.

CHAPTER 24: PALLADINO: CHEATING IN AMERICA

1. Carrington (1909b) was born on the British-owned Jersey Island as Hubert Lavington. Whaley (1990) describes him as a "self-serving credulous investigator of psychic phenomena." Shepard (1984) identifies him as a "distinguished" psychical investigator.

2. Sinclair (1906).

3. Adams (1906).

4. Carrington (1907).

5. Seventy-five years later, Dan Aykroyd and Bill Murray filmed *Ghostbusters* on the same campus.

6. "Awesome Wonders by the New Medium," *New York Times*, November 11, 1909.

7. Palladino (1910).

8. Female mediums were often reluctant to accept personal responsibility for their vocational choice. This complex dilemma of the medium is well reviewed in Moore (1977, 107).

9. "Ghostly Hands Seen after Writers Left," *New York Times*, November 15, 1909.

10. "Eusapia Palladino," *New York Times*, October 6, 1909.

11. Carrington (1908).

12. "Paladino Meets a Firm Skeptic," *New York Times*, November 17, 1909.

13. "Paladino Used Phosphorus," *New York Times*, November 19, 1909. "Illuminated paint" was also available from W. D. Devoe at the corner of William and Fulton streets in New York (Garrett 1892, 43). Roterberg's magic shop in Chicago provided many supplies for spiritualists. August Roterberg said that "without luminous paint the producer of ghosts would hardly be able to continue in business" (Hagy and Hagy 2019, 84). Maskelyne and Cook were among the first to use luminous paint on stage for magical effects (Garenne [1886], 305). One medium put luminous paint on his breast-pocket handkerchief, which he used to create the appearance of ectoplasm in the dark. He had been searched beforehand, but no one thought to consider anything hidden in plain

sight (*Pallbearers Review*, May 1968). Luminous powder was also available to order from catalogs (Grant ca. 1930).

14. A seventeenth-century Italian manuscript detailed the process for manufacturing phosphorus from urine, although it came with a warning about its unspeakably bad smell (Pieper 2013). This phosphorus, the manuscript informs us, can be mixed with pomade to rub on the face for a luminous appearance in the dark that will frighten the uninformed. (Kids: don't try this at home.) When Robert Boyle first manufactured phosphorus, he also rubbed it on his face for display in the dark. Although Boyle was the first to publish the method for making phosphorus, the process was undoubtedly discovered by others in previous times.

15. James H. Hyslop, "Paladino and Science," *New York Times*, November 21, 1909.

16. Hyslop (1910).

17. Moore (1977, 159).

18. Carrington (1909b).

19. W. S. Davis, "Sidelights on the Paladino Delusion," *New York Times*, November 21, 1909. Also see: Davis's (1910) and Hyslop's (1910) critiques, which are a master of blithering philosophical nonsense.

20. "Hyslop Hits out at Munsterberg," *New York Times*, April 3, 1910.

21. "Psychic Phenomena," *New York Times*, April 11, 1910. Hyslop was obsessed with the possibility of resurrection and life after death (Hyslop 1908). After his passing, his secretary, Gertrude Tubby (1929), channeled the return of his spirit. These messages contained only moral platitudes, so maybe it really was him.

22. Jastrow (1918).

23. Jastrow (1910).

24. Rinn (1950, 281).

25. Sargent was a professional magician, and Miller was the chair of philosophy at Columbia University.

26. "Difficult to Detect Her," *New York Times*, May 12, 1910.

27. Houdini (1924).

28. "Paladino Defiant; Again Gives Tests," *New York Times*, May 12, 1910.

29. "Palladino Test Won't Take Place," *New York Times*, May 27, 1910.

30. "Would Bolster up Palladino's Case," *New York Times*, May 13, 1910.

31. "Palladino Ready for a Strict Test," *New York Times*, May 14, 1910.

32. "Carrington Blames Rinn," *New York Times*, May 23, 1910. Palladino refused to perform séances "on condition that she gets paid only for genuine stuff" (Carrington ca. 1964, 15).

33. If spirits can rap on tables, then place a block of wood on the table and let the spirit pick it up and rap. These mysterious raps of mediums are a ridiculous way of interacting if spirits wanted to communicate with living humans.

34. "All Mediums Trick, Mme. Palladino Says," *New York Times*, May 19, 1910.

35. "Carrington Exposes Tricks of Mediums," *New York Times*, August 13, 1910.

36. Tabori (1972, 70). Magician Harry Kellar was amazed that Carrington was confused by a clumsy psychic. In a letter, Kellar said, "you are the greatest psychological puzzle I know" (Carrington ca. 1964, p. 83).

37. "Withdraws Libel Suit," *New York Times*, January 4, 1911.

38. Mann (1919, 115).

39. "Statements of Investigators," *New York Times*, May 12, 1910.

40. Krebs (1910).

41. For Krebs's exposure of Slade, see Stein (1996, 705). The Bangs sisters (Bangs and Bangs 1905) were well-known mediums active in Chicago for many years (Hunter 1894; Carrington [1913]; Abbott 1913; Swan 1991). W. U. Moore (1911) supported the sisters until he discovered their method (Moore 1913).

42. Perovsky-Petrovo-Solovovo (1911, 61).

43. Eusapia "was never still for a moment" (Richet 1923, 404).

44. The square legs of the table and the enormous volume of her skirt can be clearly seen in the picture facing page 321 in *Proceedings of the Society for Psychical Research* 23 (1909).

45. Grasset (1910, 356).

46. Carrington (1907, 184). Remember that Houdini tied Pecararo with fishing line, which stopped his manifestations. Johns Hopkins physicist Robert Hood had several suggestions for testing Palladino. He wanted to see her levitate a table while seated on a rocking chair with her feet on the chair (Carrington ca. 1964, 65). Also see Seabrook's (1941) biography of Hood.

47. Grasset (1910, 51).

48. "Miller, Dickinson, Paladino Tricks All Laid Bare," *New York Times*, May 12, 1910.

49. Brower (2010, 120). Also see his many attempts to explain away obvious fraud (Schrenck-Notzing 1920b, 292). For additional pictures of ectoplasm, see Schrenck-Notzing (1920a).

50. The book by Lodge (1916) was extensively criticized in the press, and the following books were devoted to his problematic interpretations: A Plain Citizen ([1918]), Cook (1917), Hookham (1917), Mercier (1919), Mercier [1919], and Wilson (1918).

51. Dallas (ca. 1922). Greenwood (1869) lists drunkenness and prostitution as two of the seven curses of London. Alcohol, narcotics, and prostitution were rampant in Victorian England (Diamond 2004). See Tromp (2006) for an extensive discussion of sexuality in the séance and spiritualism movement. One rare brochure (Anonymous n.d.) describes an American medium retching during a séance because her stomach was "disordered by potations of alcohol." At another time, a sitter confessed that his presumed materialized daughter had whiskey on her breath, which convinced him that fraud was more likely than a return from the hereafter.

52. Brandon (1983).

53. Kalush and Sloman (2006, 419); Kalush and Sloman (2007, 257).

54. For a review of the sexual misadventures of Myers, see Pearsall (1972, 53), Oppenheim (1985, 429), Kripal (2010, 89), and Hall (1980, xvii). Lots of people would have lined up to testify against Myers.

55. Rose Mackenberg investigated more than three hundred psychics during the two years she worked for Houdini and many more after that. She reported being repeatedly groped and propositioned (Jaher 2015, 336; Edwards 2019).

56. Brower (2010, 120). Schrenck-Notzing was a physician trained in hypnosis who was known for his treatment of sexual pathology (Morris 2006, 63). Carrington (ca. 1964) said Eva C. was the first medium to produce ectoplasm. It was generally acknowledged that she allowed searches of her orifices for hidden material by her investigators, which primarily included Schrenck-Notzing, Charles Richet, and Dr. Gustave Geley (1927). For additional improprieties of Schrenck-Notzing, see Wolffram (2012). For the attractiveness of certain mediums, see Davies (1867).

57. Brandon (1983, 122); Hall (1963). Medhurst and Goldney (1964) offered weak support for the denials of sexual improprieties of Crookes. The reputation of Crookes as a psychic investigator was damaged by his endorsement of Anna Eva Fay, whose methods were exposed by Truesdell (1883), Burlingame (1891), and others (Wiley, 2005). By 1925, Anna Eva Fay was a friend of Houdini and confessed her trickery to him (Jaher 2015, 299).

CHAPTER 25: SLATE WRITING

1. Podmore (1902, 205).

2. For séances of Slade in Australia, see Curtis (1902). For secrets of Slade's slate writing, see Truesdell (1883), Evans (1897), and Mann (1919). For the earliest "independent spirit writing," see Vizetelly ([1853a], 92), Ballou (1852, 250), and Mattison (1853, 74). For Houdini's "impossible" slate writing trick that fooled Conan Doyle, see Ernst and Carrington (1932, 240ff). Houdini never revealed the method of this amazing trick, but a compelling solution was suggested by Loomis (2016).

3. Writing on slates first occurred in England at the London home of Mr. and Mrs. Slater in 1872 (Slater 1872).

4. Ray Lankester and Horatio Donkin, who abruptly confronted Slade, were students of Charles Darwin. Slade died penniless because of alcoholism (Mann 1919, 32). See extensive support for Slade in Oxon (1882).

5. Maskelyne (1876, 65) suggests that the Davenport brothers similarly obtained enormous popularity *because* of their exposures. William Carpenter also testified against Slade (Knaggs 1879, 7).

6. Zollner (1880).

7. Hyman (1985b, 92–95). Hyman (2002, 83–89) showed he could untie knots while the ends of the rope were held. In a note to me (June 18, 2013), Hyman acknowledged in his typical modest manner: "My method almost exactly replicated an

earlier method by Milbourne Christopher which he described in the June 1949 issue of *Hugard's Magic Monthly* (p. 551, 556). Christopher called his effect *Slade's Knots Updated.* Although my method, which was published 36 years after Christopher's, differs in some details, it is close enough that it is possible that it is the result of unconscious plagiarism on my part." You can find my own method of untying a knot when the ends are secure at Pankratz (1990). Sidgwick (1886a) suggested that Slade may have switched ropes on Zollner when he was distracted.

8. Farmer (1886, 93).

9. Cholmondeley-Pennell (1884, 11).

10. Wallace (1875, 204).

11. Shapin and Schaffer (1985).

12. Dear (1997). Emsley (2000, 36) says that Boyle turned from his secret alchemical practice to work openly as a scientist, writing down all his findings, when he began his investigation of phosphorus.

13. Report (1885). Blavatsky and Eglinton created duplicate letters, one of which was on a ship and the other in Calcutta where it "mysteriously" dropped from the ceiling as if transported by astral projection. The whole sordid story of duplicity is reviewed by Hare and Hare (1936).

14. Lewis (1886).

15. Wiley (2005, 138). See this episode with Eva Fay described in Sidgwick (1886a).

16. Sidgwick (1886b).

17. Massey (1886).

18. Even Harry Houdini admitted that there was actually very little to watch because everything about slate writing appeared so straightforward. Nevertheless, he noted that slates had "infinite grafting possibilities" (Houdini 1924, 79).

19. Hoffmann (1876). Lewis wrote under the pseudonym Professor Hoffmann as a way of projecting Germanic thoroughness and authority. His book, and three that followed (*More Magic, Later Magic,* and *Latest Magic*), exposed most of the known magic at that time. Thus, magicians believed their craft had been destroyed. However, Hoffmann's books stimulated so much interest in conjuring that magic subsequently exploded in popularity. Books written before *Modern Magic* can be found in the two-volume bibliography by Toole-Stott (1976, 1978).

20. Lewis (1886).

21. Baird (1949).

22. Davey's report appears in both the *Journal* and the *Proceedings* in similar form. All references here are from the *Proceedings* because it contains the extensive reports of his subjects (Davey 1887a; Davey 1887b).

23. Typhoid fever is usually caused by bacterial contamination of water, but Londoners of that era also contracted typhoid from milk because contaminated water was often used to wash storage pans.

24. Hodgson died of a heart attack while playing handball at the age of fifty (Baird 1949). Professor Hoffmann died at seventy-nine years of age (Findlay 1977). Nora

Sidgwick was still writing, editing, and intellectually curious until the last days of her ninety-one years (Johnson 1936).

25. Wallace (1891). Spiritualists often accused magicians of hiding their mediumistic powers (Frost 1881, 334; Maskelyne 1876, 67; Ewer 1855, 30).

26. Hodgson (1892).

27. Baxter (2009) has written a historical review of secret writing devices, like graphite attached to the end of a thimble—a thimble pencil.

28. A quick review of the *Master Index to Magic in Print* suggests that more than 150 magicians have written routines for slate writing (Potter 1967). Some of the better slate secrets can be found in Grant and Smith (1931), Hull (1929), Lonergan (1930), Mann (n.d.), Warlock (1942), and Robinson (1898).

29. Hodgson (1892).

30. I have made minor edits of this description for brevity.

31. Neher (1990).

32. In 1910, a three-minute drama was constructed involving only four actors. None of the witnesses correctly reported the events, which was especially startling because almost all were attorneys or judges. They accused the wrong man of theft because they misremembered the conversation after an interruption, and they malobserved the crucial action (Research Officer, 1930). The question of eyewitness testimony is extremely complex. Some features of a situation can make the task more difficult and some make it easier (Yuille and Cutshall 1986).

33. The importance of missing information is usually studied under the heading of "omission neglect" (Sanbonmatsu et al. 2003). When an individual has incomplete or limited knowledge, the value of known information is often overestimated.

34. Robertson (1905, 7).

CONCLUSION

1. Prince (1921, 590).
2. Barnum (1866).
3. Fawkes (1920, 36).

BIBLIOGRAPHY

A. B. 1853. *Letters on Table-moving, or the Recent Miracle at Tramutola, and on the Influence of Animal Motion upon Attraction.* London: Hippolyte Bailliere.

A Converted Infidel. 1855. *The Fate of Infidelity; or the Dealings of Providence with Modern Infidels.* New York: Z. Paten Hatch.

A Plain Citizen. [1918]. *Some Revelations As to "Raymond:" An Authoritative Statement.* London: Kegan Paul, Trench, Trubner.

A Priest of the Church of England [Jean François Baltus]. 1709. *An Answer to Mr. de Fontenelle's History of Oracles . . . with Some Reflections upon the Remarks of Mr. Le Clerc.* London: printed by W. B. for Henry Clements.

Abbott, David P. 1908. *Behind the Scenes with the Mediums.* Chicago: Open Court.

———. 1913. *The Spirit Portrait Mystery: Its Final Solution.* Chicago: Open Court.

Abrams, Max. 1992. *Annemann: The Life and Times of a Legend.* Tahoma, CA: L & L Publishing.

Aczel, Amir D. 2005. *Descartes' Secret Notebook: A True Tale of Mathematics, Mysticism, and the Quest to Understand the Universe.* New York: Broadway Books.

Adamczewski, Jan. ca. 1973. *Nicolaus Copernicus and His Epoch.* Philadelphia, PA: Copernicus Society of America.

Adams, Samuel Hopkins. 1906. *The Great American Fraud.* N.p.: Collier & Son.

Adams, W. H. Davenport. 1864. *Dwellers on the Threshold; or Magic and Magicians with Some Illustrations of Human Error and Imposture.* 2 vols. London: Maxwell.

Adkins, Gregory Matthew. 2000. "The Moral Philosophy of Fontenelle." *Journal of the History of Ideas* 61: 433–52.

Africa, Thomas W. 1968. *Science and the State in Greece and Rome.* New York: Wiley & Sons.

Agnew, Robin A. L. 2002. *The Life of Sir John Forbes (1787–1861).* Bramber, West Sussex: Bernard Durnford Publishing.

Agrippa, Henrie Cornelius. 1575. *Of the Vanitie and Uncertaintie of Artes and Sciences.* London: Henrie Bynneman.

———. 1898. *Natural Magic.* Chicago: Hahn & Whitehead.

Alcock, James E. 2018. *Belief: What It Means to Believe and Why Our Convictions Are So Compelling.* Amherst, NY: Prometheus Books.

Alder, Ken. 2002. *The Measure of All Things: The Seven-year Odyssey and Hidden Error That Transformed the World.* New York: Free Press, 2002.

Alexander, Claude. n.d. *Crystal Gazing.* Los Angeles: Alexander Publishing.

———. 1921. *The Life and Mysteries of the Celebrated Dr. "Q."* Los Angeles: Alexander.

———. 1922. *Alexander's Book of Mystery.* Los Angeles: C. Alexander.

———. 1924a. *Personal Lessons, Codes, Formulas and Instructions for Members of the Crystal Silence League.* Los Angeles: Alexander Publishing.

———. 1924b. *The Inner Secrets of Psychology.* 5 vols. Los Angeles: C. Alexander Publishing.

———. 1924c. *Oriental Wisdom: Its Principles and Practice.* Los Angeles: C. Alexander Publishing.

Al-Khalili, Jim. 2011. *The House of Wisdom: How Arabic Science Saved Ancient Knowledge and Gave Us the Renaissance.* New York: Penguin.

Alpheus, A. 1902. *Complete Hypnotism: Mesmerism, Mind Reading and Spiritualism.* Chicago: M. A. Donohue.

American Psychological Association. 1992. "Ethical Principles of Psychologists and Code of Conduct." *American Psychologist* 47: 1597–1611.

An Anxious Enquirer after Truth. 1854. *The Table and the Turner; Which of the Two Is Possessed?* London: Piper, Stephenson, and Spence.

An Old Boy. 1868. *The Marvels of Optical and Chemical Magic, and How to Accomplish Them.* London: Houlston & Wright.

Andriopoulos, Stefan. 2008. *Possessed: Hypnotic Crimes, Corporate Fiction, and the Invention of Cinema.* Chicago: University of Chicago Press.

Andrus, Jerry. 1978. *Andrus Deals You In.* Albany, OR: JA Enterprises.

Angus, S. 1975. *The Mystery-religions: A Study of the Religious Background of Early Christianity.* New York: Dover.

Anonymous. n.d. *Mrs. Etta Roberts, the So-called Cage Medium.* N.p.: n.p.

———. 1661. *Humane Industry: Or a History of Most Manual Arts, Deducing the Original, Progress, and Improvement of Them. Furnished with Variety of Instances and Examples, Shewing Forth the Excellency of Humane Wit.* London: printed for Henry Herringman.

———. 1678. *Strange and Remarkable Prophesies and Predictions of the Holy, Learned, and Excellent James Usher, Late L. Arch-Bishop of Armagh, and Late Primate of Ireland.* London: printed for R. G.

———. 1680. *The Knavery of Astrology Discover'd.* London: printed for T. B. and R. E.

———. 1707. *Clavis Prophetica; or, a Key to the Prophecies of Mons. Marion, and the Other Camisars.* London: F. Morphew.

———. [1803]. *The Conjuror's Repository; or The Whole Art and Mystery of Magic Displayed*. London: T. and R. Hughes.

———. 1810. *A Faithful Account of Catharine Mewis, of Barton-under Needwood, Staffordshire, Aged Seven Years, Who for More Than Twelve Months, Has, and Still Continues, to Be Deprived of Her Eye Sight, Six Days out of Seven, and Can Only See, on the Sabbath-Day*. Derby: printed at the office of G. Wilkins.

———. 1821. *Curiosities for the Ingenious: Selected from the Most Authentic Treasures of Nature, Science and Art, Biography, History, and General Literature*. London: Thomas Boys.

———. 1824. *Oracle of the Arts; or, Entertaining Expounder of the Wonders of Science*. London: printed for John Bumpus and Richard Griffin.

———. 1825. *Catholic Miracles; Illustrated with Seven Designs, Including a Characteristic Portrait of Prince Hohenlohe, by George Cruikshank*. London: printed for Knight and Lacey.

———. 1826. *Rational Recreations: Midsummer MDCCCXXIV*. London: Knight and Lacey.

———. ca. 1827. *The Universal Conjuror, or, the Whole Art of Legerdemain, As Practiced by the Famous Breslaw, Katterfelto, Jonas, Flockton, Comas, and the Greatest Adepts in London and Paris*. London: Orlando Hodgson.

———. 1835. *Endless Amusement. The Largest Collection Ever Heretofore Published, of the Most Interesting and Instructive Experiments in Various Branches of Science*. Halifax: William Milner.

———. 1837. *Journeys into the Moon, Several Planets and the Sun. History of a Female Somnambulist, of Weilheim on the Teck, in the Kingdom of Wuektemberg, in the Years 1832 and 1833*. Philadelphia: Vollmer and Haggenmacher.

———. ca. 1840. *The Art of Conjuring Made Easy; or, Instructions for Performing the Most Astonishing Slight-of-hand Feats, with Directions for Making Fireworks*. Derby: Thomas Richardson.

———. 1847a. *Endless Amusement: A Collection of Nearly 400 Entertaining Experiments in Various Branches of Science*. Philadelphia: Lea & Blanchard.

———. 1847b. *The Book of Entertainment, of Curiosities and Wonders in Nature, Art, and Mind. Drawn from the Most Authentic Sources, Carefully Revised, and Illustrated by More Than One Hundred Engravings*. New York: Charles S. Francis.

———. 1849. *The Prophecy of Orval; Containing a Prediction of All the Remarkable Events from the First French Revolution down to the Present Time, with Proofs of Its Authenticity*. London: James Burns.

———. 1850. *Singular Revelations. Explanation and History of the Mysterious Communion with Spirits, Comprehending the Rise and Progress of the Mysterious Noises in Western New-York, Generally Received As Spiritual Communications*. Auburn, NY: Finn & Rockwell.

———. 1851. *Rochester Knockings! Discovery and Explanation of the Source of the Phenomena Generally Known As the Rochester Knockings*. Buffalo: George H. Derby.

———. 1853. *Table-turning and Table-talking Considered, in Connection with the Dictates of Reason and Common Sense.* Bath, UK: Samuel Gobbs.

———. 1855. *The Little Boy's Own Book of Sports and Pastimes.* London: David Bogue.

———. 1858a. *An Angel's Message.* London: John Wesley.

———. 1858b. *The Sociable; or One Thousand and One Home Amusements.* New York: Dick & Fitzgerald.

———. 1859. *The Boy's Own Conjuring Book: Being a Complete Hand-book of Parlour Magic.* New York: Dick & Fitzgerald.

———. 1860. *The Confessions of a Medium.* Boston: Ticknor and Fields.

———. [1865]. "What Is Clairvoyance." *The Boys' Journal: A Magazine of Literature, Science, Adventure, and Amusement.* London: Henry Vickers.

———. ca. 1866. *Prophecies of Robert Nixon, Mother Shipton, and Martha, the Gipsy.* London: Published for the Booksellers.

———. ca. 1870. *Spirit Mysteries Exposed, Being a Full and Plain Explanation of the Wonderful Feats of the Davenport Brothers and Other "Mediums," with a History of "Spirit Rapping," and Explanation of the Means by Which Its Manifestations Are Produced, Etc.* New York: Jesse Haney.

———. 1871. *The Magician's Instructor Containing Complete Details and Instructions in Performing Feats of the Black Art.* Baltimore: Fisher & Denison.

———. 1874. *How to Become a Clairvoyant, Containing Full Instructions for Producing the Psychologic, Mesmeric, and Clairvoyant Conditions; with a Comprehensive Statement of the Phenomena and Uses of the Same.* Newark, NJ: Union Publishing.

———. 1884. *The Secrets of Clairvoyance and How to Become an Operator. Mesmerism and Psychology, and How to Become a Mesmerizer and Psychologist, Being a Complete Embodiment of All the Curious Facts Connected with the above Strange Sciences, with Instructions How to Become a Medium.* New York: M. Young.

———. 1889. *Parlor Magic.* New York: Worthington.

———. 1893. "A Puzzling Problem: Clairvoyance, Telepathy, or Spirits? A Report on Professor Baldwin and His Wife." *Borderland* 1, no. 2: 142–48.

———. [ca. 1895]. *Spiritualism Exposed.* Boston: A. B. Courtney.

———. 1899. *A Complete Course in the Art of Mind Reading.* Chicago: Psychic Research Company.

———. 1901. *The Perfect Course of Instruction in Hypnotism, Mesmerism, Clairvoyance, Suggestive Therapeutics, and the Sleep Cure, Giving Best Methods of Hypnotizing by Masters of the Science.* Chicago: Psychic Research Company.

———. 1909. *The Work of the Open Court Publishing Co: An Illustrated Catalogue of Its Publications Covering a Period of Twenty-one Years (1887–1907).* Chicago: Open Court.

———. 1943. "Rich As Croesus." *Bulletin of the Business Historical Society* 17, no. 2: 47–48.

Aradi, Zsolt. 1956. *The Book of Miracles.* New York: Farrar, Straus and Cudahy.

Archdale, F. A. 1961. *Elementary Radiesthesia and the Use of the Pendulum.* McKelume Hill, CA: Health Research.

Arendt, M., and A. Elklit. 2001. "Effectiveness of Psychological Debriefing." *Acta Psychiatrica Scandinavica,* 104: 423–27.

Ariely, Dan. 2008. *Predictably Irrational: The Hidden Forces That Shape Our Decisions.* New York: HarperCollins.

Arikha, Noga. 2007. *Passions and Tempers: A History of the Humours.* New York: HarperCollins.

Arnold, George. 1862. *The Magician's Own Book, or the Whole Art of Conjuring.* New York: Dick and Fitzgerald.

Arons, Harry. 1948. *The Master Course in Hypnotism.* Newark, NJ: Power Publishers.

———. 1953. *Techniques of Speed Hypnosis.* South Orange, NJ: Power Publishers.

Aronson, Simon. 1990. *The Aronson Approach.* N.p.: privately published.

Ashburner, John. 1867. *Notes and Studies in the Philosophy of Animal Magnetism and Spiritualism with Observations upon Catarrh, Bronchitis, Rheumatism, Gout, Scrofula, and Cognate Diseases.* London: H. Bailliere.

Aubrey, John. 1857. *Miscellanies upon Various Subjects.* London: John Russell Smith.

Augustine. 1958. *The City of God.* Translated by G. G. Walsh, D. B. Zema, G. Monahan, and D. J. Honan. Garden City, NY: Image Books.

Austen, A. W. 1938. *The Teachings of Silver Birch.* London: Psychic Press.

Bacon, Roger. 1659/2011. *Friar Bacon: His Discovery of the Miracles of Art, Nature, and Magick Faithfully Translated out of Dr. Dees Own Copy, by T. M. and Never before in English.* N.p.: EEOB.

———. 1928. *The Opus Majus of Roger Bacon.* Philadelphia: University of Pennsylvania Press.

Badcock, John. 1828. *Philosophical Recreations, or Winter Amusements.* London: printed for T. Hughes.

Baggally, W. W. 1917. *Telepathy: Genuine and Fraudulent.* London: Methuen.

Bailey, Michael D. 2007. *Magic and Superstition in Europe: A Concise History from Antiquity to the Present.* Lanham, MD: Rowman & Littlefield.

———. 2013. *Fearful Spirits, Reasoned Follies: The Boundaries of Superstition in Late Medieval Europe.* Ithaca, NY: Cornell University Press.

Bailly, Christian. 1987. *Automata: The Golden Age 1848–1914.* London: Sotheby's Publications.

Baird, Alex. 1949. *The Life of Richard Hodgson.* London: Psychic Press.

Baldini, Ugo. 2001. "The Roman Inquisition's Condemnation of Astrology: Antecedents, Reasons and Consequences." In *Church, Censorship and Culture in Early Modern Italy.* Edited by G. Frangnito. Cambridge: Cambridge University Press.

Baldinucci, Filippo. 1682/1966. *The Life of Bernini.* University Park: Pennsylvania State University Press.

Baldwin, Eugene F., and Maurice Eisenberg. 1897. *Leaves from the Note-book of a Hypnotist.* Boston: Baldwin, Eisenberg & Flint.

Baldwin, S. S. 1879. *Spirit Mediums Exposed*. Melbourne: McCarron, Bird.

Baldwin, Samri S. 1895. *The Secrets of Mahatma Land Explained*. Brooklyn, NY: Dyson.

Ballou, Adin. 1852. *An Exposition of Views Respecting the Principal Facts, Causes and Peculiarities Involved in Spirit Manifestations*. Boston: Bela Marsh.

Bangs, M., and L. Bangs. 1905. *Bangs Sisters: Phenomenal Psychics*. Chicago: privately printed.

Barfield, Owen. 1985. *History in English Words*. Great Barrington, MA: Lindisfarne Press.

Bargh, J. A., and Chartrand, T. L. 1999. "The Unbearable Automaticity of Being." *American Psychologist* 54: 462–79.

Barkas, Thomas P. *Outlines of Ten Years' Investigations into the Phenomena of Modern Spiritualism, Embracing Letters, Lectures, &c.* London: Frederick Pitman, 1862.

Barker, A. T. 1948. *The Mahatma Letters to A. P. Sinnett from the Mahatmas M. & K. H.* London: Rider.

Barnello [Edward A. Barnwell]. ca. 1880. *Barnello's Voodoo Incantations or How to Eat Fire*. New York: Popular Pub.

Barnett, S. J. 1999. *Idol Temples and Crafty Priests: The Origins of Enlightenment Anti-clericalism*. London: Macmillan.

Barnum, P. T. 1866. *The Humbugs of the World: An Account of Humbugs, Delusions, Impositions, Quackeries, Deceits and Deceivers Generally, in All Ages*. New York: Carleton.

Barnum, Vance. 1919. *Joe Strong the Boy Fire-eater*. New York: Sully.

Barone, T. 1995. "Persuasive Writings, Vigilant Readings, and Reconstructed Characters: The Paradox of Trust in Educational Storysharing." In *Life History and Narrative*. Edited by J. A. Hatch and R. Wisniewski. London: Falmer Press.

Barraclough, G. 1979. *The Medieval Papacy*. New York: Norton.

Barrett, W. F. 1911. *Seeing without Eyes*. Halifax: Spiritualists' National Union.

Basterfield, Candice, Scot O. Lilienfeld, Shauna M. Bowes, and Thomas H. Costello. 2020. "The Nobel Disease: When Intelligence Fails to Protect against Irrationality." *Skeptical Inquirer* 44, no. 3: 32–37.

Bate, John. 1654. *The Mysteries of Nature and Art*. London: printed by R: Bishop for Andrew Crook.

Baxter, Thomas. 2009. *The Nail Writer Anthology*. N.p.: privately printed.

Beard, George M. 1877a. *The Scientific Basis of Delusions: A New Theory of Trance, and Its Bearings on Human Testimony*. New York: G. P. Putnam's Sons.

———. 1877b. "Physiology of Mind-reading." *Popular Science Monthly* 10: 459–73.

———. 1879a. "The Psychology of Spiritism." *North American Review* 129: 65–80.

———. 1879b. "The Delusions of Clairvoyance." *Scribner's Monthly* 18: 433–40.

———. 1882. *The Study of Trance, Muscle-reading and Allied Nervous Phenomena in Europe and America, with a Letter on the Moral Character of Trance Subjects, and a Defence of Dr. Charcot*. New York: privately printed.

Beaumont, John. 1724. *Gleanings of Antiquities*. London: printed for J. Roberts.

Beckmann, John. 1817. *A History of Inventions and Discoveries*. 4 vols. London: printed for Longman, Hurst, Rees, Orme, and Brown.

Beckmann, Petr. 1971. *A History of π (pi)*. New York: Dorset Press.

Beeching, Jack. 1983. *The Galleys at Lepanto*. New York: Charles Scribner's Sons.

Begbie, P. I. 1851. *Supernatural Illusions*. London: T. C. Newby.

Behnke, Leo. 1990. *The Collected Mental Secrets of C. A. George Newmann*. South Pasadena, CA: Daniel's Den.

Benedict, Barbara M. 2001. *Curiosity: A Cultural History of Early Modern Inquiry*. Chicago: University of Chicago Press.

Beretta, Marco. 2009. *The Alchemy of Glass: Counterfeiting, Imitation, and Transmutation in Ancient Glassmaking*. Sagamore Beach, MA: Watson.

Beretta, Marco, Antonio Clericuzio, and Lawrence M. Principe, eds. 2009. *The Accademia del Cimento and Its European Context*. Sagamore Beach, MA: Science History Publications.

[Bernard, Jean Frederic]. [1760]. *The Praise of Hell: or, a Discovery of the Infernal World. Describing the Advantages of That Place, with Regard to Its Situation, Antiquity, and Duration. With a Particular Account of Its Inhabitants; Their Dresses, Customs, Manners, Occupations and Diversions in Which Are Included, the Laws Constitution and Government of Hell, with Notes Historical and Critical to Explain the Whole*. London: M. Cooper.

Bernheim, H. 1884/1965. *Hypnosis and Suggestion in Psychotherapy*. New Hyde Park, NY: University Books.

Bernstein, Bruce. 2012. "Presentation Excuses the Method." *Magic* 22, no. 3: 44–45.

Berve, Helmut, and Gottfried Gruben. [1963]. *Greek Temples, Theatres and Shrines*. New York: Harry N. Abrams.

Bevan E. 1929. *Sibyls and Seers: A Survey of Some Ancient Theories of Revelation and Inspiration*. Cambridge: Harvard University Press.

Biondi, Massimo. 2013. "Parapsychologist Massimo Biondi Comment." *Psypioneer Journal* 9: 225.

Bishop, Washington Irving. 1880. *Second Sight Explained: A Complete Exposition of Clairvoyance or Second Sight, As Exhibited by the Late Robert Houdin and Robert Heller*. Edinburgh: John Menzies.

Black, Christopher F. 2009. *The Italian Inquisition*. New Haven, CT: Yale University Press.

Blackburn, Gabriele. 1983. *The Science and Art of the Pendulum: A Complete Course in Radiesthesia*. Ojai, CA: Idylwild Books.

Blackmore, Susan. 2018. "Daryl Bem and Psi in the Ganzfeld." *Skeptical Inquirer* 42: 44–45.

Blackwelder, Eilot. 1909. "Mars As the Abode of Life." *Science* 29: 659–61.

Blackwell, Richard J. 1991. *Galileo, Bellarmine, and the Bible*. Notre Dame, IN: University of Notre Dame Press.

Blondel, David. 1661. *A Treatise of the Sibyls, So Highly Celebrated, As Well by the Ancient Heathens, As the Holy Fathers of the Church; Giving an Accompt [sic] of the Names, and Number of the Sibyls, of the Qualities, the Form and Matter of Their Verses; As Also of the Books Now Extant under Their Names, and the Errors Crept into Christian Religion, from the Impostures Contained Therein, Particularly, Concerning the State of the Just, and Unjust after Death.* London: printed by T. R. for the author.

Boas, Marie. 1966. *The Scientific Renaissance: 1450–1630.* New York: Harper Torchbooks.

Bodie, Walford. 1905. *The Bodie Book: Hypnotism, Electricity, Mental Suggestion, Magnetic Touch, Clairvoyance, Telepathy.* London: Caxton Press.

Bok, Sissela. 1978. *Lying: Moral Choice in Public and Private Life.* New York: Pantheon Books.

Bond, William. 1728. *The Supernatural Philosopher: Or, the Mysteries of Magick, in All Its Branches, Clearly Unfolded. Containing, . . . The History of the Life and Surprising Adventures of Mr. Duncan Campbell.* London: E. Curll.

Bonomo, Joe. 1952. *Feats of Strength: Strongmen's Tricks & Secrets.* New York: Strongmen's Club of America.

Booth, John. 1931. *Magical Mentalism.* Hamilton, Ontario: Seagers Press.

Borch-Jacobsen, Mikkel. 2000. "How to Predict the Past: From Trauma to Repression." *History of Psychiatry* 15–35.

Bordelon, Laurent. 1711. *A History of the Ridiculous Extravagancies of Monsieur Oufle.* London: J. Morphew.

Borel, T. 1879. *The Count Agenor De Gasparin: A Biography.* London: Hodder & Straughton.

Bossy, John. 1991. *Giordano Bruno and the Embassy Affair.* New Haven, CT: Yale University Press.

Bowden, Hugh. 2005. *Classical Athens and the Delphic Oracle: Divination and Democracy.* Cambridge: Cambridge University Press.

Bowman, Marilyn. 1997. *Individual Differences in Posttraumatic Response: Problems with the Adversity-Distress Connection.* Mahwah, NJ: Erlbaum.

Boyle, Robert. n.d. *The Sceptical Chymist.* London: Dent & Sons.

Bracciolini, Poggio. [1928]. *The Facetiae of Poggio and Other Medieval Story-tellers.* Translated by Edward Storer. London: George Routledge & Sons.

Bradley, H. Dennis. [1937]. *Towards the Stars.* London: T. Werner Laurie.

Braid, James. 1899. *Braid on Hypnotism: Neurypnology or the Rationale of Nervous Sleep Considered in Relation to Animal Magnetism or Mesmerism and Illustrated by Numerous Cases of Its Successful Application in the Relief and Cure of Disease.* London: George Redway.

Bramwell, J. Milne. 1903. *Hypnotism: Its History, Practice and Theory.* London: Grant Richards.

Brandon, Joan. 1956. *Successful Hypnotism.* New York: Stravon Publishers.

Brandon, Ruth. 1983. *The Spiritualists: The Passion for the Occult in the Nineteenth and Twentieth Centuries*. New York: Knopf.

Braude, Ann. 1990. "News from the Spirit World: A Checklist of American Spiritualist Periodicals, 1847–1900." *Proceedings of the American Antiquarian Society* 99, no. 2: 399–462.

Brewster, David. 1839. *Letters on Natural Magic*. New York: Harper & Brothers.

———. 1882. *Letters on Natural Magic*. London: Sonnenschein.

Bridgett, T. E. 1890. *Blunders and Forgeries*. London: Kegan Paul, Trench, Trubner.

Brigham, Josiah. 1859. *Twelve Messages from the Spirit John Quincy Adams, through Joseph D. Stiles, Medium to Josiah Brigham*. Boston: Bela Marsh.

Brill, A. 1975. *A. Brill's Bible of Building Plans*. Peoria, IL: A. B. Enterprises.

Brittan, S. B. 1868. *Man and His Relations: Illustrating the Influence of the Mind on the Body*. New York: Townsend & Adams.

Britten, Emma Hardinge. 1884. *Nineteenth Century Miracles; or, Spirits and Their Work in Every Country of the Earth. A Complete Historical Compendium of the Great Movement Known As "Modern Spiritualism."* New York: William Britten, Lovell & Co.

Broad, William J. 2006. *The Oracle: The Lost Secrets and Hidden Messages of Ancient Delphi*. New York: Penguin.

Brockman, John, ed. 2017. *Today's Most Visionary Thinkers Reveal the Cutting-edge Scientific Ideas and Breakthroughs You Must Understand*. New York: Harper Perennial.

Bromhall, Thomas. 1658. *A Treatise of Specters. Or, an History of Apparitions, Oracles, Prophecies and the Cunning Delusions of the Devil, to Strengthen the Idolatry of the Gentiles*. London: John Streater.

Bronowski, Jacob. 1975. "Copernicus As a Humanist." In *The Nature of Scientific Discovery: A Symposium Commemorating the 500th Anniversary of the Birth of Nicolaus Copernicus*. Edited by Owen Gingerich, 170–88. Washington, DC: Smithsonian Institution Press.

Bronowski, J., and Bruce Mazlish. 1962. *The Western Intellectual Tradition: From Leonardo to Hegel*. New York: Harper & Row.

Brooke, Christopher. 1971. *The Structure of Medieval Society*. New York: McGraw-Hill.

Brooke-Hitching, Edward. 2018. *The Phantom Atlas: The Greatest Myths, Lies and Blunders on Maps*. San Francisco: Chronicle Books.

Brower, M. Brady. 2010. *Unruly Spirits: The Science of Psychic Phenomena in Modern France*. Urbana: University of Illinois Press.

Brown, B. G. 1938. *A Report on Three Experimental Fire-walks by Ahmed Hussain and Others*. London: University of London Council for Psychical Investigation.

Brown, Gary. 2017. "Henry Slade and His Slates." *Gibeciere* 12, no. 1: 10–112.

Browne, Thomas. 1646. *Pseudodoxia Epidemica: or, Enquiries into Very Many Received Tenents, and Commonly Presumed Truths*. London: Edward Dod.

Bruce, Colin. 2001. *Conned Again, Watson: Cautionary Tales of Logic, Math, and Probability*. Cambridge, MA: Perseus.

Bruce, F. F. 1978. *History of the Bible in English*. New York: Oxford University Press.

Brugger, Peter, and Christine Mohr. 2008. "The Paranormal Mind: How the Study of Anomalous Experiences and Beliefs May Inform Cognitive Neuroscience." *Cortex* 44: 1291–98.

Brugger, Peter, and Kirsten I. Taylor. 2003. "ESP: Extrasensory Perception or Effects of Subjective Probability?" *Journal of Consciousness Studies* 10: 221–46.

Bruno, Giordano. 1964. *Five Dialogues by Giordano Bruno: Cause, Principle and Unity*. Translated by Jack Lindsay. New York: International Publishers.

Brunvand, H. J. 1981. *The Vanishing Hitchhiker: American Urban Legends and Their Meanings*. New York: W. W. Norton.

Bungener, L. F. 1852. *History of the Council of Trent*. Edinburgh: Thomas Constable.

Burckhardt, Jacob. 1929. *The Civilization of the Renaissance in Italy*. London: George G. Harrap.

Burke, Peter. 1981. *Popular Culture in Early Modern Europe*. New York: Harper & Row.

Burkett, B. G., and Glenna Whitley. 1998. *Stolen Valor*. Dallas, TX: Verity Press.

Burlingame, H. J. 1891. *Leaves from Conjurers' Scrap Books or, Modern Magicians and Their Works*. Chicago: Honohue, Henneberry.

———. 1895. *Tricks in Magic, Illusions, and Mental Phenomena*. Chicago: Clyde Publishing.

———. [1905]. *How to Read People's Minds*. Chicago: Clyde.

Burnett, C. M. 1850. *The Philosophy of Spirits in Relation to Matter: Shewing the Real Existence of Two Very Distinct Kinds of Entity Which Unite to Form the Different Bodies That Compose the Universe, Organic and Inorganic, by Which the Phenomena of Light, Heat, Electricity, Motion, Life, Mind, Etc. Are Reconciled and Explained*. London: Samuel Highley.

Burr, Alex C. 1933. "Notes on the History of the Concept of Thermal Conductivity." *Isis* 20: 246–59.

Burtt, Edwin Arthur. 1955. *The Metaphysical Foundations of Modern Physical Science*. Garden City, NY: Doubleday.

Butterfield, Herbert. 1962. *The Statecraft of Machiavelli*. New York: Collier Books.

———. 1965. *The Origins of Modern Science 1300–1800*. New York: The Free Press.

Butterworth, Philip. 2005. *Magic on the Early English Stage*. Cambridge: Cambridge University Press.

C. D. [Sophia Elizabeth de Morgan]. 1863. *From Matter to Spirit. The Result of Ten Years' Experience in Spirit Manifestations Intended As a Guide to Enquirers*. London: Longman, Green, Longman, Roberts, & Green.

C. E. D. [Charlotte E. Dresser]. 1922. *Spirit World and Spirit Life: Descriptions Received through Automatic Writing by C.E.D.* Los Angeles: J. F. Rowny Press.

Cadwallader, M. E. 1922. *Hydesville in History*. Chicago: Progressive Thinker Publishing House.

Cadwell, J. W. 1883. *Full and Comprehensive Instructions How to Mesmerize*. Boston: published by the author.

Cahagnet, L. Alph. 1851. *The Celestial Telegraph; or, Secrets of the Life to Come, Revealed through Magnetism*. New York: Redfield.

Calmet, Augustine. 1850. *The Phantom World: The History and Philosophy of Spirits, Apparitions, Etc*. Philadelphia: Hart.

Calvert, John. n.d. *Easy to Learn Stage Hypnosis*. N.p.: privately printed.

Cameron, Euan. 2011. *Enchanted Europe: Superstition, Reason, and Religion, 1260–1750*. New York: Oxford University Press.

Campbell, E. Z. 1853. *The Spiritual Telegraphic Opposition Line; or, Science and Divine Revelation against Spiritual Manifestations*. Springfield, MA: Taylor.

Cannon, Alexander. 1953. *The Power Within*. New York: Dutton.

Capron, E. W. 1855. *Modern Spiritualism; Its Facts and Fanaticisms, Its Consistencies and Contradictions*. Boston: Bella Marsh.

Cardano, Girolamo. 2002. *The Book of My Life*. Translated by Jean Stoner. New York: New York Review Books.

Carey, George W. 1918. *The Wonders of the Human Body, Physical Regeneration According to the Laws of Chemistry and Physiology*. Los Angeles: Chemistry of Life Co.

———. 1932. *The Biochemic System of Medicine Comprising the Theory, Pathological Action, Therapeutical Application, Materia Medica, and Repertory of the Twelve Biochemic Remedies Based on the Discoveries of Dr. W. H. Schuessler*. St. Louis, MO: Luyties Pharmacal Co.

Carey, George W., and Inez Eudora Perry. 1948. *The Zodiac and the Salts of Salvation*. Los Angeles, CA: Carey-Perry School of the Chemistry of Life.

Carlson, Eric T. 1960a. "Charles Poyen Brings Mesmerism to America." *Journal of the History of Medicine* 15: 121–32.

———. 1960b. "Addenda to the Early History of Hypnotic Anesthesia." *Journal of the History of Medicine* 33: 81–83.

Caron, M., and S. Hutin. 1961. *The Alchemists*. New York: Grove Press.

Carpenter, William. 1852. "On the Influence of Suggestion in Modifying and Directing Muscular Movement, Independently of Volition." *Proceedings of the Royal Institution of Great Britain* 1: 147–53.

Carpenter, William. B. 1877. *Mesmerism, Spiritualism, Etc., Historically and Scientifically Considered*. New York: Appleton.

———. 1886. *Principles of Mental Physiology, with Their Applications to the Training and Discipline of the Mind, and the Study of Its Morbid Conditions*. New York: D. Appleton.

———. 1889. *Nature and Man: Essays Scientific and Philosophical*. New York: D. Appleton.

Carrington, Hereward. 1907. *The Physical Phenomena of Spiritualism*. Boston: Turner.

———. 1908. *Vitality, Fasting and Nutrition: A Physiological Study of the Curative Power of Fasting, Together with a New Theory of the Relation of Food to Human Vitality*. New York: Rebman.

———. 1909a. "Eusapia Palladino: The Despair of Science." *Collier's Magazine* 33: 660–75.

———. 1909b. *Eusapia Palladino and Her Phenomena*. New York: Dodge.

———. 1909c. *Hindu Magic*. London: Annals of Psychical Science.

———. [1913]. *Personal Experiences in Spiritualism*. London: T. Werner Laurie.

———. 1913. *Side-show and Animal Tricks*. Kansas City: A. M. Wilson.

———. [1928]. *Side-show Tricks Explained*. Girard, KS: Haldeman-Julius Publications.

———. 1932. *A Primer of Psychical Research*. London: Rider.

———. ca. 1964. *Letters to Hereward Carrington from Famous Psychical Researchers, Scientists, Mediums, and Magicians*. New York: Fieldcrest.

Carswell, John. 1950a. *The Prospector: Being the Life and Times of Rudolf Erich Raspe (1737–1794)*. London: Cresset Press.

———. 1950b. *The Romantic Rogue: Being the Singular Life and Adventures of Rudolph Eric Raspe, Creator of Baron Munchausen*. New York: Dutton.

Casey, Charles. 1876. *Philitis: Being a Condensed Account of the Recently Discovered Solution of the Use and Meaning of the Great Pyramid . . . to Which Is Added a Review of Professor Piazzi Smyth's Second Edition of "Our Inheritance in the Great Pyramid."* Dublin: Carson Brothers.

Cassirer, Ernst, Paul Oskar Kristeller, and John Herman Randall, eds. 1967. *The Renaissance Philosophy of Man*. Chicago: University of Chicago Press.

Ceci, Stephen J., and Maggie Bruck. 1995. *Jeopardy in the Courtroom: A Scientific Analysis of Children's Testimony*. Washington, DC: American Psychological Association.

Celenza, C. S. "Lorenzo Valla, 2004 'Paganism' and Orthodoxy." In "Studia Humanitatis: Essays in Honor of Salvatore Camporeale." Supplement, *MLN* 119, no. 1: S66–S87.

Celsus et al. 1978. *Arguments of Celsus, Porphyry, and the Emperor Julian, against the Christians*. Ann Arbor, MI: University Microfilms International.

Chandler, F. W. 1907. *The Literature of Roguery*. New York: Burt Franklin.

Changeux, Jean Pierre. 2004. *The Physiology of Truth*. Cambridge, MA: Belknap Press.

Chapin, David. 2004. *Exploring Other Worlds: Margaret Fox, Elisha Kent Kane, and the Antebellum Culture of Curiosity*. Amherst: University of Massachusetts Press.

Charcot, J-M. 1987. *Charcot the Clinician: The Tuesday Lessons*. New York: Raven Press.

Charvet, David, and John Pomeroy. 2004. *Alexander: The Man Who Knows*. Pasadena, CA: Mike Caveney's Magic Works.

Cheesman, Clive, and Johnathan Williams. 2000. *Rebels Pretenders & Impostors*. New York: St. Martin's Press.

Cheiro. 1928. *Cheiro's World Predictions*. Los Angeles: London Publishing.

Chislett, T. H. 1949. *Spirits in the House*. Birmingham, AL: Goodliffe.

Cholmondeley-Pennell, H. 1884. *"Bringing It to book:" Facts of Slate-writing through Mr. W. Eglinton.* London: Psychological Press Association and E. W. Allen.

Christie-Murray, David. 1989. *A History of Heresy.* Oxford: Oxford University Press.

Christopher, Milbourne. 1970. *ESP, Seers & Psychics.* New York: Crowell.

Cicero, Marcus Tullius. 1997. *The Nature of the Gods and on Divination.* Amherst, NY: Prometheus.

Clanny, W. Reid. 1841. *A Faithful Record of the Miraculous Case of Mary Jobson.* London: Newcastle-upon-Tyne.

Clark, Stuart. 1997. *Thinking with Demons: The Idea of Witchcraft in Early Modern Europe.* Oxford: Oxford University Press.

Clarke, E. H. 1880. *Visions: A Study of False Sight (Pseudopia).* Boston: Houghton, Osgood.

Clavelin, Maurice. 1974. *The Natural Philosophy of Galileo: Essay on the Origins and Formation of Classical Mechanics.* Cambridge, MA: MIT Press.

Clay, Edmund. [1853]. *The Certain Danger, Folly, and Sin, of Table-talking and Table-turning.* Leamington: G. C. Liebenrood.

Clement of Alexandria. 1960. *Clement of Alexandria.* Translated by G. W. Butterworth. London: William Heinemann.

Clericuzio, Antonia. 2009. "The Other Side of the Accademia del Cimento: Borelli's Chemical Investigations." In *The Accademia del Cimento and Its European Context.* Edited by Marco Beretta, Antonio Clericuzion, and Lawrence M. Principe. Sagamore Beach, MA: Science History Publications.

Close, Francis. 1853. *The Testers Tested, or, Table Moving, Turning, and Talking Not Diabolical: A Review of the Publications of the Rev. Messrs. Godfrey, Gillson, Vincent, and Dibdin.* London: T. Hatchard.

Cohen, Morris R., and I. E. Drabkin. 1958. *A Source Book in Greek Science.* Cambridge, MA: Harvard University Press.

Coleman, Christopher B. 1922. *The Treatise of Lorenzo Valla on the Donation of Constantine.* New Haven, CT: Yale University Press.

Collins, A. Frederick. 1939. *The Book of Magic.* New York: D. Appleton-Century.

Colquhoun, John Campbell. 1839. *Zoo-magnetic Journal.* Edinburgh: Adam and Charles Black.

———. 1851. *An History of Magic, Witchcraft, and Animal Magnetism.* London: Longman, Brown, Green, & Longmans.

Committee of the London Dialectical Society. 1871. *Report of Spiritualism, of the Committee of the London Dialectical Society, Together with the Evidence, Oral and Written, and a Selection from the Correspondence.* London: Longmans, Green, Reader and Dyer.

———. 1873. *Report on Spiritualism of the Committee of the London Dialectical Society.* London: Burns.

Committee on Thought-transference. 1883 "Third Report on Thought-transference." *Proceedings of the Society for Psychical Research* 1: 161–215.

Como, James. 1994. "Mere Lewis." *Wilson Quarterly* 18: 109–17.

Connor, Steven. 1999. *Dumbstruck: A Cultural History of Ventriloquism*. Oxford: Oxford University Press.

Cook, Walter. 1917. *Reflections on "Raymond:" An Appreciation and Analysis*. London: Grant Richards.

Cooke, Conrad William. 1903. *Automata Old and New*. London: Sette of Odd Volumes.

Coopland, B. W. 1952. *Nicole Oresme and the Astrologers*. Liverpool: University Press.

Coover, John Edgar. 1917. *Experiments in Psychical Research at Leland Stanford Junior University*. Stanford, CA: published by the University.

Corrington, Julian D. 1961. "Spontaneous Generation." *Bios* 32: 62–76.

Corson, Hiram. 1919. *Spirit Messages*. Boston: Christopher Publishing.

Coulomb, E. 1885. *Some Account of My Association with Madame Blavatsky from 1872 to 1884*. N.p.: n.p.

Coulter, Ottley R. 1952. *How to Perform Strong Man Stunts*. New York: Padell.

Cowdrey, Wayne L., Howard A. Davis, and Donald R. Scales. 1977. *Who Really Wrote the Book of Mormon*. Santa Ana, CA: Vision House Publishers.

Coyne, R. R. 1954. *When God Smiled on Ronald Coyne*. Sapulpa, OK: Mrs. R. R. Coyne.

Crabtree, Adam. 1988. *Animal Magnetism, Early Hypnotism, and Psychical Research, 1766–1925: An Annotated Bibliography*. White Plains, NY: Kraus International Publications.

———. 1993. *From Mesmer to Freud: Magnetic Sleep and the Roots of Psychological Healing*. New Haven, CT: Yale University Press.

Craft, A. N. 1881. *Epidemic Delusions: Containing an Exposé of the Superstitions and Frauds Which Underlie Some Ancient and Modern Delusions, Including Especial Reference to Modern Spiritualism*. Cincinnati, OH: Walden and Stowe.

Crane, Mark, Richard Raiswell, and Margaret Reeves, eds. 2004. *Shell Games: Studies in Scams, Frauds, and Deceits (1300–1650)*. Toronto: Centre for Reformation and Renaissance Studies.

Creed, F., and E. Guthrie. 1993. "Techniques for Interviewing the Somatising Patient." *British Journal of Psychiatry* 162: 467–71.

Creery, A. M. Letter. 1887. *Journal of the Society of Psychical Research* 3: 269–70.

Cremer, W. H. [ca. 1874]. *Hanky Panky: A Book of Easy and Difficult Conjuring Tricks*. Edinburgh: Grant.

———. 1892. *The Magician's Own Book*. London: Chatto & Windus.

Crichley, MacDonald. 1956. "Congenital Indifference to Pain." *Annals of Internal Medicine* 45: 737–47.

Crisciani, Chiara. 1999. "From the Laboratory to the Library: Alchemy According to Guglielmo Fabri." In *Natural Particulars: Nature and the Disciplines in Renaissance Europe*. Edited by Anthony Grafton and Nancy Siraisi. Cambridge, MA: MIT Press.

Criswell, [Jeron]. 1968. *Criswell Predicts from Now to the Year 2000*. Anderson, SC: Droke House.

Crombie, A. C. 1959. *Medieval and Early Modern Science*. Garden City, NY: Doubleday.

Crossley, Robert. 2000. "Percival Lowell and the History of Mars." *The Massachusetts Review* 41: 297–318.

———. 2008. "Mars and the Paranormal." *Science Fiction Studies* 35: 466–84.

Crowe, Michael J. 1986. *The Extraterrestrial Life Debate 1750–1900: The Idea of a Plurality of Worlds from Kant to Lowell*. Cambridge: Cambridge University Press.

Crowell, Eugene. 1879. *The Spirit World: Its Inhabitants, Nature, and Philosophy*. Boston: Colby & Rich.

Cumberland, Stuart. 1888. *Thought-reader's Thoughts: Being the Impressions and Confessions of Stuart Cumberland*. London: Sampson Low, Marston, Searle & Rivington.

———. 1905. *People I Have Read*. London: C. Arthur Pearson.

———. 1918. *That Other World: Personal Experiences of Mystics and Their Mysticism*. London: Grant Richards.

———. 1919. *Spiritualism—The Inside Truth*. London: Odhams.

Curnow, Trevor. 2004. *The Oracles of the Ancient World*. London: Duckworth.

Curry, Patrick. 1989. *Prophecy and Power: Astrology in Early Modern England*. Princeton, NJ: Princeton University Press.

Curtis, James. 1902. *Rustlings in the Golden City: Being a Record of Spiritualistic Experiences in Ballarat and Melbourne*. London: Office of "Light."

Dallas, Helen Alexandrina. ca. 1922. *Objections to Spiritualism Answered*. London: London Spiritualist Alliance.

Damasio, Antonio. 2006. *Descartes' Error*. London: Vintage Books.

Dance, Peter. 1976. *Animal Fakes and Frauds*. Maidenhead: Sampson Low.

Daniels, J. W. 1856. *Spiritualism versus Christianity; or, Spiritualism thoroughly Exposed*. New York: Miller, Orton & Mulligan.

Darnton, Robert, and Daniel Roche. 1989. *Revolution in Print: The Press in France 1775–1800*. Berkeley: University of California Press.

Daston, Lorraine, and Katharine Park. 1999. *Wonders and the Order of Nature: 1150–1750*. New York: Zone Books.

Davenport, Ruben Briggs. 1888. *The Death Blow to Spiritualism*. New York: G. W. Dillingham.

Davey, S. J. 1887a. "Experimental Investigation." *Proceedings of the Society for Psychical Research* 4: 403–95.

———. 1887b. "The Possibilities of Mal-observation &c., from a Practical Point of View." *Journal of the Society for Psychical Research* 3: 8–44.

Davey, William. 1889. *The Illustrated Practical Mesmerist: Curative and Scientific*. London: James Burns.

Davies, Charles Maurice. 1867. *Mystic London; or, Phases of Occult Life in the British Metropolis*. New York: Lovell.

Davies, Owen. 2017. *The Oxford Illustrated History of Witchcraft and Magic*. Oxford: Oxford University Press.

Davis, Andrew Jackson. 1855. *The Great Harmonia; Being a Philosophical Revelation of the Natural, Spiritual, and Celestial Universe*. Vol. 1, *The Physician*. Boston: Sanborn, Carter & Bazin.

———. 1866. *The Monk of the Mountains; or, a Description of the Joys of Paradise . . . with the Destiny and Conditions of the Nations of the Earth for One Hundred Years to Come, by the Hermit Himself*. Indianapolis, IN: Downey & Brouse.

Davis, W. S. 1910. "The New York Exposure of Eusapia Palladino." *Journal of the American Society for Psychical Research* 4: 401–24.

Davy, John. 1836. *Memoirs of the Life of Sir Humphry Davy, Bart*. 2 vols. London: Longman, Rees, Orme, Brown, Green, & Longman.

Dawson, Trevor. 2012. *Charles Dickens: Conjurer, Mesmerist and Showman*. N.p.: Trevor Dawson.

de Bergerac, Cyrano. 1962. *Voyages to the Moon and the Sun*. Translated by Richard Aldington. New York: Orion Press.

De Boer, J. Z. 2001. "New Evidence for the Geological Origins of the Ancient Delphic Oracle." *Geology* 29: 707–10.

De Boismont, A. Brierre. 1853. *Hallucinations: Or, the Rational History of Apparitions, Visions, Dreams, Ecstasy, Magnetism, and Somnambulism*. Philadelphia: Lindsay & Blakiston.

de Camp, L. Sprague, and Catherine C. de Camp. 1966. *Spirits, Stars, and Spells: The Profits and Perils of Magic*. New York: Canaveral Press.

de Caus, Isaac. 1982. *Wilton Garden: New and Rare Inventions of Water-works*. New York: Garland.

de Courmelles, Foveau. 1895. *Hypnotism*. London: Routledge.

de Gasparin, Count Agenor. 1857. *Science vs. Modern Spiritualism. A Treatise on Turning Tables, the Supernatural in General, and Spirits*. 2 vols. New York: Kiggins & Kellogg.

de Koven, Anna. 1920. *A Cloud of Witnesses*. New York: E. P. Dutton.

de la Motte, Bourdois. 1843. "Objections to Mesmerism." *The Mesmerist: A Journal of Vital Magnetism* 6: 44–48.

De Lawrence, George. 1927. *Answers to Questions*. N.p.: R. A. Nelson.

De Lawrence, L. W. 1902. *Practical Lessons in Hypnotism and Magnetism*. Chicago: Drake.

———. 1910. *India's Hood Unveiled*. Chicago: De Lawrence, Scott.

De Morgan, Augustus. 1872/1954. *A Budget of Paradoxes*. New York: Dover.

De Rosa, Peter. 1988. *Vicars of Christ: The Dark Side of the Papacy*. New York: Crown.

De Santillana, Giorgio. 1955. *The Crime of Galileo*. Chicago: University of Chicago Press.

———. 1961. *The Origins of Scientific Thought from Anaximander to Proclus 600 B.C.–500 A.D.* New York: New American Library.

Dean, Frank. 1937. *The Art of Knife Throwing*. San Jose: Frank Dean.

Dean, Henry. ca. 1739. *The Whole Art of Legerdemain; or Hocus Pocus in Perfection.* 3rd ed. London: printed for A. Bettsworth and C. Hitch, R. Ware, J. Osborn, & J. Hodges.

Dear, Peter. 1990. "Miracles, Experiments, and the Ordinary Course of Nature." *ISIS* 81: 663–83.

———. 1997. *The Scientific Enterprise in Early Modern Europe.* Chicago: University of Chicago Press.

———. 1998. "A Mechanical Microcosm: Bodily Passions, Good Manners, and Cartesian Mechanism." In *Science Incarnate: Historical Embodiments of Natural Knowledge.* Edited by Christopher Lawrence and Steven Shapin. Chicago: University of Chicago Press.

Debus, Allen G. 1992. "Science and History: The Birth of a New Field." In *Science in Early Modern Thought.* Edited by Stephen A. McKnight. Columbia: University of Missouri Press.

Defoe, Daniel. 1835. *A Journal of the Plague Year; Memorials of the Great Pestilence in London, in 1665.* London: Thomas Tegg and Son.

Deleuze, J. P. F. 1886. *Practical Instruction in Animal Magnetism.* New York: Fowler & Wells.

Despiau, M. L. 1801. *Select Amusements in Philosophy and Mathematics; Proper for Agreeable Exercising the Minds of Youth.* London: W. Glendinning.

Devant, David. 1931. *My Magic Life.* London: Hutchinson.

Dewey, D. M. 1850. *History of the Strange Sounds or Rappings, Heard in Rochester and Western New York, and Usually Called the Mysterious Noises! Which Are Supposed by Many to Be Communications from the Spirit World Together with All the Explanation That Can As Yet Be Given of the Matter.* Rochester, NY: D. M. Dewey.

Dexter, Will. 1956. *Sealed Vision.* London: George Armstrong.

———. 1979. *The Uncanny Power.* Alberta: Hades.

Diaconis, Perse, and Ron Graham. 2012. *Magical Mathematics: The Mathematical Ideas That Animate Great Magic Tricks.* Princeton, NJ: Princeton University Press.

Diamond, Michael. 2004. *Victorian Sensation: Or, the Spectacular, the Shocking and the Scandalous in Nineteenth-century Britain.* London: Anthem.

Dibdin, R. W. 1853. *Table-turning.* London: privately printed.

Dickens, A. G. 1967. *Reformation and Society in Sixteenth-century Europe.* New York: Harcourt, Brace & World.

———. 1969. *The Counter Reformation.* Harcourt, Brace & World.

Didier, Adolphe. 1856. *Animal Magnetism and Somnambulism.* London: T. C. Newby.

Dijksterhuis, E. J. 1964. *The Mechanization of the World Picture.* Oxford: Oxford University Press.

Dingwall, Eric J. 1962. *Very Peculiar People: Portrait Studies in the Queer, the Abnormal and the Uncanny.* New Hyde Park, NY: University Books.

———. 1967. *Abnormal Hypnotic Phenomena: A Survey of Nineteenth-century Cases [France].* New York: Barnes & Noble.

Dobbs, Betty Jo Teeter. 1992. "Alchemical Death and Resurrection." In *Science, Pseudo-science, and Utopianism in Early Modern Thought*. Edited by Stephen A. McKnight. Columbia, MS: University of Missouri Press.

Dodds, E. R. 1951. *The Greek and the Irrational*. Berkeley: University of California Press.

Doesticks, Q. K. Philander. 1859. *The Witches of New York*. New York: Rudd & Carleton.

Dolnick, Edward. 2008. *The Forger's Spell: A True Story of Vermeer, Nazis, and the Greatest Art Hoax of the Twentieth Century*. New York: Harper-Collins.

Donagan, Alan, and Barbara Donagan. *Philosophy of History*. New York: Macmillan, 1965.

Donaldson, I. M. L. 2005. "Mesmer's 1780 Proposal for a Controlled Trial to Test His Method of Treatment Using 'Animal Magnetism.'" *Journal of the Society of Medicine* 98: 572–75.

Dr. Q. n.d. *Instantaneous Method*. Buffalo, NY: privately printed.

Dr. Q [William Larsen]. 1941. *Dr. Q's Gyrating Tables*. N.p.: F. G. Thayer.

Dr. X [Gerald Heaney]. [1922]. *On the Other Side of the Footlights: An Expose of Routines, Apparatus and Deceptions Resorted to by Mediums, Clairvoyants, Fortune Tellers and Crystal Gazers in Deluding the Public*. Berlin, WI: Heaney.

Drachmann, A. G. 1948. *Ktesibios, Philon, and Heron: A Study in Ancient Pneumatics*. Copenhagen.

Drake, Stillman. 1957. *Discoveries and Opinions of Galileo*. New York: Anchor Books.

———. 1978. *Galileo at Work: His Scientific Biography*. Chicago: University of Chicago Press.

———. 2001. *Galileo: A Very Short Introduction*. Oxford: Oxford University Press.

Dresser, Horatio W. 1921. *The Quimby Manuscripts*. New York: Crowell.

Druckman, Daniel, and Robert A. Bjork, eds. 1991. *In the Mind's Eye: Enhancing Human Performance*. Washington, DC: National Academy Press.

Du Prel, Carl. 1889. *The Philosophy of Mysticism*. London: George Redway.

Duguid, David. 1914. *Hafed Prince of Persia: His Experiences in Earth-life and Spirit-life, Being Spirit Communications Received through Mr. David Duguid, the Glasgow Trance-painting Medium. With an Appendix, Containing Communications from the Spirit Artists, Ruisdal and Steen*. London: W. Foulsham.

Dumont, Theron Q. 1917. *The Advanced Course in Personal Magnetism*. Philadelphia, PA: Domino Publishing.

Dunbar, Robin. 1995. *The Trouble with Science*. Cambridge: Harvard University Press.

Duncan, David Ewing. 1998. *The Calendar: The 5000-Year Struggle to Align the Clock and the Heavens and What Happened to the Missing Ten Days*. London: Fourth Estate.

Dunninger, Joseph. 1926. *Popular Magic*. Vol. 1. New York: Experimenter Publishing.

———. 1928. *Houdini's Spirit Exposés from Houdini's Own Manuscripts, Records and Photographs and Dunninger's Psychical Investigations*. New York: Experimenter Publishing.

———. 1967. *Dunninger's Complete Encyclopedia of Magic*. New York: Lyle Stuart.

Dunston, A. J. 1973. "Pope Paul II and the Humanists." *Journal of Religious History* 7: 287–306.

Durant, C. F. 1837/1982. *Exposition, or a New Theory of Animal Magnetism*. New York: Da Capo Press.

Dyson, George B. 1997. *Darwin among the Machines: The Evolution of Global Intelligence*. Reading, MA: Addison-Wesley.

E. W. C. N. (1844) "Dr. Forbes and Alexis." *The Zoist* 2: 393–401.

Eamon, William. 1994. *Science and the Secrets of Nature: Books of Secrets in Medieval and Early Modern Culture*. Princeton, NJ: Princeton University Press.

———. 2010. *The Professor of Secrets: Mystery, Medicine, and Alchemy in Renaissance Italy*. Washington, DC: National Geographic.

Earle, Lee. 1989. *The Classic Reading*. N.p.: Binary Star.

Eco, Umberto. 1988. *Foucault's Pendulum*. San Diego: Harcourt Brace.

Eddy, Mary Baker. 1920. *Manual of the Mother Church*. Boston: Christian Science Publishing Society.

Editor. 1860. "The Manifold Phases of Spiritualism." *The Spiritual Magazine* 1: 49–55.

———. 1864. "Dr. Elliotson." *The Spiritual Magazine* 5: 215–17.

———. 1895. "Exit Eusapia!" *British Medical Journal* 2, no. 1819: 1182.

Edmonds, John W. 1858. *Spiritual Tracts*. New York: n.p.

Edmonds, John W., and George T. Dexter. 1853. *Spiritualism*. 2 vols. New York: Partridge & Brittan.

Edwards, Gavin. 2019, December 6. "Overlooked No More: Rose Mackenberg, Houdini's Secret Ghost-buster." *New York Times*.

Edwards, Paul, ed. 1976. *The Encyclopedia of Philosophy*. New York: Macmillan.

Egginton, William. 2016. *The Man Who Invented Fiction: How Cervantes Ushered in the Modern World*. New York: Bloomsbury.

Ehrman, Bart D. 2005. *Misquoting Jesus: The Story behind Who Changed the Bible and Why*. New York: HarperSanFrancisco.

———. 2010. *Jesus, Interrupted: Revealing the Hidden Contradictions in the Bible*. New York: HarperCollins.

———. 2011. *Forged: Writing in the Name of God—Why the Bible's Authors Are Not Who We Think They Are*. New York: HarperOne.

Eidinow, Esther. 2007. *Oracles, Curses, and Risk among the Ancient Greeks*. Oxford: Oxford University Press.

Eisenstein, Elizabeth L. 1980. *The Printing Press As an Agent of Change: Communications and Cultural Transformations in Early-modern Europe*. Cambridge: Cambridge University Press.

Ekstein, Nina. 1996. "Appropriation and Gender: The Case of Catherine Bernard and Bernard de Fontenelle." *Eighteenth-Century Studies* 30, no. 1: 59–80.

Elliotson, John. 1845. "Reports of Various Trials of the Clairvoyance of Alexis Didier." *The Zoist* 2: 477–529.

———. 1849. "Remarks of a Female Mesmerist in Reply to the Scurrilous Insinuations of Dr. F. Hawkins, Dr. Mayo, and Mr. Walkey." *The Zoist* 7: 44–53.

———. 1851. "Dr. Esdaile and the *London Medical Gazette*." *The Zoist* 9: 112–24.

———. 1852. "More Clairvoyance in Alexis Didier." *The Zoist* 10: 221–24.

———. 1856. "Conclusions of *The Zoist*." *The Zoist* 13: 444.

Elliott, Alfred. 1873. *Within-doors: A Book of Games and Pastimes for the Drawing-room*. London: T. Nelson.

Elliott, Charles Wyllys. 1852. *Mysteries; or, Glimpses of the Supernatural*. New York: Harper.

Emmons, S. Bulfinch. 1853. *Philosophy of Popular Superstitions and the Effects of Credulity and Imagination*. Boston: L. P. Crown.

Emsley, John. 2000. *The 13th Element: The Sordid Tale of Murder, Fire, and Phosphorus*. New York: John Wiley & Sons.

Enfield, W. 1821. *Scientific Amusements in Philosophy and Mathematics: Including Arithmetic, Acoustics, Electricity, Magnetism, Optics, Pneumatics, Together with Amusing Secrets in Various Branches of Science, the Whole Being Calculated to Form an Agreeable and Improving Exercise for the Mind*. London: printed for A. K. Newman and Co. and Simpkin and Marshall, T. Tegg, and Edwards and Knibbs; also Griffin and Co. Glasgow.

Englebert, Omer. 1951. *The Lives of the Saints*. New York: David McKay.

Erasmus. 1925. *Erasmus in Praise of Folly*. Edited by Horace Bridges. Chicago: Pascal Covici.

Erickson, Carolly. 1976. *The Medieval Vision: Essays in History and Perception*. New York: Oxford University Press.

Ernst, B. M. L., and Hereward Carrington. 1932. *Houdini and Conan Doyle: The Story of a Strange Friendship*. New York: Albert and Charles Boni.

Ernst, Germana. 2010. "The Mirror of Narcissus: Cardano Speaks of His Own Life." *Bruniana & Campanelliana* 16: 451–61.

Evans, Henry Ridgely. 1897. *Hours with the Ghosts*. Laird & Lee: Chicago.

Evans, J. A. S. 1978. "What Happened to Croesus?" *The Classical Journal* 74: 34–40.

Ewbank, Thomas. 1850. *A Descriptive and Historical Account of Hydraulic and Other Machines for Raising Water*. New York: Bangs, Platt.

Ewer, F[erdinand] C[artwright]. 1855. *The Eventful Nights of August 20th and 21st, 1854: And How Judge Edmonds Was Hocussed; or, Fallibility of "Spiritualism" Exposed*. New York: Samuel Hueston.

Faivre, Antoine. 1995. *The Eternal Hermes: From Greek God to Alchemical Magus*. Grand Rapids, MI: Phanes Press.

Falconer, John. 1685. *Information Disclosed without a Key. Containing, Plain and Demonstrative Rules, for Decyphering All Manner of Secret Writing. With Exact Methods, for Resolving Secret Intimations by Signs or Gestures, or in Speech. As Also an Inquiry into the Secret Ways of Conveying Written Messages: And the Several Mysterious Proposals for Secret Information, Mentioned by Trithemius, &c.* London: printed for Daniel Brown.

Fanger, Claire. 1998. *Conjuring Spirits: Texts and Traditions of Medieval Ritual Magic.* Gloucestershire: Sutton.

———, ed. 2012. *Invoking Angels: Theurgic Ideas and Practices, Thirteenth to Sixteenth Centuries.* University Park: Pennsylvania State University Press.

Fara, Patricia. 2002. *Newton: The Making of Genius.* New York: Columbia University Press.

Faraday, Michael. 1853. "Professor Faraday on Table-moving." *Athenæum* 1340: 801–3.

Farmer, Bob. 2004. "Rules for Fools—Hit the Mark and Strike It Rich." *Genii* 67, no. 8: 16–19.

———. 2014. *The Bammo Ten Card Deal Dossier.* Toronto: Magicana.

Farmer, J. S. 1886. *Twixt Two Worlds: A Narrative of the Life and Work of William Eglinton.* London: Psychological Press.

Farrer, James Anson. 1892. *Books Condemned to Be Burnt.* London: Elliot Stock.

———. 1907. *Literary Forgeries.* London: Longmans, Green.

Fawkes, F. A. 1920. *Spiritualism Exposed.* Bristol: J. W. Arrowsmith.

Fechner, Christian. 2002. *The Magic of Robert-Houdin.* Boulohne, France: FCF Editions.

Feilding, Everard. 1963. *Sittings with Eusapia Palladino & Other Studies.* New Hyde Park, NY: University Books.

Feilding, Everard, W. W. Baggally, and Hereward Carrington. 1909. "Report on a Series of Sittings with Eusapia Palladino." *Proceedings of the Society for Psychical Research* 23: 306–569.

Ferguson, Niall. 2008. *The Ascent of Money: A Financial History of the World.* New York: Penguin.

Ferguson, Wallace K. 1953. *The Renaissance.* New York: Henry Holt.

Ferguson, Wallace K. et al. 1962. *The Renaissance: Six Essays.* New York: Harper & Row.

Fermi, Laura, and Gilberto Bernardini. 1965. *Galileo and the Scientific Revolution.* Greenwich, CT: Fawcett.

Ferris, Timothy. 1988. *Coming of Age in the Milky Way.* New York: Doubleday.

Ferrone, Vincenzo. 1995. *The Intellectual Roots of the Italian Enlightenment: Newtonian Science, Religion, and Politics in the Early Eighteenth Century.* Amherst, NY: Humanity Books.

Festinger, L., H. W. Reicken, and S. Schachter. 1956. *When Prophecy Fails.* New York: Harper.

Findlay, James B., and Thomas A. Sawyer. 1977. *Professor Hoffmann: A Study*. Tustin, CA: private publication.

Findlen, Paula. 1990. "Jokes of Nature and Jokes of Knowledge: The Playfulness of Scientific Discourse in Early Modern Europe." *Renaissance Quarterly* 43: 292-331.

Finley, M. I. 1968. *Aspects of Antiquity: Discoveries and Controversies*. New York: Viking Press.

———. 1981. *Early Greece: The Bronze and Archaic Ages*. New York: Norton.

Finocchiaro, Maurice A. 1991. *The Galileo Affair: A Documentary History*. New York: Notable Trials Library.

Fisher, John. 1980. *Body Magic*: New York: Stein and Day.

Fiske, Bradley A. 1921. *Invention: The Master-key to Progress*. New York: Dutton.

Fitzkee, Dariel. 1935. *Contact Mind Reading Expanded*. San Rafael, CA: Saint Raphael House.

Fix, Andrew. 2010. "What Happened to Balthasar Bekker in England? A Mystery in the History of Publishing." *Church History and Religious Culture* 90: 609-31.

Fixen, L. G. 1912. *Mind Reading or Second Sight: As Performed by the Zancigs*. Chicago: Diamond Dust.

Flaceliere, R. 1965. *Greek oracles*. New York: Norton.

Flammarion, Camille. 1891. *Urania: A Romance*. London: Chatto & Windus.

———. 1900a. *L'inconnu: The Unknown*. New York: Harper & Brothers.

———. 1900b. *L'Inconnu et les problemes psychiques: Manifestations des mourants apparitions, telepathie communications, etc*. Paris: Flammarion.

———. 1907. *Mysterious Psychic Forces: An Account of the Author's Investigations in Psychical Research, Together with Those of Other European Savants*. Boston: Small, Maynard.

———. 1923. *Dreams of an Astronomer*. New York: Appleton.

Fleming, John V. 2013. *The Dark Side of the Enlightenment: Wizards, Alchemists, and Spiritual Seekers in the Age of Reason*. New York: Norton.

Flint, Herbert L. n.d. *Dr. Herbert L. Flint's Hypnotic Routine*. San Jose: Merlin Enterprises.

———. 1912a. *A Course in Hypnotism and Hypnotic Suggestion and How to Acquire and Utilize Hypnotic Power*. Cleveland: Flint's College of Hypnotism.

———. 1912b. *Flint's Lessons in Hypnotism*. Cleveland, OH: Flint's College of Hypnotism.

Flint, Valerie. 1999. "The Demonization of Magic and Sorcery in Late Antiquity: Christian Redefinitions of Pagan Religions." In *Witchcraft and Magic in Europe: Ancient Greece and Rome*. Edited by Bengt Ankarloo and Stuart Clark. Philadelphia: University of Pennsylvania Press.

Flothmann, Gerhard. ca. 1916. *Die Macht der Hypnose Praktischer Lehrkurs des Hypnotismus und Magnetismus für Jedermann*. Leipzig: F. W. Gloeckner.

Flournoy, Theodore. 1900. *From India to the Planet Mars*. New York: Harper & Brothers.

Floyer, John. 1713. *The Sibylline Oracles*. London: R. Bruges.

Fontenelle, Bernard Bovier. 1990. *Conversations on the Plurality of Worlds*. Berkeley: University of California Press.

Fontenelle, [Bernard de]. 1687. *Histoire des oracles*. Amsterdam: Mortier.

Fontenelle, Bernard de. 1753. *The History of Oracles in Two Dissertations*. N.p.: printed by R. Urie.

Fontenrose, Joseph. 1978. *The Delphic Oracle*. Berkeley: University of California Press.

Forbes, John. 1845a. *Illustrations of Modern Mesmerism from Personal Investigation*. London: John Churchill.

———. 1845b. *Mesmerism True—Mesmerism False: A Critical Examination of the Facts, Claims, and Pretensions of Animal Magnetism*. London: John Churchill.

Forrest, Derek. 1999. *Hypnotism*. London: Penguin.

Forte, Steve. 2004. *Casino Game Protection: A Comprehensive Guide*. Las Vegas, NV: SLF Publishing.

———. 2006. *Poker Protection: Cheating . . . and the World of Poker*. Las Vegas, NV: SLF Publishing, 2006.

———. 2020. *Gambling Sleight of Hand: Forte Years of Research*. Las Vegas, NV: Steve Forte.

Fournier d' Albe, Edmond Edward. 1923. *The Life of Sir William Crookes*. London: Fisher Unwin.

Fox, Matthew, ed. 1987. *Hildegard of Bingen's Book of Divine Works with Letters and Songs*. Santa Fe, NM: Bear.

Freedberg, David. 2002. *The Eye of the Lynx: Galileo, His Friends, and the Beginnings of Modern Natural History*. Chicago: University of Chicago Press.

Frost, Thomas. 1881. *The Lives of the Conjurors*. London: Chatto and Windus.

Fuller, G. N., and M. V. Gale. 1988. "Migraine Aura As Artistic Inspiration." *British Medical Journal* 297: 1670–72.

Fuller, J. G. 1968. *The Day of St. Anthony's Fire*. New York: Macmillan.

Fuller, Ronald. 1936. *The Beggars' Brotherhood*. London: George Allen & Unwin.

Fulves, Karl. 1979. *The Shamrock Code*. Teaneck, NJ: Fulves.

Gale, Leonard D. 1838. *Elements of Natural Philosophy: Embracing the General Principles of Mechanics, Hydrostatics, Hydraulics, Pneumatics, Acoustics, Optics, Electricity, Galvanism, Magnetism, and Astronomy*. New York: Collins, Keese.

Galilei, Galileo. 1914. *Dialogues Concerning Two New Sciences*. New York: Dover.

Gardner, Laurence. 2005. *The Magdalene Legacy: The Jesus and Mary Bloodline Conspiracy*. London: Element.

Gardner, Martin. 1981. *Science: Good, Bad and Bogus*. Buffalo: Prometheus.

———. 1989. *How Not to Test a Psychic: Ten Years of Remarkable Experiments with Renowned Clairvoyant Pavel Stapanik*. Buffalo, NY: Prometheus Books.

Garenne, Henri [Frank Lind]. [1886]. *The Art of Modern Conjuring*. London: Ward, Lock.

Garrett, Julia E. 1892. *Mediums Unmasked: An Expose of Modern Spiritualism.* Los Angeles: H. M. Lee & Bro.

Garrett, Teral. 1948. *Twenty-six Living and Dead Tests.* London: George Armstrong.

Garrison, Peter. 1985. "Kindling Courage." *Omni* 7: 44.

Garwood, Christine. 2007. *Flat Earth: The History of an Infamous Idea.* New York: St. Martin's Press.

Gauld, Alan. 1968. *The Founders of Psychical Research.* New York: Schocken Books.

———. 1995. *A History of Hypnotism.* Cambridge: Cambridge University Press.

Gauld, H. D. 1929. *Ghost Tales and Legends.* London: Chambers.

Geley, Gustave. 1927. *Clairvoyance and Materialisation: A Record of Experiments.* Translated by Stanley de Brath. New York: Doran.

Gentilcore, David. 2000. *Medical Charlatanism in Early Modern Italy.* Oxford: Oxford University Press.

Gibson, Walter. 1928. *Popular Card Tricks.* New York: E. I. Company.

Giglio, James. 1998. "A Comment on World War II Repression." *Professional Psychology: Research and Practice* 29: 470.

Gilbert, William. 1600/1958. *On the Magnet.* New York: Basic Books.

Gillson, E. [1853a]. *Table-talking; Disclosures of Satanic Wonders & Prophetic Signs.* London: Binns & Goodwin.

———. [1853b]. *A Watchman's Appeal, with Special Reference to the Unexplained Wonders of the Age.* London: Binns and Goodwin.

———. [1853c]. *Whose Is the Responsibility? A Letter to the Rev. F. Close, M.A., in Reply to His Pamphlet "The Testers Tested."* London: Binns and Goodwin.

Gilson, Etienne. 1938. *Reason and Revelation in the Middle Ages.* New York: Scribner's Sons.

Gimpel, Jean. 1976. *The Medieval Machine: The Industrial Revolution of the Middle Ages.* New York: Holt, Rinehart and Winston.

Gingerich, Owen. 1989. *Album of Science: The Physical Sciences in the Twentieth Century.* New York: Charles Scribner's Sons.

———. 1992. *The Great Copernicus Chase and Other Adventures in Astronomical History.* Cambridge: Sky Publications.

Ginzburg, Carlo. 1983. *The Night Battles: Witchcraft and Agrarian Cults in the Sixteenth and Seventeenth Centuries.* New York: Penguin Books.

———. 1992a. *Clues, Myths, and the Historical Method.* Baltimore: Johns Hopkins University Press.

———. 1992b. *Ecstasies: Deciphering the Witches' Sabbath.* New York: Penguin Books.

Giobbi, Roberto. 2010. *Secret Agenda.* Seattle, WA: Hermetic Press.

Glazebrook, J. K. 1854. *Table-talking a Fraud; or, "Godfrey's Cordial" for the Satanic Agency School.* Blackburn: printed by J. Walkden.

Glucklich, Ariel. 1997. *The End of Magic.* New York: Oxford University Press.

Godfrey, N. S. 1853. *Table-turning, the Devil's Modern Master-piece.* 4th ed. London: Seeleys.

———. 1854. *The Theology of Table-turning, Spirit-rapping, and Clairvoyance, in Connection with the Antichrist.* London: Seeleys.

Godwin, Joscelyn. 2009. *Athanasius Kircher's Theatre of the World: The Life and Work of the Last Man to Search for Universal Knowledge.* Rochester, VT: Inner Traditions.

Godwin, William. 1834. *Lives of the Necromancers; or, an Account of the Most Eminent Persons in Successive Ages, Who Have Claimed for Themselves, or to Whom Has Been Imputed by Others, the Exercise of Magical Power.* London: Mason.

Goldschmidt, E. P. 1951. "The First Edition of Lucian of Samosata." *Journal of the Warburg and Courtauld Institutes* 14: 7–20.

Goldsmith, Barbara. 1998. *Other Powers: The Age of Suffrage, Spiritualism, and the Scandalous Victoria Woodhull.* New York: Knopf.

Goldsmith, Margaret. 1934. *Franz Anton Mesmer: A History of Mesmerism.* Garden City, NY: Doubleday, Doran.

Goldston, Will. 1905. *Crystal Gazing: Astrology, Palmistry, Planchette, and Spiritualism.* London: Gamage.

———. 1921. *More Exclusive Magical Secrets.* London: Will Goldston.

Golinski, Jan. 1999. *Science As Public Culture: Chemistry and Enlightenment in Britain, 1760–1820.* Cambridge: Cambridge University Press.

Gordon, William R. 1856. *A Three-fold Test of Modern Spiritualism.* New York: Charles Scribner.

Gould, C. 1886. *Mythical Monsters.* London: W. H. Allen.

Gould, George M., and Walter L. Pyle. 1897. *Anomalies and Curiosities of Medicine.* Philadelphia: W. B. Saunders.

Grafton, Anthony. 1990. *Forgers and Critics: Creativity and Duplicity in Western Scholarship.* London: Juliet Gardiner Books.

———. 1992. *New Worlds, Ancient Texts: The Power of Tradition and the Shock of Discovery.* Cambridge: Belknap Press.

———. 1999. *Cardano's Cosmos: The Worlds and Works of a Renaissance Astrologer.* Cambridge, MA: Harvard University Press.

———. 2009. *Worlds Made by Words: Scholarship and Community in the Modern West.* Cambridge MA: Harvard University Press, 2009.

Granger, William. 1803. *Wonderful Museum, and Extraordinary Magazine: Being a Complete Repository of All the Wonders, Curiosities, and Rarities of Nature and Art.* 6 vols. London: Kelly and Hogg.

Grant, U. F. ca. 1930. *U. F. Grant's Popular Magical Effects.* Catalog No. 1. Columbus, OH: U. F. Grant.

Grant, U. F., and H. Adrian Smith. 1931. *Flap Slate Wrinkles.* Pittsfield, MA: privately published.

Grasset, Joseph. 1910. *The Marvels beyond Science.* New York: Funk & Wagnalls.

Greary, P. J. 1990. *Furta Sacra: Thefts of Relics in the Central Middle Ages.* Princeton, NJ: Princeton University Press.

Greenblatt, Stephen. 1991. *Marvelous Possessions: The Wonder of the New World*. Chicago: University of Chicago Press.

———. 2011. *The Swerve: How the World Became Modern*. New York: Norton.

Greenwood, James. 1869. *The Seven Curses of London*. London: Stanley Rivers.

Gregory, Anita. 1985. *The Strange Case of Rudi Schneider*. Metuchen, NJ: Scarecrow Press.

Gregory, Richard L. 2009. *Seeing through Illusion*. New York: Oxford University Press.

Gregory, William. 1877. *Animal Magnetism; or, Mesmerism and Its Phenomena*. London: Harrison.

Gresham, William L. 1946. *Nightmare Alley*. New York: Rinehart.

———. 1948. *Monster Midway*. New York: Rinehart.

———. 1959. *Houdini: The Man Who Walked through Walls*. New York: Holt.

Griggs, W. N. 1852. *The Celebrated "Moon Story," Its Origin and Incidents*. New York: Bunnell and Price.

Grimes, J. Stanley. 1857. *The Mysteries of Human Nature Explained by a New System of Nervous Physiology: To Which Is Added a Review of the Errors of Spiritualism*. Buffalo: R. M. Wanzer.

———. 1885. *The Mysteries of the Head and the Heart Explained: Including an Improved System of Phrenology; a New Theory of the Emotions, and an Explanation of the Mysteries of Mesmerism, Trance, Mind-reading, and the Spirit Delusion*. Chicago: Henry A. Sumner.

Grimm, Harold J. 1949. "Lorenzo Valla's Christianity." *Church History* 18: 75–88.

Groebner, Valentin. 2007. *Who Are You? Identification, Deception, and Surveillance in Early Modern Europe*. New York: Zone Books.

Guillain, Georges. 1959. *J.-M. Charcot 1825–1893: His Life—His Work*. New York: Paul B. Hoeber.

Guillemin, F. 2002. *An Illustrated History of White Magic before Robert-Houdin*. Brest, France: Ar Strobineller Breiz.

Hacking, Ian. 1984. *The Emergence of Probability: A Philosophical Study of Early Ideas about Probability, Induction and Statistical Inference*. Cambridge: Cambridge University Press.

Hadfield, J. A. 1940. "Treatment by Suggestion and Hypno-analysis. In *The Neuroses in War*. Edited by Imanuel Miller. London: Macmillan.

Hagen, M. A. 1997. *Whores of the Court: The Fraud of Psychiatric Testimony and the Rape of American Justice*. New York: Regan Books.

Hagy, Jim, and Sage Hagy. 2019. *Fair Tricks: The Magicians at the Columbian Exposition, Chicago 1893*. Glenview, IL: Reginald Scot Books.

Haidt, Jonathan. 2012. *The Righteous Mind: Why Good People Are Divided by Politics and Religion*. New York: Pantheon Books.

Hall, Marie Boas. 1975. "The Spirit of Innovation in the Sixteenth Century." In *The Nature of Scientific Discovery: A Symposium Commemorating the 500th Anniversary of the Birth of Nicolaus Copernicus*. Edited by Owen Gingerich, 309–21. Washington, DC: Smithsonian Institution Press.

Hall, Spencer T. 1843. *Phreno-magnet, and Mirror of Nature: A Record of Facts, Experiments and Discoveries in Phrenology, Magnetism, Etc*. London: Simpkin, Marshall.

Hall, Trevor H. 1963. *The Spiritualists: The Story of Florence Cook and William Crookes*. New York: Garrett Publications.

———. 1964. *The Strange Case of Edmund Gurney*. London: Duckworth.

———. 1980. *The Strange Story of Ada Goodrich Freer*. London: Duckworth.

Hamley Brothers. ca. 1906. *Illustrated Catalogue of Conjuring Tricks*. London: Hamley Brothers.

Hammerschlag, Heinz E. 1957. *Hypnotism and Crime*. Hollywood, CA: Wilshire Book Company.

Hammond, C. 1852. *The Pilgrimage of Thomas Paine, and Others, to the Seventh Circle in the Spirit World*. Rochester, NY: D. M. Dewey.

Hankins, Thomas L. 1991. *Science and the Enlightenment*. Cambridge: Cambridge University Press.

Hankins, Thomas L., and Robert J. Silverman. 1995. *Instruments and the Imagination*. Princeton, NJ: Princeton University Press.

Hansen, George P. 1990. "Magicians Who Endorsed Psychic Phenomena." *The Linking Ring* 70: 52–54, 63–65, 109.

Hanson, N. R. 1970. "Hypotheses Fingo." In *The Methodological Heritage of Newton*. Edited by Robert E. Butts and John W. Davis. Oxford: Basil Blackwell.

Hanson, Victor Davis. 2006. *A War Like No Other: How the Athenians and Spartans Fought the Peloponnesian War*. New York: Random House.

Harbin, Robert. 1979. *Harbincadabra: Brainwaves and Brainstorms of Robert Harbin*. N.p.: Goodliffe.

Harding, Frederic. ca. 1930. *Why Red Indians Are Spirit Guides*. N.p.: n.p.

Hardinge, Emma. 1870. *Modern American Spiritualism: A Twenty Years' Record of the Communion between Earth and the World of Spirits*. New York: published by the author.

Hare, Harold Edward, and William Loftus Hare. 1936. *Who Wrote the Mahatma Letters? The First thorough Examination of the Communication Alleged to Have Been Received by the Late A. P. Sinnett from Tibetan Mahatmas*. London: Williams & Norgate.

Hare, Robert. 1855. *Experimental Investigation of the Spirit Manifestations, Demonstrating the Existence of Spirits and Their Communion with Mortals*. New York: Partridge & Brittan.

Hare, Robert, and Harriet Clark Hare. 1901. *Christian Spiritual Bible Containing the Gospel of the Type of the Emanation and God, the Only Ubiquitous Son; Being the Gospel of our Lord in His Four Incarnations Together with the Gospel of Our Lady, His Altruistic Affinity*. Philadelphia: n.p.

Harre, R. 1970. *The Method of Science*. London: Wykeham.

Harrington, Anne. 1988. "Metals and Magnets in Medicine: Hysteria, Hypnosis and Medical Culture in *fin-de-siécle* Paris." *Psychological Medicine* 18: 21–38.

Hart, Ernest. 1894. "The Eternal Gullible: With Confessions of a Professional 'Hypnotist.'" *Century Illustrated Monthly Magazine* 48: 833–39.

———. 1896. *Hypnotism, Mesmerism and the New Witchcraft*. London: Smith, Elder & Co.

Hart, Ivor B. 1963. *The Mechanical Investigations of Leonardo da Vinci*. Berkeley: University of California Press.

Hawkins, Gerald S. 1983. *Mindsteps to the Cosmos*. New York: Harper & Row.

Headley, John M. 1990. "Tommaso Campanella and Jean de Launoy: The Controversy over Aristotle and His Reception in the West." *Renaissance Quarterly* 43: 529–50.

Heard, Mervyn. 2006. *Phantasmagoria: The Secret Life of the Magic Lantern*. Hastings: The Projection Box.

Hecker, J. F. C. ca. 1885. *The Black Death* [bound with] *The Dancing Mania of the Middle Ages*. New York: The Humboldt Library of Science.

Heninger, S. K. 1968. *A Handbook of Renaissance Meteorology*. New York: Greenwood Press.

Hensley, Marie E. 1921. *From the Lowest to the Highest or Truth Unveiled*. Tacoma, WA: Ray Printing.

Herbert, Randy S. et al. 2002. "Prominent Medical Journals Often Provide Insufficient Information to Assess the Validity of Studies with Negative Results." *Journal of Negative Results in Biomedicine* 1: 1–5.

Herodotus. 1956. *The History of Herodotus*. New York: Tudor.

Herschel, John F. W. 1849. *Outlines of Astronomy*. Philadelphia: Lea & Blanchard.

Hester, Rose, and Walt Hudson. 1977. *Psychic Character Analysis: Cold Reading Updated*. Baltimore: Magic Media.

Hippolytus. 1868. *The Refutation of All Heresies*. Edinburgh: Clark.

Hirshfeld, Alan W. 2001. *Parallax: The Race to Measure the Cosmos*. New York: W. H. Freeman.

Hockley, Fred. 1850. "Remarks upon the Rev. George Sandby's Review of M. Alphonse Cahagnet's *Arcanes de al vie future devoiles*." *The Zoist* 8: 54–64.

Hodgson, R. 1892. "Mr. Davey's Imitations by Conjuring of Phenomena Sometimes Attributed to Spirit Agency." *Proceedings of the Society for Psychical Research* 8: 253–310.

Hoffmann [Angelo Lewis]. 1876. *Modern Magic*. London: George Routledge and Sons.

———. [1887]. *Drawing-room Conjuring*. London: George Routledge and Sons.

Hoffman, Paul. 1988. *Archimedes' Revenge: The Joys and Perils of Mathematics*. New York: Norton.

Hofstadter, Dan. 2009. *The Earth Moves: Galileo and the Roman Inquisition*. New York: Norton.

Holmes, Richard. 2010. *The Age of Wonder: How the Romantic Generation Discovered the Beauty and Terror of Science*. New York: Vintage Books.

Holt, Hamilton. 1947. *Mysteries of Magic, Mind Reading and Hypnotism*. New York: Lev Gleason Publications.

Holt, Thaddeus. 2004. *The Deceivers: Allied Military Deception in the Second World War*. New York: Scribner.

Home, D. D. 1872. *Incidents in My Life: Second Series*. London: Tinsley Brothers.

———. 1877. *Lights and Shadows of Spiritualism*. London: Virtue.

Honig, William M. 1984. "Science's Miss Lonelyhearts." *Sciences* 24: 24–27.

Honorton, Charles et al. 1990. "Psi Communication in the Ganzfeld: Experiments with an Automated Testing System and a Comparison with a Meta-analysis of Earlier Studies." *Journal of Parapsychology* 54: 19–24.

Hookham, Paul. 1917. *"Raymond": A Rejoinder Questioning the Validity of Certain Evidence and of Sir Oliver Lodge's Conclusions Regarding It*. Oxford: B. H. Blackwell.

Hooper, William. 1794. *Rational Recreations in Which the Principles of Numbers and Natural Philosophy Are Clearly and Copiously Elucidated, by a Series of Easy, Entertaining, Interesting Experiments*. 4 vols. London: B. Law and Son.

Hope, Richard. 1961. *Aristotle's Physics*. Lincoln: University of Nebraska Press.

Hopkins, Albert A. 1898. *Magic: Stage Illusions and Scientific Diversions Including Trick Photography*. New York: Munn.

Hopkins, George M. 1890. *Experimental Science: Elementary Practical and Experimental Physics*. New York: Munn.

Hordern House. 2002. *Imaginary Voyages and Invented Worlds*. Sidney: Hordern House.

Horgan, John. 1986. *The End of Science: Facing the Limits of Knowledge in the Twilight of the Scientific Age*. New York: Broadway Books.

Houdini, Harry. 1908. *The Unmasking of Robert-Houdin*. New York: Publishers Printing.

———. 1920. *Miracle Mongers and Their Methods: A Complete Expose*. New York: Dutton.

———. 1924. *A Magician among the Spirits*. New York: Harper & Brothers.

Houran, J., and R. Lange, ed. 2001. *Hauntings and Poltergeists: Multidisciplinary Perspectives*. Jefferson, NC: McFarland.

Hoving, Thomas. 1996. *False Impressions: The Hunt for Big-time Art Fakes*. New York: Simon & Schuster.

Howe, Henry. 1856. *Memoirs of the Most Eminent American Mechanics*. New York: J. C. Derby.

Howgego, Raymond John. 2013. *Encyclopedia of Exploration Invented and Apocryphal Narratives of Travel*. Australia: Hordern House.

Howitt, William. 1845. *A Popular History of Priestcraft in All Ages and Nations*. London: John Chapman.

Hoyt, William Graves. 1976. *Lowell and Mars*. Tucson: University of Arizona Press.

Hubbell, G. G. 1901. *Fact and Fancy in Spiritualism, Theosophy and Psychical Research*. Cincinnati: Robert Clarke Company.

Huber, Volker. 2007. "The Educated Swan." *Gibeciére*: 12–43.

Hull, Burling. n.d. *Fifty Sealed Message Reading Methods*. Davenport: London.

———. 1920. *Twelve Sealed Message Reading Methods*. New York: Burling Hull.

———. 1929. *Original Slate Secrets*. New York: Burling Hull.

———. 1946. *The Last Word Blindfold Methods*. Woodside, NY: Hull.

Hull, Clark L. 1931. "Quantitative Methods of Investigating Hypnotic Suggestion." *Journal of Abnormal and Social Psychology* 15: 390–417.

Hume. 1886. *Thought Reading and Kindred Subjects Divested of Their Mystery and Explained*. Melbourne: William Inglis.

Hunt, Fred. ca. 1890. *How to Do Second Sight: Heller's Second Sight Explained*. New York: Frank Tousey.

Hunt, J. H. 1934. *Indian "Fakirs."* London: Adlard & Son.

Hunter, J. Paul. 1990. *Before Novels: The Cultural Contents of Eighteenth-century English Fiction*. New York: W. W. Norton.

Hunter, Marguerite. 1894. *A Narrative Descriptive of Life in the Material and Spiritual Spheres*. N.p.: published for Marguerite Hunter.

Hurst, Lulu. 1897. *Lulu Hurst, (the Georgia Wonder) Writes Her Autobiography*. Rome, GA: Psychic Publishing.

Hutchings, Emily Grant. 1917. *Jap Herron*. New York: Mitchell Kennerley.

Hutchison, Keith. 1982. "What Happened to Occult Qualities in the Scientific Revolution?" *ISIS* 73: 233–53.

Hutton, Edward. 1928. *A Glimpse of Greece*. London: The Medici Society.

Hyman, Ray. 1957. "Review of *Modern Experiments in Telepathy*." *Journal of the American Statistical Association* 52: 607–10.

Hyman, Ray. 1977. "Cold Reading: How to Convince Strangers That You Know All about Them." *The Zetetic* 1: 18–37.

———. 1981. "The Psychic Reading." In *The Clever Hans Phenomenon: Communication with Horses, Whales, Apes, and People*. Edited by T. A. Sebeok and R. Rosenthal, 169–81. New York: New York Academy of Sciences.

———. 1985a. "The Zöllner Phenomenon." In *The New York Magic Symposium IV*. Edited by Stephen Minch. New York: New York Magic Symposium.

———. 1985b. "The Ganzfeld Psi Experiment: A Critical Appraisal." *Journal of Parapsychology* 49: 3–49.

———. 1989. *The Elusive Quarry*. Buffalo: Prometheum.

———. 1994. "Anomaly or Artifact? Comments on Bem and Honorton." *Psychological Bulletin* 115: 19–23.

———. 1996. "Evaluation of a Program on Anomalous Mental Phenomena." *Journal of Scientific Exploration* 10: 31–58.

———. 2002. "The Transcendental Knot." In *Puzzlers' Tribute: A Feast for the Mind*. Edited by David Wolfe and Tom Rodgers. Natick, MA: A. K. Peters.

———. 2007. "Critical Thinking in Psychology." In *Critical Thinking in Psychology.* Edited by R. J. Sternberg, Henry R. Roediger III, and Diane F. Halpern. New York: Cambridge University Press.

Hyman, Ray, and Charles Honorton. 1986. "A Joint Communique: The Psy Ganzfeld Controversy." *Journal of Parapsychology* 50: 351–64.

Hyslop, James H. 1908. *Psychical Research and the Resurrection.* Boston: Small, Maynard.

———. 1910. "Eusapia Palladino." *Journal of the American Society for Psychical Research.* 4: 425–46.

Isaacson, Walter. 2017. *Leonardo da Vinci.* New York: Simon & Schuster.

Inge, William Ralph. 1956. *Christian Mysticism.* New York: Meridian Books.

Insulanus, Theophilus [Donald MacLeod]. 1763. *A Treatise on the Second Sight, Dreams and Apparitions.* Edinburgh: Ruddiman, Auld, and Company.

———. 1819. *A Treatise on the Second Sight, Dreams and Apparitions.* Glasgow: Chapman.

Ireland, W. W. 1885. *The Blot upon the Brain: Studies in History and Psychology.* Edinburgh: Bell & Bradfute.

J. D. 1886. "A Recent Contribution to the Discussion of Hypnotism." *Science* 8: 521–22.

Jackson, Brooks, and Kathleen Hall Jamieson. 2007. *Unspun: Finding Facts in a World of Disinformation.* New York: Random House.

Jackson, H. G. 1972. *The Spirit Rappers.* Garden City, NY: Doubleday.

Jaher, David. 2015. *The Witch of Lime Street: Science, Seduction, and Houdini in the Spirit World.* New York: Crown.

James, John. 1879. *Mesmerism, with Hints for Beginners.* London: W. H. Harrison.

James, Peter, and Nick Thorpe. 1994. *Ancient Inventions.* New York: Ballantine Books.

James, William. 1921. *The Will to Believe and Other Essays in Popular Philosophy.* London: Longmans.

James I. 1616. *The Workes of the Most High and Mightie Prince, Iames.* London: Robert Barker and Iohn Bill.

———. 1924. *Daemonologie and Newes from Scotland.* London: John Lane the Bodley Head.

Jardine, Lisa. 1996. *Worldly Goods: A New History of the Renaissance.* New York: Doubleday.

Jarrett, Bede. 1942. *Social Theories of the Middle Ages: 1200–1500.* Westminster, MD: Newman Book Shop.

Jastrow, Joseph. 1910. "The Unmasking of Paladino," *Collier's,* May 14.

———. 1918. *The Psychology of Conviction.* New York: Houghton Mifflin.

Jay, Ricky. 1986. *Learned Pigs and Fireproof Women.* New York: Villard.

———. 2001. *Jay's Journal of Anomalies.* New York: Farrar, Straus and Giroux.

Jaynes, Julian. 1976. *The Origin of Consciousness in the Breakdown of the Bicameral Mind.* Boston: Houghton Mifflin.

Jennings, Lee B. 1976. "Hoffmann's Hauntings: Notes toward a Parapsychological Approach to Literature." *The Journal of English and Germanic Philology* 75: 559–67.

Jeppson, Lawrence. 1971. *Fabulous Frauds: A Study of Great Art Forgeries.* London: Arlington Books.

Jermay, Luke. 2007. *3510.* Chicago: Penguin Magic.

Jerome, Joseph. 1965. *Montague Summers: A Memoir.* London: Cecil & Amelia Woolf.

Jewett, Pendie L. 1873. *Spiritualism and Charlatanism; or the Tricks of the Media Embodying an Expose of the Manifestations of Modern Spiritualism by a Committee of Business Men of New York.* New York: S. W. Green.

Jimenez-Martinez, Enrique. 2017. "Further Reflections on Joan Dalmau." *Gibeciére* 12, no. 2: 67–118.

Johns, Adrian. 1998. *The Nature of the Book: Print and Knowledge in the Making.* Chicago: University of Chicago Press.

Johnson, Alice. 1936. *Mrs. Henry Sidgwick's Work in Psychical Research.* London: Society for Psychical Research.

Johnson, J. 1994. "Henry Maudsley on Swedenborg's Messianic Psychosis." *British Journal of Psychiatry* 165: 690–91.

Johnson, Paul. 2000. *The Renaissance.* New York: Modern Library.

Johnson Smith. 1941. *Mammoth New Catalog of 9000 Novelties: Including . . . over 1500 Sensational New Novelties Just Added!* Detroit, MI: Johnson Smith & Co.

Jones, Amanda. 1910. *Psychic Autobiography.* New York: Greaves.

Jones, Frederick E. 1911. *Mind Reading: Secrets and Sealed Billet Reading.* N.p.: n.p.

Jones, John. 1861. *The Natural and Supernatural: Or, Man Physical, Apparitional, and Spiritual.* London: H. Bailliere.

Jones, M. 1990. *Fake? The Art of Deception.* London: British Museum.

Jones, Tom B. 1979. *The Silver-plated Age.* Lawrence, KS: Coronado Press.

Joseph, Eddie. ca. 1940. *Intuitional Sight.* N.p.: Max Andrews.

Joyce, J. 1815. *Scientific Dialogues, Intended for the Instruction and Entertainment of Young People in which the First Principles of Natural and Experimental Philosophy Are Fully Explained.* London: Printed for Baldwin, Cradock, and Joy.

Judges, A. V. 1930. *The Elizabethan Underworld.* London: George Routledge and Sons.

Jung-Stilling, Johann Heinrich. *Theory of Pneumatology; in Reply to the Question, What Ought to Be Believed or Disbelieved Concerning Presentiments, Visions, and Apparitions, According to Nature, Reason, and Scripture.* New York: J. S. Redfield, 1851.

Kagan, Donald. 1996. *On the Origins of War and the Preservation of Peace.* New York: Doubleday.

Kahill, Thomas. 1995. *How the Irish Saved Civilization: The Untold Story of Ireland's Heroic Role from the Fall of Rome to the Rise of Medieval Europe.* New York: Nan A. Talese.

Kallet, Lisa. 1993. *Money, Expense, and Naval Power in Thucydides' History 1–5.24.* Berkeley: University of California Press.

Kalush, William. 2002. "Sleight of Hand with Playing Cards prior to Scot's *Discoverie*." In *Puzzlers' Tribute: A Feast for the Mind*. Edited by David Wolfe and Tom Rodgers. Natick, MA: A. K. Peters.

Kalush, William, and Larry Sloman. 2006. *The Secret Life of Houdini: The Making of America's First Superhero*. New York: Atria Books.

———. 2007. *The Secret Life of Houdini Laid Bare: Sources, Notes and Additional Material*. Pasadena, CA: Mike Caveney's Magic Words.

Kane, Elisha K., and Margaret Fox. 1866. *The Love-life of Dr. Kane; Containing the Correspondence, and a History of the Acquaintance, Engagement, and Secret Marriage between Elisha K. Kane and Margaret Fox*. New York: Carleton.

Kanipe, Jeff. 2006. *Chasing Hubble's Shadows: The Search for Galaxies at the Edge of Time*. New York: Hill and Wang.

Kant, Immanuel. 1900. *Dreams of a Spirit-seer Illustrated by Dreams of Metaphysics*. London: Swan Sonnenschein.

Kaplan, Fred. 1974. "'The Mesmeric Mania:' The Early Victorians and Animal Magnetism." *Journal of the History of Ideas* 35, 691–702.

Karlyn [J. F. Burrows]. ca. 1912. *Secrets of Stage Hypnotism: Stage Electricity and Bloodless Surgery*. London: The Magician.

Karon, Bertram. 1998. "Repressed Memories: The Real Story." *Professional Psychology: Research and Practice* 29: 482–87.

Karon, Bertram, and Anmarie Widener. 1997. "Repressed Memories and World War II: Lest We Forget!" *Professional Psychology: Research and Practice* 28: 338–40.

Karr, Todd. 2006. *Essential Robert-Houdin*. N.p.: Miracle Factory.

———. 2019. *Annemann's Enigma*. N.p.: Miracle Factory.

Kaufman, Richard. 2011. *The Berglas Effect*. Washington, DC: Kaufman.

Keegan, J. 1994. *A History of Warfare*. New York: Vintage Books.

Kelley, D. R., and D. H. Sacks. 1997. *The Historical Imagination in Early Modern Britain: History, Rhetoric, and Fiction, 1500–1800*. Cambridge: Cambridge University Press.

Kells, Stuart. 2017. *The Library: A Catalogue of Wonders*. Berkeley, CA: Counterpoint.

Kenyon, John. 1972. *The Popish Plot*. New York: St. Martin's Press.

Kerchever, E. 1891/1970. *The History and Motives of Literary Forgeries*. New York: Burt Franklin.

Kerényi, C. 1959. *Asklepios. Archetypal Image of the Physician's Existence*. New York: Pantheon Books.

Kerner, Justinus. 1845. *The Seeress of Prevorst Being Revelations Concerning the Inner-life of Man, and the Inter-diffusion of a World of Spirits in the One We Inhabit*. London: J. C. Moore.

Kerr, H. 1973. *Mediums, and Spirit-rappers, and Roaring Radicals: Spiritualism in American Literature, 1850–1900*. Urbana: University of Illinois.

Kessler, Eckhard. 1990. "The Transformation of Aristotelianism during the Renaissance." In *New Perspectives on Renaissance Thought: Essays in the History of*

Science, Education and Philosophy. Edited by John Henry and Sarah Hutton. London: Duckworth.

Kieckhefer, Richard. 1989. *Magic in the Middle Ages.* Cambridge: Cambridge University Press.

Kindt, Julia. 2006. "Delphic Oracle Stories and the Beginning of Historiography: Herodotus' Croesus Logos." *Classical Philology* 101: 34–51.

King, Amy C., and Cecil B. Read. 1963. *Pathways to Probability.* New York: Holt, Rinehart and Winston.

Kinney, Arthur F. 1990. *Rogues, Vagabonds and Sturdy Beggars: A New Gallery of Tudor and Early Stuart Rogue Literature.* Amherst: University of Massachusetts Press.

Kinnunen, Taru, Harold S. Zamansky, and Martin L. Block. 1994. "Is the Hypnotized Subject Lying?" *Journal of Abnormal Psychology* 103: 184–91.

Kirby, R. S. 1815. *The Wonderful and Scientific Museum: Or Magazine of Remarkable Characters.* 6 vols. London: printed for R. S. Kirby.

Knaggs, Samuel T. 1879. *Mediums and Their Dupes: A Complete Exposure of the Chicaneries of Professional Mediums and Explanation of So-called Spiritual Phenomena.* Sydney: Grenville's Telegram Company.

Knowles, Elmer D. 1914. *Prof. Elmer E. Knowles' Complete System of Personal Influence and Healing.* London: The National Institute of Sciences.

Knuckles, Bobby [Bob Farmer]. 2018. *Beat 'Em, Cheat 'Em, Leave 'Em Bleedin'.* Brockville, ON: Every Trick in the Book.

Kodera, Sergius. 2014. "The Laboratory As Stage: Giovan Gattista Della Porta's Experiments." *Journal of Early Modern Studies* 1: 15–38.

Konnikova, Maria. 2016. *The Confidence Game: Why We Fall for It . . . Every Time.* New York: Viking.

Kopf, Jared. 2018. "Handsome Is As Handsome Does: Lovick." *Genii* 81: 72–87.

Kossy, Donna. 1994. *Kooks: A Guide to the Outer Limits of Human Belief.* Portland, WA: Feral House.

Krebs, Stanley L. 1910. *Trick Methods of Eusapia Paladino.* Philadelphia: privately printed.

Kreskin. 1991. *Secrets of the Amazing Kreskin.* Buffalo, NY: Prometheus Books.

Kripal, Jeffrey J. 2010. *Authors of the Impossible: The Paranormal and the Sacred.* Chicago: University of Chicago Press.

Krutch, Joseph Wood. 1969. *The Most Wonderful Animals That Never Were.* Boston: Houghton Mifflin.

Kucharski, Adam. 2016. *The Perfect Bet: How Science and Math Are Taking the Luck out of Gambling.* New York: Basic Books.

Kuhn, Thomas S. 1970. *The Structure of Scientific Revolutions.* Chicago: University of Chicago Press.

———. 1975. *The Copernican Revolution: Planetary Astronomy in the Development of Western Thought.* Cambridge, MA: Harvard University Press.

Kulka, R. A. et al. 1988. *Trauma and the Vietnam War Generation*. New York: Brunner/Mazel.

Lachapelle, Sofie. 2011. *Investigating the Supernatural: From Spiritism and Occultism to Psychical Research and Metaphysics in France, 1853–1931*. Baltimore: Johns Hopkins University Press.

Lacroix, Paul. 1878. *Science and Literature in the Middle Ages, and at the Period of the Renaissance*. New York: D. Appleton.

Laicus. 1854. *Satan's Miracles Proved Unreal: Together with the Pretensions of Table-turning. As Set Forth by Rev. Mr. Godfrey & Others, Scripturally & Rationally Considered*. Cheltenham: R. Edwards.

Lamont, Peter. 2006. *The First Psychic: The Peculiar Mystery of a Notorious Victorian Wizard*. London: Abacus.

———. 2012. "Minor Episodes in the Marvelous History of Mesmerists, Mediums and Mindreaders." *Genii* 75: 14–15.

Lamont, Peter, and Jim Steinmeyer. 2018. *The Secret History of Magic: The True Story of the Deceptive Art*. New York: TarcherPerigee.

Landels, J. G. 1978. *Engineering in the Ancient World*. Berkeley: University of California Press.

Landes, David S. 1983. *Revolution in Time: Clocks and the Making of the Modern World*. Cambridge: Harvard University Press.

Lane, Maria D. 2006. "Mapping the Mars Canal Mania: Cartographic Projection and the Creation of a Popular Icon." *Imago Mundi* 58: 198–211.

Lang, A. 1894. *Cock Lane and Common Sense*. London: Longmans, Green.

Langford, David. 1990. "Me and Whitley and the Continuum." *Sglodion* 2.

Langford, Jerome J. 1966. *Galileo, Science and the Church*. New York: Desclee.

Langham, J. 1951. *More Than Meets the Eye*. London: Evans Brothers.

Langley, S. P. 1877. "The First 'Popular Scientific Treatise.'" *Popular Science Monthly* 10: 718–25.

Lanners, Edi. 1977. *Illusions*. New York: Holt, Rinehart and Winston.

Lanning, K. V. 1992. *Investigator's Guide to Allegations of "Ritual" Child Abuse*. Quantico, VA: Federal Bureau of Investigation.

Lapponi, Joseph. 1907. *Hypnotism and Spiritism: A Critical and Medical Study*. London: George Bell & Sons.

Lattis, James M. 1994. *Between Copernicus and Galileo: Christoph Clavius and the Collapse of Ptolemaic Cosmology*. Chicago: University of Chicago Press.

LaVater, John Caspar. 1804. *Essays on Physiognomy; for the Promotion of the Knowledge and the Love of Mankind*. London: H. D. Symonds.

LaVelle, Steven. *Superhuman Feats You Can Do*. Fresno, CA: Theater Center America, 1981.

Lea, Henry Charles. 1922. *A History of the Inquisition of the Middle Ages*. 3 vols. New York: Macmillan.

Leasor, James. 1962. *The Plague and the Fire*. London: George Allen & Unwin.

Lee, Edwin. 1843. "Cases of Clairvoyance." *The Mesmerist: A Journal of Vital Magnetism* 1: 83–88.

——. 1866. *Animal Magnetism and Magnetic Lucid Somnambulism*. London: Longmans, Green.

Lee, Frederick George. 1885. *Glimpses in the Twilight*. Edinburgh: William Blackwood and Sons.

Lefaivre, Liane. 2005. *Leon Battista Alberti's Hypnerotomachia Poliphili*. Cambridge, MA: MIT Press.

Leff, A. A. 1976. *Swindling and Selling*. New York: Free Press.

Leger, Theodore. 1846. *Animal Magnetism; or Psycodunamy*. New York: D. Appleton.

Lehman, Amy. 2009. *Victorian Women and the Theatre of Trance: Mediums, Spiritualists and Mesmerists in Performance*. Jefferson, NC: McFarland.

Lehoux, Daryn. 2007. "Drugs and the Delphic Oracle." *The Classical World* 101: 41–56.

Leikind, Bernar J., and William J. McCarthy. 1985. "An Investigation of Firewalking." *Skeptical Inquirer* 10: 23–34.

LeLoyer, Peter. 1605. *A Treatise of Specters or Straunge Sight, Visions and Apparitions Appearing Sensibly unto Men*. London: Printed by Val. S. for Mathew Lownes.

Lender, J. A. 1903. "Seeds Samson." In *The Angels' Diary and Celestion Study of Man*. Denver: privately published.

Leonard, Jonathan Norton. 1930. *Crusaders of Chemistry*. Garden City, NY: Doubleday, Doran.

Leonard, Todd Jay. 2005. *Talking to the Other Side: A History of Modern Spiritualism and Mediumship*. Lincoln, NE: iUniverse.

Levi, Eliphas. 1914. *The History of Magic: Including a Clear and Precise Exposition of Its Procedure, Its Rites and Its Mysteries*. Philadelphia: David McKay.

Lewes, George Henry. 1864. *Aristotle: A Chapter from the History of Science, Including Analyses of Aristotle's Scientific Writings*. London: Smith, Elder.

Lewinsohn, Richard. 1954. *Animals, Men and Myths: An Informative and Entertaining History of Man and the Animals around Him*. New York: Harper & Brothers.

Lewis, A. J. 1886. "How and What to Observe in Relation to Slate-writing Phenomena." *Journal of Society for Psychical Research* 2:362–75.

Lewis, C. S. 1954. *English Literature in the Sixteenth Century Excluding Drama*. Oxford: Oxford University Press.

Lewis, H. C. 1886. "Account for Professor H. Cavill Lewis." *Proceedings of the Society for Psychical Research* 4: 352–77.

Lewis, Matthew. 1796/2002. *The Monk*. New York: Modern Library.

Leys, Ruth. 2000. *Trauma: A Genealogy*. Chicago: University of Chicago Press.

Lilienfeld, S. O., J. Marshall, J. T. Todd et al. 2015. "The Persistence of Fad Interventions in the Face of Negative Scientific Evidence: Facilitated Communication for Autism As a Case Example." In *Evidence-Based Communication Assessment and Intervention*. New York: Routledge.

Lilley & Co. ca. 1905. *Specialties for Secret Societies: Paraphernalia, Supplies, Masks, Bears, Wigs, Etc.* Columbus, OH: M. C. Lilley.

Lindhorst, [Will]. ca. 1930s. *Flash Magic Illusions and Circus Tricks for Exclusive Magicians Only.* St. Louis, MO: Lindhorst.

Linn, William Alexander. 1923. *The Story of the Mormons: From the Date of Their Origin to the Year 1901.* New York: Macmillan.

Linton, Charles. 1855. *The Healing of the Nations.* New York: Society for the Diffusion of Spiritual Knowledge.

Little, George H. 1899. "To Pick Out All the Court Cards Blindfolded." *Mahatma* 2, no. 8.

Litvag, Irving. 1972. *Singer in the Shadows: The Strange Story of Patience Worth.* New York: Macmillan.

Litz, B. T. et al. 2002. "Early Intervention for Trauma: Current Status and Future Directions." *Clinical Psychology: Science and Practice* 9: 112–34.

Lloyd, G. E. R. 1973. *Greek Science after Aristotle.* New York: Norton.

Locke, Richard Adams. 1835. *The History of the Moon, or an Account of the Wonderful Discoveries of Sir John Herschell.* [London]: printed by B. D. Cousins.

———. 1859. *The Moon Hoax; or, a Discovery That the Moon Has a Vast Population of Human Beings.* New York: William Gowans.

Lodge, Oliver. 1916. *Raymond of Life and Death.* New York: Doran.

Loftus, Elizabeth F., and K. Ketcham. 1991. *Witness for the Defense; the Accused, the Eyewitness, and the Expert Who Puts Memory on Trial.* New York: St. Martin's Press.

Lombroso, Cesare. 1909. *After Death What? Spiritistic Phenomena and Their Interpretation.* Boston: Small, Maynard.

Lonergan, D. J. 1930. *Slate Secrets: Consisting of Routines, Methods, Chemical Formulas, Manipulations, Subterfuges, Etc.* N.p.: privately printed.

Long, Pamela O. 2001. *Openness, Secrecy, Authorship: Technical Arts and the Culture of Knowledge from Antiquity to the Renaissance.* Baltimore: Johns Hopkins University Press.

Lonk, Adolph. F. 1940. *Manual of Hypnotism and Psycho-therapeutics.* Palatine, IL: privately printed.

———. 1947. *"The Original" Complete Seventy-two Part Manual of Hypnotism and Psycho-therapeutics and Also Mysteries of Time and Space.* Palatine, IL: privately printed.

Loomis, Bob. 2016. *Houdini's Final Incredible Secret: How Houdini Mystified Sherlock Holmes' Creator.* N.p.: Bob Loomis.

Lovell, Simon. 2003. *Billion Dollar Bunko.* Tahoma, CA: L & L Publishing.

———. 2007. *How to Cheat at Everything: A Con Man Reveals the Secrets of the Esoteric Trade of Cheating, Scams, and Hustles.* New York: Thunder's Mouth Press.

Lowell, Percival. 1895. *Mars.* Boston: Houghton Mifflin.

———. 1906. *Mars and Its Canals.* New York: Macmillan.

———. 1908. *Mars As the Abode of Life.* New York: Macmillan.

Loyd, Sam. 1927. *Tricks and Puzzles: A Selected Series of Tricks, Puzzles and Conundrums, Arranged for the Layman, Student, Puzzle-fan or Scientist.* New York: Experimenter.

Loyd-Jones, Hugh. 1976. "The Delphic Oracle." *Greece & Rome* 23: 60–73.

Lucas-Dubreton, J. 1966. *Daily Life in Florence in the Time of the Medici.* New York: Macmillan.

Lucian. 1949. *Lucian: True History, Dialogues of the Dead, Dialogues of the Heterae, and Other Selected Essays.* Chicago: Henry Regnery.

———. 1961. *Satirical Sketches.* Translated by Paul Turner. Harmondsworth, UK: Penguin.

Luck, Georg. 1985. *Arcana Mundi: Magic and the Occult in the Greek and Roman Worlds.* Baltimore: Johns Hopkins Press.

Luckhurst, Roger. 2002. "Passages in the Invention of the Psychi: Mind-reading in London, 1881–84." In *Transactions and Encounters: Science and Culture in the Nineteenth Century.* Edited by Roger Luckhurst and Josephine McDonagh, 117–50. Manchester: Manchester University Press.

Lunt, Edward D. 1903. *Mysteries of the Seance and Tricks and Traps of Bogus Mediums.* Boston: Lunt Bros. Publishers.

Lustig, David J. 1930. *Vaudeville Hypnotism.* New York: La Vellma Publication.

MacAire, Sid. 1889. *Mind Reading or Muscle Reading As Exhibited by the Late Washington Irving Bishop and Others.* Manchester, UK: John Heywood.

Macknik, Stephen L., and Susana Martinez-Conde. 2010. *Sleights of Mind: What the Neuroscience of Magic Reveals about Our Everyday Deceptions.* New York: Henry Holt.

MacWalter, J. G. 1854. *The Modern Mystery; or, Table-tapping, Its History, Philosophy, and General Attributes.* London: John Farquyhar Shaw.

Madden, R. R. 1857. *Phantasmata, or Illusions and Fanaticisms of Protean Forms Productive of Great Evils.* 2 vols. London: T. C. Newby.

Magee, W. C. 1854. *Talking to Tables: A Great Folly or a Great Sin.* Bath: R. E. Peach.

Magida, Arthur J. 2011. *The Nazi Seance: The Strange Story of the Jewish Psychic in Hitler's Circle.* New York: Palgrave Macmillan.

Malone, Michael S. 2012. *The Guardian of All Things: The Epic Story of Human Memory.* New York: St. Martin's Press.

Mangan, Michael. 2007. *Performing the Dark Arts: A Cultural History of Conjuring.* Bristol: Intellect.

Mann, Al. n.d. *Master Slate Secrets.* 4 vols. Freehold, NJ: Al Mann Exclusives.

Mann, Walter. 1919. *The Follies and Frauds of Spiritualism.* London: Watts.

Marsak, Leonard M. 1959. "Bernard de Fontenelle: The Idea of Science in the French Enlightenment." *Transactions of the American Philosophical Society* 49: 1–64.

Marsh, Leonard. 1854. *The Apocatastasis; or Progress Backwards.* Burlington, VT: Chauncey Goodrich.

Martinez-Conde, Susana, and Stephen Macknik. 2017. *Champions of Illusion: The Science behind Mind-boggling Images and Mystifying Brain Puzzles*. New York: Scientific American.

Maskelyne, James Nevil. 1876. *Modern Spiritualism: A Short Account of Its Rise and Progress, with Some Exposures of So-Called Spirit Media*. London: Frederick Warne.

Massey, C. C. 1886. "The Possibility of Mal-observation in Relation to Evidence for the Phenomena of Spiritualism." *Proceedings of the Society of Psychical Research* 4: 75–110.

Massironi, Mauro. 2016. "Galasso: On the Trail of Horatio." *Gibeciére* 11, no. 2: 15–56.

Masters, W. J. 1882. *A Key to the Mysteries of Thought-reading: Nerves and Sensations, Mater and Motion*. Landport, UK: printed by C. Annett.

Mathiesen, Robert. 1998. "A Thirteenth-century Ritual to Attain the Beatific Vision from the *Sworn Book of Honorius* of Thebes." In *Conjuring Spirits: Texts and Traditions of Medieval Ritual Magic*. Edited by Claire Fanger. Gloucestershire: Sutton.

Matlock, Jann. 1996. "The Invisible Woman and Her Secrets Unveiled." *The Yale Journal of Criticism* 9: 175–221.

Matossian, Mary Kilbourne. 1989. *Poisons of the Past: Molds, Epidemics, and History*. New Haven, CT: Yale University Press.

Mattison, H. 1853. *Spirit Rapping Unveiled! An Expose of the Origin, History, Theology and Philosophy of Certain Alleged Communications from the Spirit World, by Means of "Spirit Rapping," "Medium Writing," "Physical Demonstrations," Etc*. New York: Mason Brothers.

———. 1855. *Spirit-rapping Unveiled! An Expose of the Origin, History, Theology and Philosophy of Certain Alleged Communications from the Spirit World, by Means of "Spirit Rapping," "Medium Writing," "Physical Demonstrations," Etc*. New York: Derby.

Maudsley, Henry. 1887/2011. *Natural Causes and Supernatural Seemings*. Cambridge: Cambridge University Press.

———. 1897. *Pathology of Mind*. London: Macmillan.

Maurizio, Lisa. 1993. *Delphic Narratives: Recontextualizing the Pythia and Her Prophecies*. Princeton, NJ: Princeton University.

———. 1997. "Delphic Oracles as Oral Performances: Authenticity and Historical Evidence. *Classical Antiquity* 16: 308–34.

Maxwell, J. 1905. *Metaphysical Phenomena: Methods and Observations*. London: Duckworth.

Maxwell-Stuart, P. G. 2003. *Witch Hunters: Professional Prickers, Unwitchers and Witch Finders of the Renaissance*. Stroud, UK: Tempus.

Mayo, Herbert. 1851. *On the Truths Contained in Popular Superstitions with an Account of Mesmerism*. Edinburgh: William Blackwood.

Mayor, Adrienne. 2009. *Greek Fire, Poison Arrows, and Scorpion Bombs: Biological and Chemical Warfare in the Ancient World*. New York: Overlook.

McCartney, Eugene S. 1920. "Spontaneous Generation and Kindred Notions in Antiquity." *Transactions and Proceedings of the American Philological Association* 51: 101–15.

McConaughy, John. 1931. *From Cain to Capone: Racketeering down the Ages.* New York: Brentano's.

McCormick, Donald. 1976. *Taken for a Ride: The History of Cons and Con-men.* London: Harwood-Smart.

McEwen, P. H. 1912. *Hypnotism Made Plain.* Sydney: William Brooks.

McGlashan, Thomas H., Frederick J. Evans, and Martin T. Orne. 1969. "The Nature of Hypnotic Analgesia and Placebo Response to Experimental Pain." *Psychosomatic Medicine* 31: 227–64.

McIntyre, Lee. 2018. *Post-truth.* Cambridge: MIT Press.

McKenzie, J. Hewat. ca. 1910. *First Steps to Spirit Intercourse.* London: J. Hewat McKenzie.

McKenzie, Judith S., Sheila Gibson, and A. T. Reyes. 2004. "Reconstructing the Serapeum in Alexandria from the Archaeological Evidence." *The Journal of Roman Studies* 94: 73–121.

McKnight, Stephen A., ed. 1992. *Science in Early Modern Thought.* Columbia: University of Missouri Press.

McNally, Richard J. 2005. *Remembering Trauma.* Cambridge: Belknap Press.

McNeill, W. H. 1976. *Plagues and Peoples.* Garden City, NY: Anchor Press.

Medhurst, R. G., and K. M. Goldney. 1964. "William Crookes and the Physical Phenomena of Mediumship." *Proceedings of the Society for Psychical Research* 54: 25–157.

Méheust, Bertrand 2007. *A Historical Approach to Psychical Research: The Case of Alexis Didier (1826–1886).* Big Sur, CA: Esalen Institute.

Melton, John. 1620. *Astrologaster, or, the Figure-caster.* London: Barnard Alsop for Edward Blackmore.

Menchi, Silvana Seidel. 1994. "Characteristics of Italian Anticlericalism." In *Anticlericalism in Late Medieval and Early Modern Europe.* Edited by Peter A. Dykema and Heiko A. Oberman. New York: E. J. Brill.

Menghi, Girolamo. 2002. *The Devil's Scourge: Exorcism during the Italian Renaissance.* Translated by Gaetano Paxia. Boston: Weiser Books.

Mercier, Charles A. 1919. *Spirit Experiences.* London: Watts.

———. [1919]. *Spiritualism and Sir Oliver Lodge.* London: Watts.

Merton, Robert K. 1970. *Science, Technology and Society in Seventeenth-century England.* New York: Harper & Row.

Middleton, Conyers. 1749. *A Free Inquiry into the Miraculous Powers.* London: printed for R. Manby and H. S. Cox.

Middleton, W. E. Knowles. 1971. *The Experimenters: A Study of the Accademia Del Cimento.* Baltimore: Johns Hopkins Press.

Miles, Richard. 2011. *Carthage Must Be Destroyed: The Rise and Fall of an Ancient Civilization*. New York: Viking.

Milis, L. J. R. 1999. *Angelic Monks and Earthly Men*. Woodbridge, Suffolk: Boydell Press.

Milman, Henry Hart. 1862. *History of Latin Christianity*. New York: Sheldon.

Milmine, Georgine. 1971. *The Life of Mary Baker G. Eddy and the History of Christian Science*. Grand Rapids, MI: Baker Book House.

Minch, Stephen. 1975. *Mind and Matter: A Handbook of Parapsychokinetic Phenomena*. Calgary: Micky Hades.

Minois, Georges. 2012. *The Atheist's Bible: The Most Dangerous Book That Never Existed*. Chicago: University of Chicago Press.

Moehring, John C. 2013, April. *Journal of Magic Research*.

Moffatt, John M. 1842. *The Boy's Book of Science; a Familiar Introduction to the Principles of Natural Philosophy*. London: Thomas Tegg.

Monet, Joseph, ed. 1930. *Casanova's Memoirs*. New York: privately printed.

Montaigne, Michel. 1946. *The Essays of Michel de Montaigne*. Translated by George B. Ives. New York: Heritage Press.

Montgomery, G. H., K. N. DuHamel, and W. H. Redd. 2000. "A Meta-analysis of Hypnotically Induced Analgesia: How Effective Is Hypnosis?" *The International Journal of Clinical and Experimental Hypnosis* 48: 138–53.

Moore, Laurence. 1972. "Spiritualism and Science: Reflections on the First Decade of the Spirit Rappings." *American Quarterly* 24: 474–500.

Moore, R. L. 1977. *In Search of White Crows*. New York: Oxford University Press.

Moore, W. U. 1911. *Glimpses of the Next State*. London: Watts.

———. 1913. *The Voices: A Sequel to "Glimpses of the Next State."* London: Watts.

Morgan, Catherine. 1990. *Athletes and Oracles: The Transformation of Olympia and Delphi in the Eighth Century B.C.* Cambridge: Cambridge University Press.

Morgan, R. C. [1854]. *An Inquiry into Table-miracles, Their Cause, Character, and Consequences; Illustrating by Recent Manifestations of Spirit-writing and Spirit-music*. London: Binns & Goodwin.

Morley, Henry. 1854. *Jerome Cardan: The Life of Girolamo Cardano, of Milan, Physician*. 2 vols. London: Chapman and Hall.

Morris, Richard. 2006. *Harry Price the Psychic Detective*. England: Sutton.

Morrow, Albert. [1914]. *Thought-reading Exposed*. Manchester, UK: Daisy Bank Printing & Publishing.

Moss, H. St. L. B. 1964. *The Birth of the Middle Ages: 395–814*. New York: Oxford University Press.

Mother Shipton. 1641. *The Prophesie of Mother Shipton in the Raigne of King Henry the Eighth*. London: Richard Lownds.

Mottelay, Paul Fleury. 1922. *Bibliographical History of Electricity and Magnetism*. Mansfield Centre, CT: Martino Fine Books.

Mueller, Tom. 2013. "CSI: Italian Renaissance." *Smithsonian* 44, no. 4: 51–59.

Munsterberg, Hugo. 1910. "My Friends the Spiritualists: Some Theories and Conclusions Concerning Eusapia Palladino." *Metropolitan Magazine* 31: 559–72.

Munthe, Axel. 1930. *The Story of San Michele*. New York: Dutton.

Musson, Clettis V. 1937. *Thirty-five Weird and Psychic Effects*. N.p.: Chas. C. Eastman.

Mystic Helper. 1924. *The Evolution of the Universe or, Creation According to Science Transmitted from Michael Faraday Late Electrician and Chemist of the Royal Institution of London*. Los Angeles: Cosmos Publishing.

Nasier, Alcofribas. 1904. *De Tribus Impostoribus, A.D. 1230. The Three Impostors Translated (with Notes and Comments) from a French Manuscript of the Work Written in 1716*. N.p.: privately printed for the subscribers.

Natale, Simone. 2016. *Supernatural Entertainments: Victorian Spiritualism and the Rise of Modern Media Culture*. University Park: Pennsylvania State University Press.

National Spiritualist Association. 1948. *Centennial Book of Modern Spiritualism in America*. Chicago: National Spiritualist Association of United States of America.

Naudé, G. 1657. *The History of Magic by Way of Apology, for All the Wise Men Who Have Unjustly Been Reputed Magicians, from the Creation, to the Present Age*. London: John Streater.

Neher, Andrew. 1986. *Jewish Thought and the Scientific Revolution of the Sixteenth Century: David Gans (1541–1613) and His Times*. Oxford: Oxford University Press.

———. 1990. *Paranormal and Transcendental Experience: A Psychological Examination*. New York: Dover.

Nelson, Robert A. 1929. *Nelson Enterprises Catalog*. No. 12. Columbus, OH: Nelson Enterprises.

———. 1968. *The Art of Cold Reading*. N.p.: Nelson Enterprises.

New Testament of Our Lord and Savior Jesus Christ As Revised and Corrected by the Spirits. 1861. New York: published by the proprietors.

Newman, John B. *Fascination, or the Philosophy of Charming Illustrating the Principles of Life in Connection with Spirit and Matter*. New York: Fowlers and Wells, 1850.

Newman, William R. 2004. *Promethean Ambitions: Alchemy and the Quest to Perfect Nature*. Chicago: University of Chicago Press.

Newnham, William. 1830. *Essay on Superstition; Being an Inquiry into the Effects of Physical Influence on the Mind, in the Production of Dreams, Visions, Ghosts, and Other Supernatural Appearances*. London: Hatchard.

Nickell, Joe, ed. 1994. *Psychic Sleuths: ESP and Sensational Cases*. Amherst, NY: Prometheus Books.

Nicolson, Adam. 2005. *God's Secretaries: The Making of the King James Bible*. New York: HarperCollins.

Nisbett. R. E., and T. D. Wilson. 1977. "Telling More Than We Can Know: Verbal Reports on Mental Processes. *Psychological Review* 84: 231–59.

Nixon, Robert. ca. 1818. *The Original Predictions or Prophecies of Robert Nixon, the Cheshire Prophet, from Lady Cowper's Correct Copy, in the Reign of Queen Anne: To Which Is Prefixed, Some Particulars of the Life of This Extraordinary Character:*

Likewise, Mother Shipton's Yorkshire Prophecy; with the Explanations: Which Was Discovered among Other Valuable Manuscripts, Preserved in the Family of the P—'s for Many Years. London: J. Bailey.

Noll, Richard. 1997. *The Aryan Christ: The Secret Life of Carl Jung*. New York: Random House.

Northrup, John. 1885. *The True Testament Bible and Millennium. Containing an Abridged History of God, the Goddesses, the People, the Universe, and Many Things. The Constitution of the Earth, and the Number of Governments There Shall Be. The Laws of Marriage and Divorce. The Beauties of Heaven, and What and Where It Really Is; and Death No Longer a Terror*. Hornitos, CA: privately printed.

Norwich, John Julius. 2011. *Absolute Monarchs: A History of the Papacy*. New York: Random House.

Notestein, Wallace. 1911. *A History of Witchcraft in England from 1558 to 1718*. Washington, DC: American Historical Association.

Novella, Steven. 2012, November 8. "Facilitated Communication Persists despite Scientific Criticism." *Neurologica*. https://theness.com/neurologicablog/index.php/facilitated-communication-persists-despite-scientific-criticism.

Ofshe, Richard, and E. Watters. 1994. *Making Monsters: False Memories, Psychotherapy, and Sexual Hysteria*. New York: Scribner's Sons.

Ogden, Emily. 2018. *Credulity: A Cultural History of US Mesmerism*. Chicago: University of Chicago Press.

Oldfield, Taverse. 1852. *"To Daimonion," or the Spiritual Medium*. Boston: Gould and Lincoln.

Oldmixon, John. 1746. *Nixon's Cheshire Prophecy at Large*. N.p.: privately printed.

Olds, Katrina B. 2015. *Forging the Past: Invented Histories in Counter-reformation Spain*. New Haven, CT: Yale University Press.

Oppenheim, Janet. 1985. *The Other World: Spiritualism and Psychical Research in England, 1850–1914*. Cambridge: Cambridge University Press.

Ore, Oystein. 1953. *Cardano: The Gambling Scholar*. Princeton, NJ: Princeton University Press.

Orne, Martin T. 1951. "The Mechanisms of Hypnotic Age Regression: An Experimental Study." *Journal of Abnormal and Social Psychology* 46: 213–25.

———. 1952. "The Nature of Hypnosis: Artifact and Essence." *Journal of Abnormal and Social Psychology* 45: 277–99.

———. 1962. "On the Social Psychology of the Psychological Experiment: With Particular Reference to Demand Characteristics and Their Implications." *American Psychologist* 17: 776–83.

———. 1979. "The Use and Misuse of Hypnosis in Court." *International Journal of Clinical and Experimental Hypnosis* 27: 311–41.

Orne, Martin T. et al. 1996. "'Memories' of Anomalous and Traumatic Autobiographical Experiences: Validation and Consolidation of Fantasy through Hypnosis." *Psychological Inquiry* 7: 168–72.

Ornstein, Martha. 1938. *The Role of Scientific Societies in the Seventeenth Century*. Chicago: University of Chicago Press.

Ovette, Joseph. [1921]. *Bargain Magic*. Berlin, WI: Heaney Magic.

———. 1947. *Miraculous Hindu Feats*. Oakland: Lloyd Jones.

Owen, Alex. 1990. *The Darkened Room: Women, Power and Spiritualism in Late Victorian England*. Philadelphia: University of Pennsylvania Press.

Owen, Iris M., and Margaret Sparrow. 1976. *Conjuring up Philip: An Adventure in Psychokinesis*. Toronto: Filzhenry & Whiteside.

Owen, Robert. 1836. *The Book of the New Moral World, Containing the Rational System of Society, Founded on Demonstrable Facts, Developing the Constitution and Laws of Human Nature and of Society*. London: Effingham Wilson.

Owen, Robert Dale. 1860. *Footfalls on the Boundary of Another World*. Philadelphia: Lippincott.

Oxon, M. A. 1882. *Psychography: A Treatise on One of the Objective Forms of Psychic or Spiritual Phenomena*. London: Psychological Press Association.

Ozanam, [Jacques], [Jean Etienne] Montucla, and Charles Hutton [editor and translator]. 1803. *Recreations in Mathematics and Natural Philosophy*. 4 vols. London: printed for G. Kearsley.

Page, Charles Grafton. 1853. *Psychomancy: Spirit-rappings and Table-tippings Exposed*. New York: Appleton.

Page, Sophie. 2012. "Uplifting Souls: The *Liber de essential spirtuum* and the *Liber Razielis*." In *Invoking Angels: Theurgic Ideas and Practices, Thirteenth to Sixteenth Centuries*. Edited by Clair Fanger. University Park: The Pennsylvania State University Press.

Palladino, Eusapia. 1910. "My Own Story." *Cosmopolitan* 48: 292–300.

Palter, Robert M., ed. 1961. *Toward Modern Science*. New York: Noonday Press.

Panek, Richard. 1998. *Seeing and Believing: How the Telescope Opened Our Eyes and Minds to the Heavens*. New York: Viking.

Pankratz, Loren. 1979. "Symptom Validity Testing and Symptom Retraining: Procedures for the Assessment and Treatment of Functional Sensory Deficits." *Journal of Consulting and Clinical Psychology* 47: 409–10.

———. 1986. "Do It to Yourself Section: Surgery." *Journal of the American Medical Association* 255: 324.

———. 1988. "Fire Walking and the Persistence of Charlatans." *Perspectives in Biology and Medicine* 31: 291–98.

———. 1990. "The Scourge of Scrooge." *Genii* 54: 118.

———. 1998. *Patients Who Deceive: Assessment and Management of Risk in Providing Benefits and Care*. Springfield, IL: Charles C Thomas.

———. 2002. "Demand Characteristics and the Development of Dual, False Belief Systems." *Prevention and Treatment* 5.

———. 2003a. "More Hazards: Hypnosis, Airplanes, and Strongly Held Beliefs." *Skeptical Inquirer* 27, no. 3: 31–36.

———. 2003b. "The Misadventures of Wanderers and Victims of Trauma." In *Malingering and Illness Deception*. Edited by Peter W. Halligan, Christopher Bass, and David A. Oakley. Oxford: Oxford University Press.

———. 2006. "Persistent Problems with the Munchausen Syndrome by Proxy Label." *Journal of the American Academy of Psychiatry and Law* 34: 90–95.

———. 2010. "Persistent Problems with the 'Separation Test' in Munchausen Syndrome by Proxy." *Journal of Psychiatry and Law* 38: 307–23.

———. 2014. "The Historical Roots of the Con Game." *The Yankee Magic Collector* 16: 102–17.

Pankratz, Loren, David Hickam, and Shirley Toth. 1989. "The Identification and Management of Drug-seeking Behavior in a Medical Center." *Drug and Alcohol Dependence* 24: 115–18.

Pankratz, Loren, and Gregory McCarthy. 1986. "The Ten Least Wanted Patients." *Southern Medical Journal* 79: 613–20.

Pankratz, Loren, and James Jackson. 1994. "Habitually Wandering Patients." *New England Journal of Medicine* 331: 1752–55.

Pankratz, Loren, and John Lipkin. 1978. "The Transient Patient in a Psychiatric Ward: Summering in Oregon." *Journal of Operational Psychiatry* 9: 42–47.

Pankratz, Loren, and Landy Sparr. 1984. "Drs. Pankratz and Sparr Reply." *American Journal of Psychiatry* 141: 473.

Pankratz, Loren, and Lyle Kofoed. 1988. "The Assessment and Treatment of Geezers." *Journal of the American Medical Association* 259: 1228–29.

Parallax. 1873. *Zetetic Astronomy: Earth Not a Globe. An Experimental Inquiry into the True Figure of the Earth, Proving It a Plane, without Orbital or Axial Motion, and the Only Known Material World*. London: John B. Day.

Parish, Edmund. 1897. *Hallucinations and Illusions: A Study of the Fallacies of Perception*. London: Scott.

Park, James Allan. 1799. *A System of the Law of Marine Insurances*. Boston: Thomas and Andrews, David West, and John West.

Parke, H. W. 1943. "The Days for Consulting the Delphic Oracle." *The Classical Quarterly* 37: 19–22.

———. 1967. *Greek Oracles*. London: Hutchinson University Library.

———. 1985. *The Oracles of Apollo in Asia Minor*. London: Routledge.

———. 1988. *Sibyls and Sibylline Prophecy in Classical Antiquity*. London: Routledge.

Parkyn, H. A. 1900. *Suggestive Therapeutics and Hypnotism*. Chicago: Suggestion Publishing.

Partington, Charles F. 1825. *The Century of Inventions of the Marquis of Worcester*. London: John Murray.

Partridge, John. 1703. *Astrological Predictions of What Shall Happen throughout Europe, in the Year of Our Lord 1703*. N.p.: privately printed.

Patihis, L. et al. 2013. "False Memories in Highly Superior Autobiographical Memory Individuals." *Proceedings of the National Academy of Sciences of the United States of America* 110: 20947–52.

Pattie, Frank A. 1956. "Mesmer's Medical Dissertation and Its Debt to Mead's *De Imperio Solis ac Lunae*." *Journal of the History of Medicine* 11: 275–87.

Paul, Henry H. 1845. *The Book of Chemical Amusements: A Complete Encyclopedia of Experiments in Various Branches of Chemistry, &c.* Philadelphia: Getz & Smith.

Pearl, Jonathan L. 1999. *The Crime of Crimes: Demonology and Politics in France 1560–1620*. Wilfrid Laurier University Press.

Pearsall, Ronald. 1972. *The Table-rappers*. London: Michael Joseph.

Pendergrast, Mark. 2003. *Mirror Mirror: A History of the Human Love Affair with Reflection*. New York: Basic Books.

Pennington, Kenneth. 2003. "Innocent until Proven Guilty: The Origins of a Legal Maxim." *The Jurist* 63: 106–24.

Pepper, John Henry. 1861. *Scientific Amusements for Young People*. London: Routledge, Warne, and Routledge.

Perovsky-Petrovo-Solovovo. 1911. "Statement by Count Perovsky-Petrovo-Solovovo." *Proceedings for the Society for Psychical Research* 25: 59–63.

Persuitte, David. 2000. *Joseph Smith and the Origins of the Book of Mormon*. Jefferson, NC: McFarland.

Peters, Edward. 1978. *The Magician, the Witch, and the Law*. Sussex: Harvester Press.

Petersilea, Carlyle. 1889. *The Discovered Country*. Boston: Banner of Light Publishing.

Petrakos, Basil. 1977. *Delphi*. N.p.: Clio.

Phillips, R. P. 1950. *Modern Thomistic Philosophy: An Explanation for Students*. Westminster, MD: Newman Press.

Phillips, Will. ca. 1902. *Test Experiments in Spiritualistic Phenomena by Professor Hare. A Reply to Frank Podmore's Criticism*. Manchester: World Publishing Company.

Phin, John. 1906. *The Seven Follies of Science: A Popular Account of the Most Famous Scientific Impossibilities and the Attempts Which Have Been Made to Solve Them*. New York: Van Nostrand.

Pieper, Lori. 2013. "The Asti Manuscript." *Gibeciére* 8, no. 1: 29–235.

Pinchbeck, William Frederick. 1805. *The Expositor; or Many Mysteries Unravelled*. Boston: printed for the author.

Pinetti, Herman. 1905. *Second Sight Secrets and Mechanical Magic*. Bridgeport, CT: Dunham Press.

Piper, August. 1997. *Hoax and Reality: The Bizarre World of Multiple Personality Disorder*. Northvale, NJ: Aronson.

Pirenne, Henri. 1937. *Economic and Social History of Medieval Europe*. New York: Harcourt, Brace.

Pittinger, David J. 1993. "Measuring the MBTI . . . and Coming up Short." *Journal of Career Planning and Employment* 54: 48–52.

Plato. 1909. *The Apology*. In *The Harvard Classics*. Edited by Charles W. Elliot. Translated by Benjamin Jowett. New York: Collier & Son.

Platts, I. 1822. *The Book of Curiosities; or, Wonders of the Great World: Containing an Account of Whatever Is Most Remarkable in Nature & Art, Science & Literature*. London: Henry Fisher.

———. 1875. *Encyclopedia of Natural and Artificial Wonders and Curiosities*. New York: World Publishing House.

Plotkin, Henry. 1997. *Darwin Machines and the Nature of Knowledge*. Cambridge, MA: Harvard University Press.

Plutarch. 1993. *Moralia*. Cambridge, MA: Harvard University Press.

Podmore, Frank. 1893. "(1) Experiences de Milan; Notes de M. Charles Richet; (2) Rapport de la Commission reunie a Milan pour l'Etude des Phenomenes Psychiques. Both Being Articles in the Annales des Sciences Psychiques." *Proceedings of the Society for Psychical Research* 9: 218–25.

———. 1897. *Studies in Psychical Research*. New York: Putnam's Sons.

———. 1902. *Modern Spiritualism*. London: Methuen.

———. 1909. *From Mesmer to Christian Science: A Short History of Mental Healing*. London: Methuen.

———. 1910. *The Newer Spiritualism*. London: Unwin.

Polidoro, Massimo. 2003. *Secrets of the Psychics: Investigating Paranormal Claims*. Buffalo: Prometheus Books.

Pollard, Justin, and Howard Reid. 2006. *The Rise and Fall of Alexandria: Birthplace of the Modern Mind*. New York: Viking.

Popkin, Richard H., and Arjo Vanderjagt, eds. 1993. *Scepticism and Irreligion in the Seventeenth and Eighteenth Centuries*. New York: Brill.

Popper, Karl R. 1972. *The Logic of Scientific Discovery*. London: Hutchinson.

Popular Mechanics Press. 1915. *The Boy Mechanic: Book 2*. Chicago: Popular Mechanics.

Porta, John Baptist. 1658. *Natural Magick*. London: Thomas Young and Samuel Speed.

———. 1957. *Natural Magick*. New York: Basic Books.

Porter, Roy. 2000. *The Creation of the Modern World: The Untold Story of the British Enlightenment*. New York: Norton.

Post, Isaac. 1852. *Voices from the Spirit World, with Communications from Many Spirits, by the Hand of Isaac Post, Medium*. Rochester, NY: Charles H. McDonell.

Potter, J. 1967. *The Master Index to Magic in Print; Covering Books and Magazines in the English Language Published up to and Including December 1964*. Calgary: M. Hades Enterprises.

Poulsen, Fredrick. 1920. *Delphi*. London: Gyldenal.

———. 1945. "Talking, Weeping, and Bleeding Sculptures: A Chapter of the History of Religious Fraud." *Acta Archaeologica* 16: 178–95.

Powers, Melvin. 1961. *Advanced Techniques of Hypnosis*. Hollywood: Wilshire Book Company.

———. 1964. *Hypnotism Revealed*. Hollywood: Wilshire Book Company.

Poyen, Charles. 1837. *A Letter to Col. Wm. L. Stone, of New York, on the Facts Related in His Letter to Dr. Brigham, and a Plain Refutation of Durant's Exposition of Animal Magnetism, &c. with Remarks on the Manner in Which the Claims of Animal Magnetism Should Be Met and Discussed. By a Member of the Massachusetts Bench*. Boston: Seeks, Jordan and Company.

Prater, Horatio. 1846. *On the Injurious Effects of Mineral Poisons in the Practice of Medicine*. London: Sherwood.

———. 1851. *Lectures on True and False Hypnotism, or Mesmerism*. London: Piper Brothers.

Prevost, E. 1851–1852. *The Zoist* 9: 413–14.

Price, David. 1985. *Magic: A Pictorial History of Conjurers in the Theater*. New York: Cornwall Books.

Price, Harry. 1930. *Rudi Schneider: A Scientific Examination of His Mediumship*. London: Methuen.

———. 1933. *An Account of Some Further Experiments with Rudi Schneider*. London: Council at the Rooms of the National Laboratory of Psychical Research.

———. 1934. *Exhibition of Rare Works from the Research Library of the University of London Council for Psychical Investigation: From 1490 to the Present Day*. London: University of London Council for Psychical Investigation.

———. 1936a. *Confessions of a Ghost-hunter*. London: Putnam.

———. 1936b. *A Report of Two Experimental Fire-walks*. London: University of London Council for Psychical Investigation.

———. 1939. *Fifty Years of Psychical Research*. London: Longmans, Green.

———. 1942. *Search for Truth*. London: Collins.

Prince, Walter Franklin. 1921. "A Survey of American Slate Writing Mediumship." *Proceedings of the American Society for Psychical Research* 15: 315–592.

———. 1964. *The Case of Patience Worth*. New Hyde Park, NY: University Books.

Princess Mary's Gift Book. [ca. 1915]. London: Hodder & Stoughton.

Principe, Lawrence M. 1992. Robert Boyle's Alchemical Secrecy: Codes, Ciphers and Concealments. *Ambix* 39, no. 2: 63–74.

Pritchard, Maxwell. 2018. "The Magic of the Pont Neuf." *Gibeciére* 13, no. 1: 87–171.

Procopius. 1981. *The Secret History*. Translated by G. A. Williamson. London: Penguin Books.

Ptolemy, Claudius. 1991. *The Geography*. New York: Dover.

Putnam, Allen. [1874]. *Agassiz and Spiritualism: Involving the Investigation of Harvard College Professors in 1857*. Boston: Colby & Rich.

Putnam, George Haven. 1962. *Books and Their Makers during the Middle Ages*. New York: Hillary House.

Pybus, William. 1810. *A Manual of Useful Knowledge, Being a Collection of Valuable Miscellaneous Receipts and Philosophical Experiments, Selected from Various Authors*. London: William Rawson.

Ralph, Philip Lee. 1974. *The Renaissance in Perspective*. New York: St. Martin's Press.

Randall, John Herman. 1963. *Aristotle*. New York: Columbia University Press.

Randi, James. 1980. *Flim-Flam! The Truth about Unicorns, Parapsychology and Other Delusions*. New York: Lippincott & Crowell, 1980.

———. 1982. *Flim-Flam! Psychics, ESP, Unicorns and Other Delusions*. Buffalo: Prometheus.

———. 1983a. "The Project Alpha Experiment: Part l. The First Two Years." *Skeptical Inquirer* 7: 24–33.

———. 1983b. "The Project Alpha Experiment: Part 2. Beyond the Laboratory." *Skeptical Inquirer* 9: 36–45.

Ranke, Leopold. 1847–1848. *The History of the Popes: Their Church and State, and Especially of the Conflicts with Protestantism in the Sixteenth & Seventeenth Centuries*. 3 vols. London: Henry G. Bohn.

Raspe, Rudolph. 1811. *Surprising Adventures of the Renowned Baron Munchausen, Containing Singular Travels, Campaigns, Voyages, and Adventures. Also an Account of a Voyage to the Moon and Dog Star*. London: Thomas Tegg.

Raue, C. G. 1889. *Psychology As a Natural Science Applied to the Solution of Occult Psychic Phenomena*. Philadelphia: Porter & Coates.

Rauscher, William. 2011. *Encyclopedia: Mentalism & Mentalists*. N.p.: 1878 Press.

Raymond, Professor. 1875. *Parlor Pastimes; or, the Whole Art of Amusing*. New York: Hurst.

Rea, W. T. 1982. *The White Lie*. Turlock, CA: M & R Publications.

Read, John. 1937. *Prelude to Chemistry: An Outline of Alchemy, Its Literature and Relationships*. New York: Macmillan.

Rebmuh, R. ca. 1889. *Prof. R. Rebmuh's Descriptive Catalogue of New and Original Anglo-American Illusions*. Leichester: Flavel Barker.

Rees, Terence, and David Wilmore. 1996. *British Theatrical Patents: 1801–1900*. London: Society for Theatre Research.

Reichenbach, Charles von. 1851. *Physico-physiological Researches on the Dynamics of Magnetism, Electricity, Heat, Light, Crystallization, and Chemism, in Their Relation to Vital Force*. London: Hippolyte Bailliere.

Reilly, S. W. 1957. *Table Lifting Method Used by Fake Mediums*. Chicago: Ireland Magic Company.

Report of the Committee Appointed to Investigate Phenomena Connected with the Theosophical Society. 1885. *Proceedings of the Society for Psychical Research* 3: 254.

Research Officer. 1930. *Pseudo-prophecies and Pseudo-sciences by the Research Officer. Also "A Test of the Accuracy of the Testimony of Bystanders" by George Glover Crocker*. Boston: Boston Society for Psychic Research.

Rice, N. L. 1849. *Phrenology Examined, and Shown to Be Inconsistent with the Principles of Phisiology [Sic], Mental and Moral Science, and the Doctrines of Christianity. Also an Examination of the Claims of Mesmerism*. New York: Robert Carter & Brothers.

Richards, J. T. 1982. *SORRAT: A History of the Neihardt Psychokinesis Experiments, 1961–1981*. Metuchen, NJ: Scarecrow Press.

Richet, Charles. 1923. *Thirty Years of Psychical Research Being a Treatise on Metaphysics*. London: W. Collins Sons.

———. [1929]. *Our Sixth Sense*. London: Rider.

Riding, Joe. 1989. *Joe Riding's Modern Technique of Cold Reading*. N.p.: Joe Riding.

Rinn, Joseph F. 1950. *Sixty Years of Psychical Research*. New York: Truth Seeker.

Robert-Houdin, J. n.d. *Card-sharpers: Their Tricks Exposed, or the Art of Always Winning*. Chicago: Charles Powner.

Robert-Houdin, Jean-Eugene. ca. 1878. *The Secrets of Conjuring and Magic or How to Become a Wizard*. London: Routledge.

———. 1900. *The Secrets of Stage Conjuring*. London: Routledge and Sons.

———. ca. 1903. *Life of Robert Houdin, the King of the Conjurers*. Philadelphia: Henry T. Coates.

Roberts, David. 1982. *Great Exploration Hoaxes*. San Francisco: Sierra Club Books.

Robertson, James. 1905. *Mediums and the Spiritual Press*. Glasgow: John Rutherford.

Robinson, Henry Morton. 1935. *Science Catches the Criminal*. New York: Blue Ribbon Books.

Robinson, J. M., ed. 1972. *Religion and the Humanizing of Man*. Waterloo, ON: Council on the Study of Religion.

Robinson, James M., ed. 1981. *The Nag Hammadi Library in English*. San Francisco: Harper & Row.

Robinson, William E. 1898. *Spirit Slate Writing and Kindred Phenomena*. New York: Munn.

Robison, John. 1822. *A System of Mechanical Philosophy*. 2 vols. Edinburgh: John Murray.

Romains, Jules. 1978. *Eyeless Sight*. Secaucus, NJ: Citadel Press.

Romano, Chuck. 2006. *The Mechanics of Marvels*. Aurora, IL: self-published.

Rosen, George. 1946. "A Strange Chapter in the History of Anesthesia. *Journal of the History of Medicine and Allied Sciences* 1: 527–50.

———. 1965. *Kepler's Conversation with Galileo's Sidereal Messenger*. New York: Johnson Reprint Corporation.

Rothman, M. A. 1988. *A Physicist's Guide to Skepticism*. Buffalo, NY: Prometheus.

Rousseau, Jean-Jacques. 1979. *Emile: Or, On Education*. New York: Basic Books.

Rowland, Ingrid D. 2004. *The Scarith of Scornello: A Tale of Renaissance Forgery*. Chicago: University of Chicago Press.

Rue, Loyal. 1994. *By the Grace of Guile*. New York: Oxford University Press.

Rummel, Erika. 1998. *The Humanist-scholastic Debate*. Cambridge, MA: Harvard University Press.

Russell, Bertrand. 1950. *Unpopular Essays*. New York: Simon and Schuster.

Russell, Jeffrey Burton. 1991. *Inventing the Flat Earth: Columbus and Modern Historians*. New York: Praeger.

Saberi, Reza. 2016. "One Thousand and One Years of Persian Magic." *Gibeciére* 11: 129–61.

Sacks, Oliver. 1986. *The Man Who Mistook His Wife for a Hat.* New York: Summit Books.

Salerno, Steve. 2005. *SHAM: How the Self-help Movement Made America Helpless.* New York: Crown.

Salverte, Eusebe. 1846. *The Philosophy of Magic, Prodigies and Apparent Miracles.* London: Richard Bentley.

Sambursky, Samuel. 1963. *The Physical World of the Greeks.* London: Routledge.

Samson, G. W. 1869. *Physical Media in Spiritual Manifestations: The Phenomena of Responding Tables and the Planchette.* Philadelphia: J. B. Lippincott.

Sanbonmatsu, David M. et al. 2003. "Overestimating the Importance of the Given Information in Multiattribute Consumer Judgment." *Journal of Consumer Psychology* 13: 289–300.

Sanbonmatsu, David M., and Katie K. Sanbonmatsu. 2017. "The Structure of Scientific Revolutions: Kuhn's Misconceptions of (Normal) Science." *Journal of Theoretical and Philosophical Psychology* 37: 133–51.

Sandars, Joseph. ca. 1817. *Hints to Credulity! Or, An Examination of the Pretensions of Miss M. McAvoy; Occasioned by Dr. Renwick's "Narrative" of Her Case.* Liverpool: sold by William Robinson.

Sandby, George. 1848. *Mesmerism and Its Opponents.* London: Longman, Brown, Green, and Longmans.

———. 1850. "Review of M. Alphonse Cahagnet's *Arcanes de la vie future devoiles,* &c." *The Zoist.*

———. 1853. "Can Professor Faraday Never Be Wrong? Or, Is Table Turning All a Delusion?" *The Zoist* 11: 320–24.

Santanelli. 1902. *The Law of Suggestion.* Lansing: Santanelli.

Sargent, Epes. 1881. *The Scientific Basis of Spiritualism.* Boston: Colby and Rich.

Saville, Thomas K., and Herb Dewey. 1984. *The Professional Pseudo Psychic: Red Hot Cold Reading.* Denver, CO: In Visible Print.

Schaffer, Simon. 1998. "Regeneration: The Body of Natural Philosophers in Restoration England." In *Science Incarnate: Historical Embodiments of Natural Knowledge.* Edited by Christopher Lawrence and Steven Shapin. Chicago: University of Chicago Press.

Schatz, Edward R. 1974. *Practical Contact Mind Reading.* Calgary, AB: Micky Hades.

Scheflin, Alan W., and Jerrold Lee Shapiro. 1989. *Trance on Trial.* New York: Guilford Press.

Schlesinger, Julie. 1886. "Robert Hare, M.D." *The Carrier Dove* 2: 100–104.

Schmidt, Leigh Eric. 2000. *Hearing Things: Religion, Illusion, and the American Enlightenment.* Cambridge, MA: Harvard University Press.

Schrenck-Notzing, A. Freiherrn von. 1920a. *Physikalische phaenomene des mediumismus.* Munchen: Von Ernst Reinhardt.

Schrenck-Notzing, Albert von. 1920b. *Phenomena of Materialisation: A Contribution to the Investigation of Mediumistic Teleplastics*. London: Kegan Paul.

"Science and the Spirits." 1893. *Scientific American Supplement* 35, no. 897.

Sconce, Jeffrey. 2000. *Haunted Media: Electronic Presence from Telegraphy to Television*. Durham, NC: Duke University Press.

Scot, Reginald. 1584/1930. *The Discoverie of Witchcraft*. London: John Rodker.

———. 1651. *The Discovery of Witchcraft*. London: Printed by R. C.

Seabrook, William. 1941. *Doctor Wood: Modern Wizard of the Laboratory*. New York: Harcourt, Brace.

Seltman, Charles. 1957. *Wine in the Ancient World*. London: Routledge & Kegan Paul.

Seybert Commission. 1887. *Preliminary Report of the Commission Appointed by the University of Pennsylvania to Investigate Modern Spiritualism in Accordance with the Request of the Late Henry Seybert*. Philadelphia: Lippincott.

Shapin, Steven. 1996. *The Scientific Revolution*. Chicago: University of Chicago Press.

Shapin, Steven, and Simon Schaffer. 1985. *Leviathan and the Air-pump: Hobbes, Boyle, and the Experimental Life*. Princeton, NJ: Princeton University Press.

Sharpe, S. H. 1985. *Conjurers' Optical Secrets*. Calgary: Micky Hades.

———. 1991. *Conjurers' Hydraulic and Pneumatic Secrets*. Alberta: Alberta Foundation for Literary Arts.

———. 1992. *Conjurors' Mechanical Secrets*. N.p.: Tannen's Magic.

Sharps, Matthew J. 2018. "Percival Lowell and the Canals of Mars." *Skeptical Inquirer* 42, no. 3: 41–46.

Sharratt, Michael. 1999. *Galileo: Decisive Innovator*. Cambridge: Cambridge University Press.

Shaw, W. H. J. 1893. *Magic and Its Mysteries*. Blue Island, IL: M. N. Smith Printers and Binders.

———. 1999. *Book of Acts for Carnival, Side Show Museum and Circus*. Victoria, Missouri: Shaw.

Shea, William R., and Mariano Artigas. 2003. *Galileo in Rome: The Rise and Fall of a Troublesome Genius*. New York: Oxford University Press.

Shepard, Leslie A. 1984. *Encyclopedia of Occultism and Parapsychology*. 2nd ed. Detroit, MI: Gale Research.

Shepard, Odell. 1982. *The Lore of the Unicorn*. New York: Avenel Books.

Shimron, Binyamin. 1989. *Politics and Belief in Herodotus*. Stuttgart: F. Steiner Verlag.

Shorto, Russell. 2008. *Descartes' Bones: A Skeletal History of the Conflict between Faith and Reason*. New York: Doubleday.

Sibley, Brian. 2001. *Three Cheers for Pooh: A Celebration of the Best Bear in All the World*. London: Methuen.

Sidgwick, Eleanor Mildred. 1886a. "The Charges against Mr. Eglinton." *Journal of Society for Psychical Research* 2: 467–69.

Sidgwick, Mrs. Henry. 1886b. "Results of a Personal Investigation into the Physical Phenomena of Spiritualism." *Proceedings of the Society for Psychical Research* 4: 45–74.

———. 1909a. "Psicologia e Spiritismo: Impressioni e note critiche sui fenomeni medianici di Eusapia Paladino, by Morselli." *Proceedings of the Society for Psychical Research* 21: 425–516.

———. 1909b. "Introductory Note to the 'Report on Sittings with Eusapia Palladino.'" *Proceedings of the Society for Psychical Research* 23: 306–8.

Sigmund, Karl. 2017. *Exact Thinking in Demented Times: The Vienna Circle and the Epic Quest for the Foundations of Science.* New York: Basic Books.

Signor Blitz. 1889. *The Parlor Book of Magic and Drawing-room Entertainments.* New York: Hurst.

Simonetta, Marcello. 2008. *The Montefeltro Conspiracy: A Renaissance Mystery Decoded.* Doubleday: New York.

Sims, Henry B. 1845. Letter. *The Zoist* 2: 518–21.

Sinclair, Upton. 1906. *The Jungle.* New York: Doubleday.

Singer, Charles. 1928/1958. *From Magic to Science: Essays on the Scientific Twilight.* New York: Dover.

Singer, Charles et al., eds. 1957. *A History of Technology.* Vol. 3, *From the Renaissance to the Industrial Revolution c. 1500–c. 1750.* New York: Oxford University Press.

Siraisi, Nancy G. 1997. *The Clock and the Mirror: Girolamo Cardano and Renaissance Medicine.* Princeton, NJ: Princeton University Press.

Skinner, W. E., ed. [1895]. *Wehman's Wizards' Manual: A Practical Treatise on Mind Reading, According to Stuart Cumberland and the Late Washington Irving.* New York: Wehman Bros.

Slater, Mrs. 1872. "Direct Spirit-writing." *The Spiritual Magazine* 7: 568–9.

Smith, Francis H. 1860. *My Experience, or Foot-prints of a Presbyterian to Spiritualism.* Baltimore: N.p.

Smith, H. Adrian. 1953. *The Compleat Practical Joker.* Garden City, NY: Doubleday.

———. 1987. *Books at Brown.* Providence, RI: Friends of the Library of Brown University.

Smith, Horatio. 1831. *Festivals, Games, and Amusements: Ancient and Modern.* London: Colburn and Bentley.

Smoller, Laura Ackerman. 1994. *History, Prophecy, and the Stars: The Christian Astrology of Pierre d'Ailly (1350–1420).* Princeton, NJ: Princeton University Press.

Smullyan, Raymond. 1983. *5000 B.C. and Other Philosophical Fantasies.* New York: St. Martin's Press.

Snyder, Laura J. 2011. *The Philosophical Breakfast Club: Four Remarkable Friends Who Transformed Science and Changed the World.* New York: Broadway Books.

Soal, S. G. 1937. *Preliminary Studies of a Vaudeville Telepathist.* London: University of London Council for Psychical Investigation.

Sobel, Dava. 2000. *Galileo's Daughter: A Historical Memoir of Science, Faith, and Love*. New York: Penguin Books.

Southwick, S. M., C. A. Morgan, A. L. Nicolaou, and D. S. Charney. 1997. "Consistency of Memory for Combat-related Traumatic Events in Veterans of Operation Desert Storm. *American Journal of Psychiatry* 154: 173–77.

Spanos, N. P. 1996. *Multiple Identities and False Memories: A Sociocognitive Perspective*. Washington, DC: American Psychological Association.

Sparr, Landy, and Loren Pankratz. 1983. "Factitious Posttraumatic Stress Disorder." *American Journal of Psychiatry* 140: 1016–19.

Spear, John Murray. 1857. *The Educator: Being Suggestions, Theoretical and Practical, Designed to Promote Man-culture and Integral Reform, with a View to the Ultimate Establishment of a Divine Social State on Earth*. Boston: Office of Practical Spiritualists.

Sperber, Burton S. 1982. *Miracles of My Friends*. Malibu, CA: Sperber.

Spicer, Henry. 1853. *Sights and Sounds: The Mystery of the Day: Comprising an Entire History of the American "Spirit" Manifestations*. London: Thomas Bosworth.

Spitz, Herman H., and Yves Marcuard. 2001. "Chevreul's Report on the Mysterious Oscillations of the Hand-held Pendulum: A French Chemist's 1833 Open Letter to Ampere." *Skeptical Inquirer* 25, no. 4: 35–39.

Stafford, Tom, and Matt Webb. *Mind Hacks*. Sebastopol, CA: O'Reilly Media, 2005.

Stagnaro, Angelo. 2004. *Conspiracy: An Original Collection of Mentalist Codes and Signals*. N.p.: published by the author.

Stanovich, Keith E. 1998. *How to Think Straight about Psychology*. New York: Longman.

Stein, Gordon. 1996. *The Encyclopedia of the Paranormal*. Buffalo: Prometheus Books.

Steinmeyer, J. H. 1981. *Jarrett*. Chicago: Magic.

Stillman, P., and B. Gordon. 1972. *Roman Rulers and Rebels*. Wellesley Hills, MA: Independent School Press.

Stock, Armin, and Claudia Stock. 2004. "A Short History of Ideo-motor Action." *Psychological Research* 68: 76–188.

Stone, Alex. 2012. *Fooling Houdini: Adventures in the World of Magic*. London: William Heinemann.

Stone, Tom. n.d. "Too Perfect, Imperfect." In *Magic in Mind: Essential Essays for Magicians*. Edited by Joshua Jay, 183–91. Denver, CO: Society of American Magicians.

Stone, William L. 1835. *Matthias and His Impostures: Or, the Progress of Fanaticism. Illustrated in the Extraordinary Case of Robert Matthews and Some of His Forerunners and Disciples*. New York: Harper.

———. [1836]. *A Refutation of the Fabulous History of the Arch-impostor Maria Monk Being the Result of a Minute and Searching Inquiry to Which Are Added Other Interesting Testimonies, &c*. [New York]: n.p.

———. 1837. *Letter to Doctor A. Brigham, on Animal Magnetism: Being an Account of a Remarkable Interview between the Author and Miss Loraina Brackett while in a State of Somnambulism*. New York: George Dearborn.

Storr, Will. 2014. *The Unpersuadables: Adventures with the Enemies of Science*. New York: Overlook Press.

Strieber, Whitley. 1988. *Communion: A True Story*. New York: Avon.

Sully, James. 1881. *Illusions: A Psychological Study*. New York: Appleton.

Summers, Montague. 1950. *The Physical Phenomena of Mysticism with Especial Reference to the Stigmata, Divine and Diabolic*. New York: Barnes & Noble.

Sumpton, Jonathan. 1975. *Pilgrimage: An Image of Mediaeval Religion*. Totowa, NJ: Rowman & Littlefield.

Swan, Irene. 1991. *The Bangs Sisters and Their Precipitated Spirit Portraits*. Chesterfield, IN: Hett Memorial Art Gallery and Museum.

Swedenborg, Emanuel. 1840. *On the Earths in Our Solar System, Which Are Called Planets; and on the Earths in the Starry Heavens; with an Account of Their Inhabitants, and Also of the Spirits and Angels There; from What Has Been Heard and Seen*. London: James S. Hodson.

———. 1875. *Heaven and Hell; also, the Intermediate State, or Worlds of Spirits; a Relation of Things Heard and Seen*. London: Swedenborg Society.

Sylvestre, Ralph E. 1901. *Gambols with the Ghosts*. Chicago: Sylvestre.

Symonds, John Addington. 1909. *Renaissance in Italy: The Revival of Learning*. London: Smith, Elder.

T. S. 1862. "Spiritualism in Biography—J. Heinrich Jung Stilling." *The Spiritual Magazine* 3: 189–312.

"Table Turning." 1854. *Metaphysical Magazine* 1: 125–43.

Tabori, Paul. 1972. *Pioneers of the Unseen*. New York: Taplinger.

Tamariz, Juan. 2014. *The Magic Way: The Method of False Solutions and the Magic Way*. Seattle: Hermetic Press.

Tarbell, Harlan E. 1954. *The Tarbell Course in Magic*. Vol. 6. New York: Louis Tannen.

Taub, Liba Chaia. 1993. *Ptolemy's Universe: The Natural Philosophical and Ethical Foundations of Ptolemy's Astronomy*. Chicago: Open Court.

Taut, Steffen. 2020. "Bosch's 'Juggler' Revisited." *Gibeciere* 15, no. 2: 12–114.

Taylor, Charles Henry. 1922. *By Wireless from Venus; or, the Primal Elements, Involving a New Theory of Creation . . . and an Explanation of the Unexplained in Physics and Metaphysics*. Los Angeles: Austin Publishing.

Taylor, Joseph. 1815. *Apparitions; or, the Mystery of Ghosts, Hobgoblins, and Haunted Houses*. London: Lackington, Allen.

Tegg, Thomas. ca. 1860. *The Young Man's Book of Knowledge*. New York: E. Kearny.

Telano, Rolf. 1960. *A Spacewoman Speaks*. El Monte, CA: Understanding Publishing.

Telepathist. 1926, August 9. "Thought-transference." *Chamber's Journal*: 493–96.

Temkin, Owsei. 1975. "Science and Society in the Age of Copernicus." In *The Nature of Scientific Discovery: A Symposium Commemorating the 500th Anniversary of the Birth of Nicolaus Copernicus*. Edited by Owen Gingerich, 106–33. Washington, DC: Smithsonian Institution Press.

Teresi, Dick. 2002. *Lost Discoveries: The Ancient Roots of Modern Science—from the Babylonians to the Maya*. New York: Simon & Schuster.

Thomas, Keith. 1971. *Religion and the Decline of Magic*. New York: Charles Scribner's Sons.

Thompson, E. A. 1946. "The Last Delphic Oracle." *The Classical Quarterly* 40: 25–26.

Thorndike, L. 1923. *A History of Magic and Experimental Science*. 8 vols. New York: Columbia University Press.

Thucydides. 1954. *The Peloponnesian War*. Translated by Rex Warner. Baltimore: Penguin.

Thurston, Herbert. 1935. *The Church and Spiritualism*. Milwaukee: Bruch Publishing.

———. 1951. *The Physical Phenomena of Mysticism*. Fort Collins, CO: Roman Catholic Books.

Timbs, John. 1856. *Things Not Generally Known, Familiarly Explained*. London: David Bogue.

———. 1860. *Stories of Inventors and Discoverers in Science and the Useful Arts*. London: Kent.

Toland, John. 1747. *The Miscellaneous Works of Mr. John Toland*. London: printed for J. Whiston and J. Robinson.

Toole-Stott, R. 1976–1978. *A Bibliography of English Conjuring: 1581–1876*. 2 vols. Derby, England: Harpur & Sons of Derby.

Tout, Thomas Frederick. 1934. *Mediaeval Forgers and Forgeries*. Manchester: Manchester University Press.

Townshend, Chauncy Hare. 1844. *Facts in Mesmerism, with Reasons for a Dispassionate Inquiry into It*. London: Hippolyte Bailliere.

———. 1852. "Recent Clairvoyance of Alexis Didier." *The Zoist* 9: 402–10.

———. 1854. *The Quarterly Reviewer Reviewed: Or, Mesmeric Agency Proved by Facts*. London: Thomas Bosworth.

Tracy, David F. 1952. *How to Use Hypnosis*. New York: Sterling.

Trivers, Robert. 2011. *The Folly of Fools: The Logic of Deceit and Self-deception in Human Life*. New York: Basic Books.

Trobridge, George. 1928. *Life of Emanuel Swedenborg*. New York: New-Church Press.

Tromp, Marlene. 2006. *Altered States: Sex, Nation, Drugs, and Self-transformation in Victorian Spiritualism*. Albany, NY: State University of New York Press.

Trudel, J. H. 1919. *Side Show Tricks Exposed*. Lowell, MA: Trudel.

Truesdell, John W. 1883. *The Bottom Facts Concerning the Science of Spiritualism*. New York: G. W. Carleton.

Tubby, Gertrude Ogden. 1929. *James H. Hyslop-X: His Book*. York, PA: York Printing.

Tuchman, Barbara. 1978. *A Distant Mirror: The Calamitous 14th Century*. New York: Alfred A. Knopf.

Turnbull, Herbert Westren. *The Great Mathematicians*. New York: Barnes & Noble, 1993.

Tuttle, Hudson, and J. M. Peebles. 1871. *The Year-book of Spiritualism for 1871*. Boston: William White.

Underhill, A. Leah. 1885. *The Missing Link in Modern Spiritualism*. New York: Knox.

Underhill, Evelyn. 1971. *The Mystics of the Church*. New York: Schocken Books.

Underhill, Samuel. 1868. *Underhill on Mesmerism with Criticisms on Its Opposers, with a Review of Humbugs and Humbuggers*. Chicago: Religio-Philosophical Publishing Association.

Valla, Lorenzo. 2008. *Lorenzo Valla: On the Donation of Constantine*. Translated by B. W. Bowersock. Cambridge, MA: Harvard University Press.

Vallee, Jacques. 1979. *Messengers of Deception: UFO Contacts and Cults*. Berkeley: And/Or Press.

Van der Waerden, B. L. 1963. *Science Awakening: Egyptian, Babylonian and Greek Mathematics*. New York: Wiley & Sons.

Van Helden, Albert. 1977. "The Invention of the Telescope." *Transactions of the American Philosophical Society* 67, no. 4: 1–67.

Vandenberg, Philipp. 1982. *The Mystery of the Oracles*. New York: Macmillan.

Vere, Arprey. 1879. *Ancient and Modern Magic with Explanations of Some of the Best Known Tricks Performed by Messrs. Maskelyne and Cooke*. London: Routledge and Sons.

Vergil, Polydore. 2002. *On Discovery*. Cambridge, MA: Harvard University Press.

Verneule, Cornelius C. 1950. "Lydian Gold Heavy Third-stater of Croesus." *Bulletin of the Fogg Art Museum* 11, no. 2: 52–53.

Vickers, Brian, ed. 1984/1986. *Occult and Scientific Mentalities in the Renaissance*. Cambridge: Cambridge University Press.

Victor, Jeffrey S. 1994. *Satanic Panic*. Peru, IL: Open Court.

Vincent, William. 1853. *Satanic Influence: Its Probable Connection with Table-talking*. London: J. H. Jackson.

Vizetelly, Henry. [1853a]. *Spirit Rapping in England and America: Its Origin and History*. London: Henry Vizetelly.

———. [1853b]. *Table Turning and Table Talking*. London: Henry Vizetelly.

Volkman, Ernest. 2002. *Science Goes to War: The Search for the Ultimate Weapon, from Greek Fire to Star Wars*. New York: John Wiley & Sons.

von Bothmer, Dietrich, and Joseph V. Noble. 1961. *An Inquiry into the Forgery of the Etruscan Terracotta Warriors in the Metropolitan Museum of Art*. New York: The Metropolitan Museum of Art.

Wallace, Alfred R. 1891. "Mr. S. J. Davey's Experiments." *Journal of the Society for Psychical Research* 5: 43.

———. 1898. "Extract from De Mirville's 'Des espirits et de leurs manifestations fluidiques.'" *Proceedings of the Society for Psychical Research* 14: 373–81.

Wallace, Alfred R. et al. 1878. *The Psycho-physiological Sciences, and Their Assailants. Being a Response by Alfred R. Wallace, of England, Professor J. R. Buchanan, of New*

York; Darius Lyman, of Washington; Epes Sargent, of Boston; to the Attacks of Prof. W. B. Carpenter, of England, and Others. Boston: Colby & Rich.

Wallace, Alfred Russel. 1875. *On Miracles and Modern Spiritualism.* London: James Burns.

Wallace, William A. 1981. *Prelude to Galileo: Essays on Medieval and Sixteenth-century Sources of Galileo's Thought.* Dordrecht, Holland: D. Reidel.

Wang, W. B. et al. 2018. "Identification of a Novel Mutation of the NTRK1 Gene in Patients with Congenital Insensitivity to Pain with Anhidrosis (CIPA)." *Gene* 679: 253–59.

Ward, Artemus. 1865. *Artemus Ward (His Travels) among the Mormons.* Edited by E. P. Hingston. London: Hotten.

Warlock, Peter. 1942. *The Best Tricks with Slates.* New York: Holden.

Watt, Ian. 1965. *The Rise of the Novel: Studies in Defoe, Richardson, and Fielding.* Berkeley: University of California Press.

Watts, Anna Mary Howitt. 1883. *Pioneers of the Spiritual Reformation: Life and Works of Dr. Justinus Kerner [and] William Howitt and His Work for Spiritualism.* London: Psychological Press Association.

Weatherly, Lionel A., and J. N. Maskelyne. 1891. *The Supernatural?* Bristol: J. W. Arrowsmith.

Webb, W. A. 1957. "On the Rejection of the Martian Canal Hypothesis." *Scientific Monthly* 85: 23–28.

Webster, Charles. 1982. *From Paracelsus to Newton: Magic and the Making of Modern Science.* New York: Barnes & Noble.

Webster, John. 1677. *The Displaying of Supposed Witchcraft. Wherein Is Affirmed That There Are Many Sorts of Deceivers and Impostors, and Divers Persons under a Passive Delusion of Melancholy and Fancy.* London: Jonas Moore.

Webster, Richard. 1986. *How to Build up a Psychic Practice with Full Length Cold Readings.* Aukland: Brookfield Press.

Webster, Richard. 1995. *Why Freud Was Wrong: Sin, Science, and Psychoanalysis.* New York: Basic Books.

Wegner, Daniel M., and Thalia Wheatley. 1999. "Apparent Mental Causation: Sources of the Experience of Will." *American Psychologist* 54: 480–92.

Wegner, Daniel M., V. A. Fuller, and B. Sparrow. 2003. "Clever Hands: Uncontrolled Intelligence in Facilitated Communication." *Journal of Personal and Social Psychology* 85: 5–19.

Weill-Parot, Nicolas. 2012. "Antonio da Montolomo's *De occultis et manifestis* or *Liber intelligentiarum*: An Annotated Critical Edition with English Translation and Introduction." In *Invoking Angels: Theurgic Ideas and Practices, Thirteenth to Sixteenth Centuries.* Edited by Clair Fanger. University Park: Pennsylvania State University Press.

Weinberg, Steven. 2015. *To Explain the World: The Discovery of Modern Science.* New York: HarperCollins.

Wier, Johann. 1583/1964. *Concerning the Deceptive Tricks of Demons, Enchantments and Works in Poisoning.* New York: Psychoanalytic Clinic for Training and Research.

Weisberg, Barbara. 2004. *Talking to the Dead: Kate and Maggie Fox and the Rise of Spiritualism.* New York: HarperSanFrancisco.

Wells, H. G. 1920/1971. *The Outline of History: Being Plain History of Life and Mankind.* 2 vols. Garden City, NY: Doubleday.

Wenegrat, Brant. 2001. *Theater of Disorder: Patients, Doctors, and the Construction of Illness.* New York: Oxford University Press.

Werner, H. 1847. *Guardian Spirits, a Case of Vision into the Spiritual World Translated from the German of H. Werner, with Parallels from Emanuel Swedenborg, by A. E. Ford.* New York: John Allen.

Weyer, Johann. 1998. *Witches, Devils, and Doctors in the Renaissance: Johann Weyer, De praestigiis Daemonum.* Tempe, AZ: Medieval & Renaissance Texts & Studies.

Whaley, Bart. 1989. *The Encyclopedic Dictionary of Magic: 1584–1988.* Oakland: Jeff Busby Magic.

———. 1990. *Who's Who in Magic.* Oakland, CA: Jeff Busby Magic.

Whewell, William. 1837. *History of the Inductive Sciences, from the Earliest to the Present Times.* 3 vols. London: John W. Parker.

———. 2001. *Of the Plurality of Worlds: A Facsimile of the First Edition of 1853; Plus Previously Unpublished Material Excised by the Author Just before the Book Went to Press; and Whewell's Dialogue Rebutting His Critics.* Chicago: University of Chicago Press.

Whiston, William. 1737. *A New Theory of the Earth, From Its Original, to the Consummation of All Things. Wherein the Creation of the World in Six Days, the Universal Deluge, and the Conflagration, As Laid Down in the Holy Scriptures, Are Shewn to Be Perfectly Agreeable to Reason and Philosophy.* London: printed for John Whiston.

White, Andrew Dixon. 1896. *A History of the Warfare of Science with Theology in Christendom.* New York: Appleton.

White, John. 1677. *A Rich Cabinet, with Variety of Inventions: Unlock'd and Open'd for the Recreation of Ingenious Spirits at Their Vacant Hours.* London: printed for William Witwood.

White, Nathan Francis. 1854. *Voices from Spirit-land.* New York: Partridge and Brittan.

Wicker, Christine. 2003. *Lily Dale: The True Story of the Town That Talks to the Dead.* New York: HarperSanFrancisco.

Wicks, Frederick. 1907. *Thought Reading: Second Sight & "Spiritual" Manifestations Explained.* London: Simpkin, Marshall, Hamilton, Kent & Co.

Wilberforce, Archibald. 1898. *The Great Battles of All Nations: From Marathon to Santiago.* New York: Collier.

Wilcox, Donald J. 1987. *The Measure of Times Past: Pre-Newtonian Chronologies and the Rhetoric of Relative Time.* Chicago: University of Chicago Press.

Wiley, Barry H. 2004. *The Georgia Wonder: Lulu Hurst and the Secret that Shook America.* Seattle: Hermetic Press.

———. 2005. *The Indescribable Phenomenon: The Life and Mysteries of Anna Eva Fay*. Seattle, WA: Hermetic Press.

———. 2009. "The Thought Reader Craze." *Gibeciére* 4: 9–134.

———. 2012. *The Thought Reader Craze: Victorian Science at the Enchanted Boundary*. Jefferson, NC: McFarland.

Wilkins, J[ohn]. 1691. *Mathematical Magic: Or, the Wonders That May Be Performed by Mathematical Geometry*. London: printed for Ric. Baldwin.

Wilkins, John. 1641. *Mercury: Or the Secret and Swift Messenger: Shewing How a Man May with Privacy and Speed Communicate His Thoughts to a Friend at Any Distance*. N.p.: printed by I. Norton for John Maynard and Timothy Wilkins.

Willey, Basil. 1949. *The Seventeenth Century Background: Studies in the Thought of the Age in Relation to Poetry and Religion*. London: Chatto & Windus.

Williams, Harley. 1947. *Doctors Differ: Five Studies in Contrast*. London: Johnathan Cape.

Williams, Schafer. 1954. "The Pseudo-Isidorian Problem Today." *Speculum* 29: 702–7.

Wilson, C. H. 1877/2015. *The 52 Wonders or Cards Manipulated by Science*. San Francisco: Alta California Book and Job Printing House.

Wilson, Catherine. 1995. *The Invisible World: Early Modern Philosophy and the Invention of the Microscope*. Princeton, NJ: Princeton University Press.

Wilson, P. 1918. *Sir Oliver Lodge and His Son Raymond; or What Comes after Death*. London: William R. Duff.

Wilson, Tyler. 2015. Wilson on Wilson. *Gibeciére* 10, no 1: 117–57.

Winer, Gerald A. et al. 2002. "Fundamentally Misunderstanding Visual Perception; Adults' Belief in Visual Emissions." *American Psychologist* 57: 417–24.

Winter, Alison. 1998. *Mesmerized: Powers of Mind in Victorian Britain*. Chicago: University of Chicago Press.

Wiseman, Richard. 1992. "The Feilding Report: A Reconsideration." *Journal of the Society for Psychical Research* 58: 129–52.

———. 2011. *Paranormality: Why We See What Isn't There*. London: Macmillan.

———. 2016. *101 Bets You Will Always Win: Jaw-Dropping Illusions, Remarkable Riddles, Scintillating Science Stunts, and Cunning Conundrums That Will Astound and Amaze Everyone You Know*. New York: St. Martin's Griffin.

Wiseman, Richard, and Robert L. Morris. 1995. *Guidelines for Testing Psychic Claimants*. Amherst, NY: Prometheus Books.

Wolf, Hubert. 2015. *The Nuns of Sant' Ambrogio: The True Story of a Convent in Scandal*. New York: Alfred A. Knopf.

Wolf, Tony. 2017. *Houdini's "Girl Detective": The Real-life Ghost-busting Adventures of Rose Mackenberg*. San Bernardino, CA: privately printed.

Wolfe, David W. 2001. *Tales from the Underground: A Natural History of Subterranean Life*. Cambridge, MA: Perseus.

Wolfe, N. B. 1875. *Startling Facts in Modern Spiritualism.* Chicago: Religio-Philosophical Publishing House.

Wolffram, Heather. 2012. "An Object of Vulgar Curiosity: Legitimizing Medical Hypnosis in Imperial Germany." *Journal of the History of Medicine and Allied Sciences* 67: 149–76.

Wolpert, Lewis. 1996. "The Unnatural Nature of Science." In *Unveiling the Microcosmos: Essays on Science and Technology.* Edited by Peter Day, 143–55. Oxford: Oxford University Press.

Wood, J. G. et al. ca. 1866. *The Boy's Own Treasury of Sports and Pastimes.* London: George Routledge and Sons.

Woodcroft, Bennet, trans. 1851. *The Pneumatics of Hero of Alexandria.* London: Charles Whittingham.

Woods, Roger, and Brian Lead. 2005. *Showmen or Charlatans? The Stories of "Dr." Walford Bodie and "Sir" Alexander Cannon.* Accrington, England: Caxton Printing.

Wootton, David. 1983. *Paolo Sarpi: Between Renaissance and Enlightenment.* Cambridge: Cambridge University Press.

———. 1985. "Unbelief in Early Modern Europe." *History Workshop* 20: 82–100.

———. 2007. *Bad Medicine: Doctors Doing Harm since Hippocrates.* Oxford: Oxford University Press.

———. 2010. *Galileo: Watcher of the Skies.* New Haven, CT: Yale University Press.

———. 2015. *The Invention of Science: A New History of the Scientific Revolution.* New York: HarperCollins.

Wormell, D. E. W. 1963. "Croesus and the Delphic Oracle's Omniscience." *Hermathena* 97: 2–22.

Worth, Patience. 1917. *The Sorry Tale.* New York: Henry Holt.

———. 1918. *Hope Trueblood.* New York: Henry Holt.

———. 1923. *Light from beyond: Poems of Patience Worth.* New York: Patience Worth Publishing.

———. 1928. *Telka, an Idyll of Medieval England.* New York: Worth.

Wright, Alex. 2007. *Glut: Mastering Information through the Ages.* Washington, DC: Joseph Henry Press.

Wrightsman, Bruce. 1975. "Andreas Osiander's Contribution to the Copernican Achievement." In *The Copernican Achievement.* Edited by Robert S. Westman. Berkeley: University of California Press.

Wylde, James. 1861. *The Magic of Science: A Manual of Easy and Instructive Scientific Experiments.* London: Richard Griffin.

Wyndham, Horace. 1937. *Mr. Sludge, the Medium.* London: Bles.

Yates, Frances. 1979a. *The Occult Philosophy in the Elizabethan Age.* London: Routledge.

———. 1979b. *Giordano Bruno and the Hermetic Tradition.* Chicago: University of Chicago Press.

Yonge, C. D. 1909. *The Deipnosophists or Banquet of the Learned of Athenaeus*. London: George Bell.

Yost, Casper S. 1916. *Patience Worth: A Psychic Mystery*. New York: Holt.

Young, Allan. 1995. *The Harmony of Illusions: Inventing Post-traumatic Stress Disorder*. Princeton, NJ: Princeton University Press.

Young, J. H. 1967. *The Medical Messiahs: A Social History of Health Quackery in Twentieth-century America*. Princeton, NJ: Princeton University Press.

Young, Kevin. 2017. *Bunk: The Rise of Hoaxes, Humbug, Plagiarists, Phonies, Post-facts, and Fake News*. Minneapolis, MN: Graywolf Press.

Young, L. E. 1928. *The Science of Hypnotism*. Chicago: Franklin Publishing.

Youngson, Robert M. 1998. *Scientific Blunders: A Brief History of How Wrong Scientists Can Sometimes Be*. New York: Carroll & Graf.

Yuille, John C., and Judith L. Cutshall. 1986. "A Case Study of Eyewitness Memory of a Crime." *Journal of Applied Psychology* 71: 291–301.

Zancigs. 1907. *Two Minds with but a Single Thought*. London: Paul Naumann.

Zagorin, Perez. 1990. *Ways of Lying: Dissimulation, Persecution, and Conformity in Early Modern Europe*. Cambridge, MA: Harvard University Press.

Zerffi, G. G. 1871. *Spiritualism and Animal Magnetism. A Treatise on Dreams, Second Sight, Somnambulism, Magnetic Sleep, Spiritual Manifestations, Hallucinations, and Special Visions*. London: Robert Hardwicke.

Zinsser, Hans. 1967. *Rats, Lice, and History*. New York: Bantam.

Zollner, Johann Carl Friedrich. 1880. *Transcendental Physics: An Account of Experimental Investigations*. London: Harrison.

INDEX

Abbott, David, 25, 247

academies, Italian, 66–68; of Experiment, 108–18; of Lynxes, 81, 94–96, 102–5, 350n60; of Secrets of Nature, 66–68, 71, 77, 79, 353n3

aeolipile, 333n6

Aesculapius, 37

Agrippa, Cornelius, 343n5

Ahlers, Cyriacus, 120

air pump, 110

alchemy, 70, 343n15, 347n17, 392n12

Alcock, Jim, 120

Alexander, the Man Who Knows, 267, 331n35, 383n11

Alexander of Abonuteichos, 24, 27, 330n27

Alexander the Great, 330n30, 333n29

algebra, Cardano and, 63, 342n51

Alphonse of Aragon and Naples, 40–41

American Society for Psychical Research, 293

Ammon, shrine of, 7, 33

anagrams, 96, 113, 350n53

Andros family, 201–2

Andrus, Jerry, 287, 388n32

anesthesia, approaches to, 161, 168–69

angels, 20, 44, 129

animal magnetism. *See* hypnosis; mesmerism

Annemann, Theodore, 287

Annius of Viterbo, 107, 352n16

antidotes, 78

anvil trick, 156

Apollo, 3–4, 6, 9, 13, 35, 325n6

aqua regia, 111

Aristotle, 14, 30, 57, 67, 117, 354n32; Academy of Experiment and, 110; on astronomy, 92, 96; and botany, 102; on changeless universe, 89–90; Church and, 99; on defense of ideas, 84; experiments and, 112, 114; Fontenelle on, 123; Galileo and, 85, 88, 98; and megaphone, 330n30; on new world, 58; and physics, 68; scholastics and, 45; and spontaneous generation, 116; Valla on, 50–51

Arons, Harry, 231

Aronson, Simon, 286

Ashburner, John, 184, 261–62, 367n25, 383n25

astrology, 14, 59–60, 89, 341n44; decline of, 60–62

astronomy, 58; Augustine on, 44,
341n31; Galileo and, 85, 87, 89–92;
heliocentrism, 86–87, 98–99, 120–26,
348n29; rings of Saturn, 113–14
Athens, 10, 12, 14
audience: Didier and, 171–72, 180,
183–84; Fox sisters and, 244;
uncooperative, 259–60; willing,
376n42
Augustine, saint, 20, 23, 33, 43–45,
341n31; on astrology, 59; on
magnets, 72; on new world, 58
aura, 219
automatic writing, 239, 255–56
automatons, 31, 74, 190
autosuggestion, 6, 209, 268

Babinet, Jacques, 275
Bacon, Francis, 345n46
Bacon, Roger, 91, 338n48, 343n5
Baggally, W. W., 285–87, 388n32
Baldwin, Samri S., 176–78
Balsamo, Giuseppe, 75
Baltus, Jean François, 23–24, 27
Bancroft, Richard, 345n41
Bangs sisters, 298, 390n41
Barberini family, 104, 351n77
bar bets, types of, 342n57
Barbieri, Nicolo, 76
Barnum, P. T., 190, 321
barometer, 111
Beard, George, 174–75, 185, 218,
373n37
Beaumont, Henry, 27
Beckmann, John, 36, 156, 164
bees, 104–5
Bekker, Balthasar, 329n2
belief(s): contradictory, 60; examination
of, vii; and experience, 210, 219–21,
265; Fontenelle on, 123–24; and
investigation, 304; justification of,
11, 279; and mesmerism, 162; nature

of, 11; in paranormal, 65; power of,
10–17; Redi on, 116
belief systems, 11–12, 120; defenses
of, 57, 84; dual false, 223–24;
explanations and, 69
Bell, doctor, 258
Bellarmine, Robert, 97, 99–100, 121,
338n53, 351n67–68
Bergerac, Cyrano de, 127–28
Berglass, David, 164
Berna, doctor, 160, 362n40
Bernard, Prudence, 362n51
Bernheim, Hippolyte, 219–21
Bernstein, Bruce, 170
Betson, Thomas, 76
bias, 113
Bishop, Washington Irving, 174–75,
178, 365n45, 365n48
Blackburn, Douglas, 175, 365n49
Blackwelder, E., 144
Blaine, David, 229
Blavatsky, madame, 309, 392n13
blindfold sight, 165f, 362n50; Didier
and, 152, 154, 195; mechanism of,
157–59, 165–66
blind obedience, doctrine of, 99, 350n67
Blondel, David, 329n14
Bodin, Jean, 77, 345n44, 347n7
body, humanists on, 44–45
Boethius, 51
Bologna stone, 95
book tricks, 150–51; Didier and, 152,
167; mechanism of, 167–68
Borch-Jacobson, professor, 220
Borelli, Giovanni, 116
Boswell, James, 367n9
botany, 102–8
Bowden, Hugh, 10
Boyle, Robert, 115, 118, 308, 389n14,
392n12
Bracciolini, Poggio, 45–46, 51, 336n17,
337n26–27, 338n52

Brackett, Loraina, 366n2
Braid, James, 215–17
brain: and pain insensitivity, 225–26; and paranormal experiences, 209, 357n10
Brandon, Ruth, 304
Braude, Ann, 370n33
Breslaw, Philip, 191
Brewster, David, 38, 40, 356n22
Brown, investigator, 259
Brown, John Randall, 173–75, 178
Browne, Thomas, 75, 80, 346n58
Bruno, Giordano, 110, 121, 348n27
Burkett, B. G., 374n15
Burlingame, H. J., 156, 165f, 204, 228, 366n55
Burq, Victor Jean-Marie, 219
Burr, Chauncey, 241, 245
Butterfield, Herbert, 81
Bux, Kuda, 35, 158–59, 361n28
Byrne, W. P., Mrs., 171–72

Cadwallader, M. E., 235
Cagliostro, count, 75
Cahagnet, Alphonse, 135–37
calcinated barium sulfide, 95
Calvin, John, 57, 63, 87
camera obscura, 71
Campanella, Tommaso, 20, 110, 353n5
Campbell, Duncan, 180–81
Campbell, E. Z., 205, 263
canals, on Mars, 143–44, 359n41
Capron, E. W., 237, 241–42, 244–45
Cardano, Gerolamo, 54–65, 74, 164, 340n10, 342n48, 342n51
card tricks, 76, 287, 363n6; Cardano and, 62; Didier and, 150, 152, 154, 165–66, 195; mechanisms of, 164–78
Carnegie Institute, 293
Carpenter, William, 169, 172, 185, 228, 268–69
Carrington, Hereward, 285–89, 292–94, 296–98, 299f, 301, 303; background of, 388n1; and Kellar, 390n36

Carson, Johnny, 165
cartography, 58
Casaubon, Isaac, 340n13
Casciorolo, Vincenzio, 350n50
Castellio, Sebastian, 351n72
catalepsy, 155–57
Catholics: Donation of Constantine and, 40–53; and heliocentrism, 348n29; Royal Society and, 115. See also Church; Inquisition
Cecco, 339n4
censorship, 115, 118, 341n37, 352n13
Cesi, Federico, 81, 91, 94–98, 100, 102–5
challenges to beliefs, reactions to, 16–17, 57, 320–22; Charcot and, 219; versus experience, 210; Faraday and, 273–74; Fontenelle and, 22–24; Fox sisters and, 244–45; Godfrey and, 275–78; Hare and, 257; Hayden and, 260; Inquisition and, 49–50; Palladino and, 289, 296–97; Phillips and, 382n10; Townshend and, 182
Charcot, Jean-Martin, 219–21, 223, 373n3, 377n51
charlatans, 24, 75–76
cheating. See tricks
chemistry, 343n15
Chevreul, Michel-Eugene, 265, 268, 383n7
chloral, 230
chloroform, 225, 230, 375n32
Chomondeley-Pennell, ambassador, 307
Chopra, Deepak, 370n32
Christianity: Constantine and, 43, 48; crisis in, 276; defenses of, 22–24; and demons, 20; Fontenelle and, 18, 20–21; and table turning, 275–78. See also Church
Christopher, Milbourne, 392n7
Church: and astrology, 59–60; and curiosity, 20, 43–44; Fontenelle and, 125–26; and Galileo, 90–91, 97–100;

and magnets, 207; and mesmerism, 208–10; and microscopes, 105; and oracles, 16; in Renaissance, 67, 75. *See also* Catholics; Christianity; Inquisition

Cicero, 9, 15

cimento: term, 110. *See also* experiment

Cinniger, Ron, 330n17

clairvoyance, 25, 197, 367n27; Jung-Stilling and, 131; mechanisms of, 148–63, 196–97; term, 135, 179. *See also* mind reading; remote viewing

Clark, Stuart, 44

class, and credibility, English on, 260–61

classification, 102–8

Clavius, Christoph, 347n1

Clement IX, pope, 116

Clement XII, pope, 101

Close, Francis, 278

codes, 175, 191–92, 196–97, 369n11

cold reading, 174, 252, 381n5; Beard and, 185; Fox sisters and, 244

Colombe, Ludovico, 351n70

Colquhoun, John C., 181, 210

Columbus, Christopher, 58

concave mirrors, 36

confederates, 228–29

confirmation bias, 11, 114, 327n4

Conlin, Claude, 267

Constantine, emperor of Rome, 16; and Donation, 40–53

controls, and tricks, 159–60, 362n40; Didier and, 162, 166, 183, 185–86; Eglinton and, 312; Fox sisters and, 243; Hayden and, 259–60; Palladino and, 285–86, 288, 291, 295–97, 300–303; and slate writing, 308

Cook, Florence, 261, 305

Copernicus, Nicolaus, 61, 86–87, 92, 95–96, 338n48, 356n13; Fontenelle on, 120–26

Corrington, Julian, 116

Corson, Hiram, 140

Cosimo I (de' Medici), grand duke, 110

cosmology, 348n21; Bruno and, 121; Galileo and, 85, 87, 92–93; Ptolemy and, 72

Cottrell, Suzie, 165

Coyne, Roscoe Ronald, 159, 361n32

Crandon, Margery, 304

Creery sisters, 196, 365n48, 370n22

Cremonini, Cesare, 95, 350n49

Croesus, 3–4, 9, 12–14, 325n2, 327n33, 328n14

Crookes, William, 166, 260–61, 305

Culver, Norman, Mrs., 244–45

Cumberland, Stuart, 166, 175, 178

curiousity, 20, 43–44, 50; Fontenelle on, 122; Porta and, 67–68, 80; punishment for, 329n9; Valla and, 52–53

Curran, Pearl Lenore, 379n24

Cuse, Nicolas de, 337n35

Cyrus, king of Persia, 13, 327n33, 328n13

Daniels, J. W., 262*f*

Dante Alighieri, 336n17

d'Argenson, marquis, 125

Dark Ages, term, 43

d'Ascoli, Cecco, 55

Davenport, Reuben, 249

Davenport brothers, 176, 391n5

Davey, S. J., 312–17

Davidson, Joy, 376n47

Davies, Charles, 274

Davis, Andrew Jackson, 137, 236

Davis, W. S., 293–96, 322

Davy, Humphry, 7, 140, 375n32

debunking: of astrology, Swift and, 61–62; of Donation of Constantine, 47–49; Hippolytus and, 28; Oenomaus and, 15; Robertson and, 26. *See also* tricks, mechanisms of

Defoe, Daniel, 181

Deleuse, Philippe, 210

Delphi, 5*f,* 320, 330n17; oracle of, 3–9, 14–17, 176–78; pilgrims and, 21–22

Del Rio, Martin, 76

demons, 19–21; Cardano on, 55; Daniels and, 262*f;* defenses of, 22–24; Fontenelle and, 18–19; James I on, 76; Summers and, 209. *See also* devil

demonstration: experiments, 271; Galileo and, 88

de Morgan household, 260–61, 382n18–19

Descartes, René, 64, 355n3

Despiau, M. L., 192

devil, 386n62; Hare and, 254; Jung-Stilling and, 132; Porta and, 74; rapping and, 236. *See also* demons

Dewey, D. M., 234, 236

Diana, 35

dice, 64

Dick, Thomas, 142

Dickens, Charles, 150, 259

Dickinson, Emily, 102

Didbin, R. W., 277

Didier, Adolphe, 214

Didier, Alexis, 149*f;* background of, 151–52; and blindfold tricks, 165–66; and book tricks, 167–68; and clairvoyance, 148–63; Gautier on, 193; influence of, 148–50; and mind reading, 173–76; and remote viewing, 179–80, 182–86; Robert-Houdin and, 188–89, 194–96; and sealed messages, 170–73

Dingwall, Eric, 387n13

Dini, Piero, 349n48

discomfort: lack of explanations and, 286–88; questioning and, vii

disputation, 8, 46, 54–55, 70, 337n29

dissembling, 354n36

Dodds, E. R., 14

Dodona, 7, 16, 31

Dominicans, 99

Donation of Constantine, 40–53, 42*f;* analysis of, 47–49

Donkin, Horatio, 391n4

double-entry bookkeeping, Cardano and, 63

doubt: academies and, 67; and oracles, 10–17; punishment for, 8, 327n26. *See also* investigation

Doyle, Arthur Conan, 261

Drake, Stillman, 347n4

Drury, Edward, 237

Dupotet, baron, 210

Durant, C. F., 201–3, 322

ear trumpet, 330n30

Eckeberg, John von, 156

Eco, Umberto, 338n50

Eddy, Mary Baker, 201, 375n27

Edison, Thomas, 192

Edmonds, John W., 270

education: and beliefs, 11; Faraday on, 273

Eglinton, William, 307–9, 311–13, 392n13

electro-biology, 268

electroscope, 295

Elliotson, John, 152–53, 183, 199–200, 216, 224, 373n1; and mesmerism, 210–14

Elliott, Charles, 245

Elzevier, Louis, 101

empiric, term, 339n10

Ephesus, 9

Epicurus, Epicureanism, 24, 45, 78

epilepsy, Swedenborg and, 129

Erasmus, Desiderius, 338n53

errors, 368n36; Cahagnet and, 136; Didier and, 154, 166, 169, 183–85; Godfrey and, 279; ignoring, 14, 88, 328n20; Lombroso and, 281; Page on, 253

Esdaile, James, 224–25, 228, 375n30

ether, 161, 225, 358n27, 369n2, 375n32; Forbes and, 168–69; Jung-Stilling and, 131

ethylene, 6

Etruscans, 106–7

Eugenius, pope, 53, 335n1

Eusebius, 24

examination. *See* investigation

excommunication, 47

experience: versus belief, 210, 219–21; of impossibility, 286–88; of mesmerism, 228

experiment: academy of, 108–18; designs for, 111–12, 116; Galileo and, 88–89; on observation, 311–18; and O'Key sisters, 213; and parapsychology, 197–98; term, 270, 385n28. *See also* investigation

extrasensory perception. *See* clairvoyance

extraterrestrial life, 121, 356n13; beliefs about, 127–45; Fontenelle on, 122–24

eyewitness testimony, issues with, 196, 393n32

Fabri, Guglielmo, 70

Fabri, Honoré, 116

Faraday, Michael, 182, 215, 253, 271*f*, 276, 321; and Home, 385n31; spiritualists and, 382n11; and table turning, 264, 271–73

Farmer, Bob, 164, 342n57

Fay, Anna Eva, 166, 174, 309

Fechner, Christian, 194

Feilding, Everard, 285–88

Felix V, pope, 70, 343n12

Ferdinand II, grand duke, 101, 106, 108, 110, 115

Festinger, L., 84

Finocchiaro, Maurice, 101

fire effects, 33–34; walking on fire, 35–36, 334n44

Fish, Leah, 237, 242–43, 246

Fishbough, William, 236

fishing, 171; term, 364n34

Flammarion, Camille, 143, 282, 304, 359n36

Flint, Herbert, 230–31, 360n15

Flint, Valerie, 20

Florence, 92

Flournoy, Theodore, 139–40

Fludd, Robert, 348n18

Fontenelle, Bernard de, 10, 19*f*, 236–37; on Delphi, 17–18, 20–21, 27; on heliocentrism, 120–26; success of, 37–38

Forbes, John, 156–57, 166, 168–71, 200; and mesmerism, 216–18, 228; and observation, 184–87

forcing, Davey and, 317

forgeries: Donation of Constantine, 40–53; Etruscan artifacts, 106–7; Porta and, 71

Forte, Steve, 164

fossils, 103

fountains, pneumatic, 31

Fox sisters, 234–37, 244–48; confession of, 249–50; investigation of, 239–44, 248; and table turning, 264

Franklin, Ben, 208, 358n27, 372n7

Freeman, mesmerist, 200

French Academy of Sciences, 268, 335n60

French Royal Academy (Society) of Medicine, 159, 207

French Royal Commission, 205

Freud, Sigmund, 220–21, 225, 374n10

Friedland, Ron, 348n31

Frikell, Wiljalb, 368n1

fumes, and Delphic oracle, 5–7

Gaffarel, Jacques, 207

Galasso, Horatio, 76

Galen, 57

Galilei, Vincenzo, 85–86, 347n4
Galileo Galilei, 50, 84–101, 86*f*,
    102–4, 110; Fontenelle and, 120; and
    investigation, 210; on nature, 109;
    and pendulum, 265; and rings of
    Saturn, 113–14, 114*f*
gambling, 62–65, 342n48, 363n10
Ganzfeld experiments, 198
Gardner, Martin, 158–59, 165, 303,
    363n9, 386n63
garlic, 72
Gasparin, Agénor de, 274–75
Gassendi, Pierre, 110
Gauld, Alan, 148, 160, 201, 208
Gautier, Theophie, 193
Geller, Uri, 370n25
Gibbon, Edward, 43
Gibson, Walter, 287
Gilbert, William, 79–80, 166, 261,
    346n57
Gillson, Edward, 276–77
Ginzburg, Carlo, 50, 223–24
Giobbi, Robert, 164
Glanvil, Joseph, 340n14
glass, 110
glass harmonica, 207, 372n7
Glazebrook, minister, 278
goats, at Delphi, 4–7
Goble, George, 364n33, 365n39
Godfrey, Nathaniel, 263–64, 275–79
Godwin, William, 189
Goldsmith, Margaret, 160
Goldston, Will, 360n6
Gourlay, Mrs., 255–56
Gratian, 20
gravity, 88
Greece, ancient, 3–12, 14
Greek fire, 333n30
Greeley, Horace, 367n13
Green, Lennart, 363n6
Greenblatt, Stephen, 45
Gregory, Richard, 353n17

Gregory, William, 181
Gresham, William, 229, 376n47
Griggs, William, 142
Grimes, J. Stanley, 320
Grosseteste, Robert, 348n19
guilds, 69
Guillain, Georges, 220, 373n3
Guyot, Edme-Gilles, 163

Haddock, Joseph, 236
Halley's comet, 297
Hammond, Charles, 239
hand holding, 365n50
Hankins, Thomas, 271
Hardinge, Emma, 249, 381n59
Hare, Robert, 253–57
Hart, Ernest, 284
Hauffe, Frederika, 134–35, 137–38, 236,
    278
Hayden, Maria, 258–60
Healy, Piers, 171
heliocentrism, 86–87, 98–99, 348n29;
    Fontenelle on, 120–26
Helmholtz, Herman, 196
Henry VIII, king of England, 41
heresy, 45
Hermas, 330n14
Hermes Trismegistus, 57, 340n13
Hermon, Harry, 240*f*
Herodotus, 12–14, 327n31, 328n11–12
Hero of Alexandria, 30–33, 36
Herschel, John, 141–43, 142*f*, 358n28
Herschel, William, 143
Herzig, C. S., 30
Hildegard of Bingen, 209, 372n17
Hippocrates, 56, 330n30
Hippolytus, 25, 28–29, 34, 37, 59
hoaxes, 107, 141–43, 142*f*, 358n29,
    359n33
Hockley, Fred, 136–37
Hodgson, Richard, 312–15, 392n24
Hogg, R. W., 312

Holmes, Oliver Wendell, 184
Home, D. D., 258, 260–61, 373n1, 381n57, 385n31
Honorton, Charles, 198
Hood, Robert, 390n46
Hooke, Robert, 104
Hooper, William, 74
Horace, 34
horoscopes, 55, 60
horses, 228–29; term, 368n40
Houblier, mesmerist, 200
Houdini, Harry, 150, 170, 189, 381n1, 392n18; and Fox sisters, 234–35; and spirits, 261
Hoving, Thomas, 107
Howes, mesmerist, 200
Hull, Burling, 29
humanists, 43–47, 53, 58, 339n63
human plank, 155–57
humbugometer, 321
Hume, David, 370n21
Hurkos, Peter, 368n36
Hurn, John, 241
Hurst, Lulu, 157, 361n22
Huygens, Christiaan, 65, 113–14, 353n15
Hyman, Ray, 197–98, 287, 307, 363n9, 370n25, 391n7
hypnosis, 218–31, 371n1; accounts of, 221–27; evaluation of, 216–17; term, 215–16, 218. See also mesmerism
hypothesis, 112, 348n20
Hyslop, James, 293–94, 379n24, 389n21
hysteria, 6, 138, 211, 216, 373n40

illusions, 227–28; malobservation and, 318. See also tricks
impossibility, discomfort of, 286–88
The Index of Prohibited Books, 45, 53, 77, 100, 118, 121
informants, 177–78
Inghirami family, 106–7

innocence, Fox sisters and, 234–36
Innocent IV, pope, 337n28
Inquisition, 337n28; and Academy of Experiment, 116; and Cardano, 55; and Cecco, 339n4; and dual false belief systems, 223–24; and Galileo, 95, 98–100; and heliocentrism, 125; and Helmont, 207; and Porta, 66, 75–79; Spanish, 47; and Valla, 49–51
insects, 105
Institut General Psychologique, 284–85
Insulanus, 181
investigation: Academy of Experiment and, 109–18; Ahlers and, 120; Cardano and, 56–59; Church and, 43–44; versus credulity, 199–201; English versus French, 284; Fontenelle and, 37–38; of Fox sisters, 239–44, 248; Galileo and, 84–85, 210; Greeks and, 8; Helmholtz and, 196; of hypnosis, 221–27; medicine and, 184; of oracles, 10–17; of Palladino, 281–85, 288–90, 294–96, 298–302; Porta and, 71–72, 80; Prater and, 161–62; Rhine and, 197; of slate writing, 308–9, 313–14; Socrates and, 10; of spiritualism, 251–62; of table turning, 271–75; Townshend and, 182. See also tricks, mechanisms of
invisible college, 118
Invisible Girl trick, 25–27, 26f
invisible ink, 28, 71

James, John, 364n18, 371n36
James, William, 7, 11
James I, king of England, 76, 345n40–42
Jastrow, Joseph, 295–96, 303
Jay, Ricky, 360n16
Jerome, saint, 49–50, 338n46
Jesuits, 97, 99
Joachim, Georg, 87, 339n6

Johnson, Samuel, 367n9
John XXII, pope, 20, 329n11
Jones, Amanda, 368n39
Jordan, David Starr, 382n11
Josephus, 327n26
Jung, Carl, 137–39
Jung-Stilling, Johann Heinrich, 131–32
Jupiter, 92
Jusserand, J. J., 133

kabbala, 57; term, 340n17
Kane, Elisha, 245–46
Kant, Immanuel, 356n9
Keeler, William, 320
Kellar, Harry, 307, 390n36
Kellogg, James L., 295–96
Kepler, Johannes, 93–94
Kerner, Justinus, 133–35, 137–38
Kibler, Austin, 370n25
kinesiology, applied, 157
Kircher, Athanasius, 72, 73f
kook, term, 359n46
Kossy, Donna, 359n46
Krauss, Lawrence, 370n32
Krebs, Stanley, 294, 298–302, 322
Kreskin, 267
Kuhn, Thomas, 109, 353n2

Langford, David, 359n43
Langley, Samuel, 355n2
language: Flournoy and, 139–40; Luther
    and, 98; Porta and, 80; Robert-
    Houdin and, 193; Stelluti and, 106
Langworthy, doctor, 239
Lankester, Ray, 307, 391n4
Larsen, William, 158
Lee, Edwin, 181, 199
Lefaivre, Liane, 353n8
Leff, A. A., 16
legal issues, 37; eyewitness testimony,
    196; and hypnosis, 222; Lewis and,

311; seventeenth-century, 351n73;
    Wright on, 306
Leikind, Bernard, 334n44
LeLoyer, Pierre, 378n11
lenses, 104
Leopold de' Medici, 110, 113, 115–16
Leo X, pope, 59
Lewes, editor, 259–60
Lewis, Angelo, 311–12, 392n19,
    392n24
Lewis, Carvill, 309
Lewis, C. S., 369n7, 376n47
Lewis, E. E., 235
lighting, Palladino and, 291, 300–301
lightning effects, 34
Lilly, William, 61, 342n45
Lily Dale, 289
Linn, W. A., 179
Linnaeus, Carl, 105
Locke, Richard Adams, 142, 142f
lodestone: Porta and, 71–75. See also
    magnets
Lodge, Oliver, 283, 304
Lombroso, Cesare, 281–83
Lonk, Adolph, 230
Loomis, professor, 240
lost knowledge, 57–58, 134, 340n14
Louis XVI, king of France, 125
Lovick, John, 287
Lowell, Percival, 143–44
Lucian of Samosata, 3, 24, 29, 328n12,
    332n52
Lucius III, pope, 337n28
luck, 63–64
luminous paint, 292, 292f, 388n13
Luther, Martin, 46, 51, 53, 57, 87, 98,
    130, 338n53
Luys, Jules Bernard, 284, 387n18
lycopodium powder, 34
Lynxes, Academy of, 81, 94–96, 102–5,
    350n60

Mach, Ernst, 355n10

Mackenberg, Rose, 391n55

Macknik, Stephen, 227

MacLeod, Donald, 181

MacWalter, J. G., 238

Maelcotes, Odo van, 97

Magalotti, Lorenzo, 113

Magellan, Ferdinand, 58

magic, 20, 44, 87, 343n5; Porta on, 68–70

magic lantern, 37

magnets, 73f, 79–81; Mesmer and, 206–7; Porta and, 71–75; and rapping, 241

malobservation, 309, 318; categories of, 312

Malone, Michael, 43

Malpighi, Marcello, 117

Marcellini, Carlo, 86f

Marchetti, Alessandro, 110

Marcillet, Jean, 152, 156, 160, 162–63, 166, 187; as informant, 177–78; and mistakes, 183

marine insurance, 60–61, 341n39

Mars, 139–40, 143–44, 359n38, 359n41

Martinez-Conde, Susana, 227

Maskelyne, John Nevil, 284, 307, 321, 379n31, 391n5

Massey, C. C., 311

mathematics, 58, 339n1, 346n1; Cardano and, 54, 63; experiments and, 112; Galileo and, 88

Maudsley, Henry, 129

M'Avoy, Margaret, 158

Mayo, Herbert, 265–66

McClure, S. S., 289, 292, 296

McGee, minister, 277

Mead, Richard, 206

Medici family, 92, 109–11, 113, 115–16

medicine, 362n45; Cardano and, 56; Forbes and, 168; hypnosis and, 206–7, 210–12, 220, 224–25; and investigation, 184; magnets and, 207;

microscope and, 105; Porta and, 78; Prater and, 161

mediums, 388n8; Jung and, 138; Palladino, 280–305; and pendulum, 267; and rapping, 237; and slate writing, 306–19; term, 237. *See also* Fox sisters; spiritualism

Méheust, Bertrand, 149–51, 159–60, 163, 167–69, 362n35

Melanchthon, Philipp, 50

Melton, John, 61

memory: implanted, 221;performances, 369n7; recovered, 220–23, 372n20, 374n17, 375n23

Mesmer, Franz Anton, 205–9

mesmerism, 133, 204–17; accounts of, 214–17; Cahagnet and, 135–37; Didier and, 151–52; Forbes and, 168–69; Jung and, 137–39; Jung-Stilling and, 131–32; Kerner and, 134–35; mechanisms of, 200; occult and, 206–7; Prater and, 161–62; term, 215; Townshend and, 182. *See also* hypnosis

metallotherapy, 219

metaphors: alchemists and, 70; Copernicus and, 87; Gilbert on, 79; Mesmer and, 206

meteorology, 343n7

methylsulfonal, 230

Mewis, Catherine, 158

microscope, 102, 104–6, 110, 352n7

Middleton, Conyers, 14

Miller, Dickson, 303

Miller, Dixon S., 296, 389n25

mind reading: Didier and, 154, 173–76; mechanism of, 173–76

Minnock, Tommy, 229–30, 377n51

Mirandola, Pico della, 343n5

mirror effects, 36–37, 71

Mirville, marquis de, 194–95

Monachus, Johannes, 337n37

Montaigne, Michel de, vii
moon: beliefs about, 127–32; Fontenelle
    on, 123; Galileo and, 92
Moore, W. U., 390n41
Morgan, R. C., 277
Morland, Samuel, 24
Morley, Henry, 63
Morse, Samuel, 181–82
Mount Parnassus, 3, 5*f*, 21
Moynagh, Digby, 35
Mozart, W. A., 372n7
Munchausen, baron von, 130, 357n12
Munsterberg, Hugo, 288, 294
Munthe, Axel, 220
muscle reading, 175, 269, 365n50;
    Didier and, 179–80; term, 174
Musson, Clettis, 166
Myers, Frederic, 283–84, 304
Myers-Briggs Type Indicator, 138–39
myths, 13–14

names, 103
naphtha, 34
Naples, 125
narcotics, and pain tolerance, 230
naturalists, 102–8
natural philosophy, 122, 343n5
nature, Porta and, 68–70, 77
Naudé, Gabriel, 55, 345n37, 354n36
Nero, emperor of Rome, 16
Newton, Isaac, 31, 87, 125, 347n15–16,
    349n39
New World, 58, 103, 121
Nicaea, council of, 48
Nicholas V, pope, 53
nitrohydrochloric acid, 111
nitrous oxide, 7, 375n32
Noll, Richard, 138
Nostradamus, 339n9
not-a-prayer bets, 342n57
novelty, 50, 338n48; term, 80

observation: Academy of Experiment
    and, 115; experiment on, 311–18;
    Forbes and, 184–87; Galileo and,
    84–85, 89, 96–97, 103; Lewis and,
    311; microscope and, 105
occhialino, 352n7
occult, term, 104
Ochino, Bernardino, 336n18
odds bets, 342n57
odic force, 215, 266, 361n23, 373n34
Odoacer, 43
Oehler, Andrew, 289
Oenomaus of Gadara, 15, 320, 328n25
O'Key sisters, 211–13, 284
Onassis, Aristotle, 18–29
optics, 348n30
oracles, 11–12; of Delphi, 3–9, 176–78;
    doubting, 10–17; end of, 14–17; and
    fire effects, 33–34; Jesus and, 18–20;
    term, 3
Ore, Oystein, 63
Oresme, Nicole, 54, 341n26, 341n36
Orne, Martin, 221–23, 227, 374n16–17
Orvieto, 107
Ottley family, 172
Ovette, Joseph, 229
Owen, Robert Dale, 236, 261
Ozanam, Jacques, 33–34

Padua, 89–91
Page, Charles Grafton, 242–44, 252–53,
    268, 269
pain control: hypnotic, 226–27; surgeons
    and, 375n26, 375n29
Paine, Thomas, 239
pain insensitivity, 224–25; congenital,
    376n50; mechanisms of, 228–30
Palingenio, Marcello, 118
Palladino, Eusapia, 280–305, 290*f*,
    387n13

pamphlet wars, 45–46, 275–78

Paracelsus, 161

paradigm shift, 109

parallax, 89, 93, 348n24

parapsychology, term, 197

Parke, H. W., 27

Parkyn, 230

Parr, Henry, 278

Partridge, John, 61–62

Pascal, Blaise, 64–65, 355n3

Pasteur, Louis, 117

Patrick, saint, 45, 336n18

Paul, saint, 50

Paul II, pope, 339n63

Paul III, pope, 59, 121

Paul V, pope, 99, 350n65

peace, spiritualists and, 257, 261–62

Pecararo, Nino, 381n1

Peloponnesian War, 12, 14

pencil reading, 172

pendulum, 84, 264–67

Penn and Teller, 287, 369n15

perspicillum, 352n7

Phillips, Will, 382n10

philosophy: Galileo and, 88; Greek, 8, 10

Phin, John, 34

phosphorus, 34, 333n30, 389n14;
    Bolognian, 95

phrenology, 184, 199, 214, 252–53,
    367n25, 371n36

physics: Galileo and, 88; Porta and,
    68–70

physiognomy, 77

Pinetti, Joseph, 191

Pisa, 85

placebo, versus hypnosis, 226

plants, 102–8

Plato, 7–8, 19–20

pleasure, humanists on, 45

Plutarch, 6, 15–16, 72

Podmore, Frank, 135, 150, 273, 306,
    318, 382n10

Polidoro, Massimo, 288

poltergeist, 237, 378n10

Porphyry, 23

Porta, Giovanni Battista della, 28–29,
    66–81, 91–93, 353n3, 369n11

Post, Isaac, 239

posttraumatic stress disorder, 374n15,
    375n24

Potts, reverend, 241

Powers, Melvin, 231

Poyen, Charles, 201–2, 210, 322

Prater, Horatio, 161–63, 170–71, 183,
    185–86

precognition, term, 197

Prevorst, Seeress of, 134–35, 137–38,
    236, 278

Price, Harry, 35, 361n28

priests: of Delphi, 8, 12, 16, 21–22,
    27–29; and tricks, 33–34, 36–37

Prince, Walter, 320

printing press, 56, 87, 90

probability, 54–65, 342n57

projection lens, 37

Prometheus, 329n9

proposition bets, 342n57

Proppe, Hal, 342n56

Protestants, 57, 90, 98, 125, 348n29,
    356n17

pseudo-Isidorian Decretals, 329n14,
    337n34

Ptolemy, 72, 350n54

Pullein, Thomas, 348n22

pumping, 171

Puritans, 355n41

Puységur, marquis de, 208–10

Pybus, William, 29

Pythagoras, 123

Pythia, 6, 8

Pytho, term, 325n6

quantum physics, 370n32

Quimby, Phineas, 201, 375n27

Randi, James, 165, 363n8, 368n36

Raphael, 42*f*

rapping, 234–37; Hare and, 254–57; mechanisms of, 239–41, 240*f*, 244, 249–50, 379n31; Palladino and, 291

Raspe, Rudolph, 130, 357n11

reading through obstacles: Didier and, 150–52, 154, 165, 167, 170–73; mechanism of, 165–68, 170–73. *See also* blindfold sight; sealed messages

recovered memories, 220–23, 372n20, 374n17, 375n23

Redfield, Mrs., 234, 236

Redi, Francesco, 116–17

Reichenbach, Charles von, 215, 266, 361n23, 373n34

religion: ancient Greeks and, 7, 10; crisis in, 276; Fontenelle and, 125; Swedenborg and, 130; and table turning, 275–78. *See also* Christianity; Church

remote viewing, 179–88; Didier and, 152–54, 179–80; history of, 180–82; Swedenborg and, 181; term, 197

Renaissance, 43–47, 53, 58, 67, 75, 339n63. *See also* academies

republic of letters, 118

research experiments, 271

Rheticus, 87, 339n6

Rhine, J. B., 197

Richelieu, Armand du Plessis, duke, cardinal, 207

Richet, Charles, 219, 283–84, 304

Rinaldini, Carlo, 110

Rinn, Joseph, 248, 295–97

Robbins, Tony, 36, 334n44, 334n47

Robert-Houdin, Emile, 188*f*, 190–92

Robert-Houdin, Jean-Eugene, 31, 188*f*, 368n1; background of, 189–94; and Didier, 149–50, 188, 194–96; and rapping, 241

Robertson, Etienne-Gaspard, 26, 331n41

Robinson, Henry Mansfield, 359n38

Romains, Jules, 361n28

Rome, 94, 107; fall of, 43; Vatican, 45–46, 52*f*, 53

rosin, 34

Roterberg, August, 388n13

Rowland, Ingrid, 107

Royal Society, 115, 118, 308, 355n41

Rummel, Erika, 45

Russell, Bertrand, 14, 88

saints, versus mesmerism, 208–10

salting, term, 30

Salverte, Eusebe, 37–38, 148

Samson, G. W., 273–74

Sandby, George, 136, 184, 264, 274

Santanelli, professor, 229–30, 377n52

Sargent, John W., 296, 389n25

Sarpi, Paolo, 91, 99, 126, 350n64, 351n76

Satan. *See* devil

Saturn, rings of, 113–14, 114*f*

Savonarola, Girolamo, 67

Schiaparelli, Giovanni, 143

Schlessenger, doctor, 247–48

Schneider, Rudi, 288

scholastics, 23, 43–45, 354n39

Schrenck-Notzing, baron von, 304–5, 391n56

science: Carpenter and, 185; debunkers and, 26; development of, 111–13, 117–18, 308, 347n17; Fontenelle and, 122; Galileo and, 96; paradigm shifts in, 109; and religion and magic, 87; requirements for, 343n2

Scot, Reginald, 170, 191, 345n44

scripture: Catholics versus Protestants on, 115, 350n66; Galileo and, 98–100; Hare and, 256–57; Jung-Stilling and, 132; and rapping, 240; Swedenborg and, 129; and table turning, 277 scrying, 36

sealed messages, reading: Didier and, 170–73; mechanism of, 27–29, 170–73

séances: Palladino and, 283–85; term, 237–38

second sight, 135, 137, 150, 180–81, 366n4; Robert-Houdin and, 190–91; term, 180

Secrets of Nature, Academy of, 66–68, 71, 77, 79, 353n3

SETI Institute, 144

Severianus, general, 330n27

sex detector, 267

sexual improprieties: hypnosis and, 208, 230–31; priests of Delphi and, 16; spiritualism and, 304–5, 373n29, 390n51

sexuality: humanists on, 45; Palladino and, 283, 304, 387n13; Victorians and, 304, 390n51

Seybert Commission, 248

Sforza, Galeazzo, 46–47

Shakespeare, William, 55, 195

Shepard, Leslie, 235

shrines, 4–5; of Bacchus, 31–32, 32f. See also oracles

sibyls, 6, 325n6

Sidgwick, Eleanor Balfour, 283–85, 309–11, 310f, 313, 380n40, 386n12, 392n24

Sidgwick, Henry, 283, 309

Sims, Henry, 153–54, 179

Siricius, pope, 338n43

sitters, term, 237

Sixtus IV, pope, 47

Sixtus V, pope, 95

Slade, Henry, 298, 306–7, 386n12, 391n4

slate writing, 306–19, 392n18; mechanism of, 315–18

Smith, Francis, 140

Smith, Helene, 139–40

Smullyan, Raymond, 341n44

Soal, S. G., 197

social justice movements, spiritualism and, 247, 380n49

Society for Psychical Research, 197, 283, 285, 319, 365n49; Proceedings, 175; Sidgwick and, 309–11; and slate writing, 308–9

Socrates, 10, 20

Solomon, 340n13

Solovovo, count, 300

somnambulism, 131–32, 208, 366n1, 372n12; Cahagnet and, 135–37; Colquhoun and, 210; Didier and, 152; Prater and, 185; Swedenborg and, 181; term, 135

sophists, term, 45

Spanos, Nicholas, 226

Sparta, 14

speaking trumpet, 24, 330n30

speaking tubes, 24–25

Sperber, Burton, 158

spies, 177–78

spirit guides, 139, 358n19

spiritoscope, 253–57

spirit portraits, 298

spirits: Cahagnet and, 135; communication with, methods of, 238–39, 253f, 254–57; Hare and, 253–57; Jung and, 137–38; Jung-Stilling and, 131–32; nature of, 238; Swedenborg and, 129; and table turning, 263–64; Turing test for, 251–62

spiritualism, 139–41; Ashburner and, 383n25; branches of, 246–47; development of, 234–50; Flammarion and, 143; Jung and, 138; mechanisms of, 176–78; and sexual improprieties, 304–5, 373n29, 390n51; and slate writing, 306–19; and table turning, 263–79; Wallace and, 308

spontaneous generation, 116–17

state control, 67, 75–76

statues: moving, 30–38; Porta and, 71; talking, 22–25

steam engines, 31

Stelluti, Francesco, 105–6

stethoscope, 168

Steven II, pope, 49

stiff-leg tricks: Didier and, 152, 154; mechanism of, 156–57

Stigliola, Nicola, 76

Stone, William, 366n2

Summers, Montague, 209, 372n15

sunspots, 96–97, 350n52

supernova, 89–91

superstition: versus conjuring, 164; gambling and, 62–63; humanists on, 44–45

swan trick, 74

Swedenborg, Emanuel, 128–30, 136, 181, 236

Swift, Jonathan, 61–62

Sylvester, bishop of Rome, 41, 49

Sylvester II, pope, 336n8

Sylvestre, Ralph E. *See* Burlingame, H. J.

Symonds, John, 46

sympathy, term, 343n13

table lifting: mechanism of, 281–82, 290*f*, 294, 299–302; Palladino and, 291; term, 264

table talking, term, 264

table tipping, 182; mechanism of, 384n25

table turning, 253, 263–79; mechanism of, 269–73

Tacitus, 354n36

Tamariz, Juan, 286

taxonomy, 102–8

telegraph, 181–82, 192

telepathy: term, 197. *See also* clairvoyance; mind reading

telescope, 85, 91, 94–95, 110, 347n2, 349n32, 349n42, 353n15, 358n28; and moon hoax, 141–43, 142*f*

Telesio, Bernardino, 354n23

Temkin, Owsei, 87

Theodosius the Great, 6, 16

theory, 348n20

thermometer, 110–11

thimble pencil, 315

Thomas, Keith, 61

Thomas Aquinas, saint, 20, 44, 59, 96, 105

Thompson, investigator, 259

Thorndike, Lynn, 87

Thucydides, 333n29

thunder effects, 34

Thurston, Herbert, 209

Toronto Society for Psychical Research, 386n63

Townshend, Chauncey, 150–51, 167, 177, 273, 360n8, 182182

trance, 239; accounts of, 214–16; Cahagnet and, 135–36; Didier and, 152; Palladino and, 291; and remote viewing, 179

trauma, 220, 375n24

travelling clairvoyance. *See* remote viewing

Trent, council of, 75, 79, 99, 350n66

tricks, mechanisms of, 76, 154; blindfold sight, 157–59; card tricks, 164–78; clairvoyance, 148–63, 196–97; Durant and, 201–3; Elliotson and, 201; fire effects, 33–36; Gibson and, 287; Invisible Girl, 25–27, 26*f*; luminous paint, 292, 292*f*; mesmerism, 200; moving statues, 30–33; pain insensitivity, 228–30; Palladino and, 292–96, 298–302; Porta and, 71–75; rapping, 239–41, 240*f*, 244, 249–50, 379n31; remote

viewing, 185–87; sealed messages, 28–29; Sidgwick and, 309; slate writing, 315–18; spiritualism, 176–78; stiff-leg tricks, 156–57; table lifting, 281–82, 290f, 294, 301–2; table tipping, 384n25; table turning, 270–73; talking statues, 24–25
trional, 230
Trivers, Robert, 13–14
Tubby, Gertrude, 389n21
Turing, Alan, 261
Turing test: nature of, 251–52; for spirits, 251–62
Tuscany, 106–8
two-person codes, 175, 191–92
Tyndall, John, 196

UFO craze, 144
Uliva, Antonio, 116
Underhill, Samuel, 204
Urban VIII, pope, 20, 100–101, 104, 107, 329n12, 351n71, 351n77, 353n5
Ussher, James, 352n7
Utts, Jessica, 198

vacuum, 112
Valla, Lorenzo, 40–53, 336n18, 337n25
Van Dale, Anthony, 18
van Helmont, Johannes, 207, 371n5
Vatican, 45–46, 52f, 53
Venice, 91–92, 99, 341n39
Venus, 98
Vesalius, Andreas, 57, 340n20, 347n8
vessels: Hero and, 33; Porta and, 71
viewing. See remote viewing
Vincent, William, 278
Vivani, Vincenzio, 101
Vizetelly, Henry, 234
Voltaire, 19, 120

Wakley, Thomas, 211–13
Walker, Byron, 366n55
walking on fire, 35–36, 334n44
Wallace, Alfred Russel, 150, 189, 274, 308, 314–15
Wallis, John, 180
Ward, Artemus, 365n50
Weatherhead, Maud, 140
Webster, Richard, 220
weevils, 105
Weir (Weyer), Johann, 77
Weiss, Eric. See Houdini, Harry
Wesley, John, 236
Whewell, William, 356n22
White, John, 265, 344n18
white magic, 44
Wicks, Frederick, 365n45
Wilkins, John, 30
Willey, Basil, 69
Winter, Alison, 148, 163
Wiseman, Richard, 288
witches, 76–77
Wolfe, David, 352n9
women, Fontenelle on, 122, 126
wonder, Augustine on, 44
Wootton, David, 67, 81, 89, 109
Wordsworth, Christopher, 333n17
Wright, Peter, 306
writing: automatic, 239, 255–56; slate, 306–19, 392n18; spirits and, 379n24

Yates, Frances, 205
Young, L. E., 231

Zachary, pope, 338n40
Zancigs, 287, 388n32
Zerffi, professor, 372n12
The Zoist, 136–37, 151–53, 163, 199–201, 360n10
Zollner, Johann, 307
Zwingli, Ulrich, 57